Microbes and Microbiom Clean and Green Environment

(Volume 1)

The Role of Microbes and Microbiomes in Ecosystem Restoration

Edited by

Shiv Prasad
&
Govindaraj Kamalam Dinesh
Division of Environment Science
ICAR-Indian Agricultural Research Institute
New Delhi-110012, India

Murugaiyan Sinduja
National Agro Foundation, Taramani, Chennai
Tamil Nadu, India

Velusamy Sathya
Tamil Nadu Pollution Control Board, Chennai
Tamil Nadu, India

Ramesh Poornima
&
Sangilidurai Karthika
Department of Environmental Sciences
Tamil Nadu Agricultural University, Coimbatore
India

Microbes and Microbiomes for Clean and Green Environment

(Volume 1)

The Role of Microbes and Microbiomes in Ecosystem Restoration

Editors: Shiv Prasad, Govindaraj Kamalam Dinesh, Murugaiyan Sinduja,

Sathya Velusamy, Ramesh Poornima, Sangilidurai Karthika

ISBN (Online): 978-981-5256-59-8

ISBN (Print): 978-981-5256-60-4

ISBN (Paperback): 978-981-5256-61-1

©2024, Bentham Books imprint.

Published by Bentham Science Publishers Pte. Ltd. Singapore. All Rights Reserved.

First published in 2024.

need for a court order if at any point you breach any terms of this License Agreement. In no event will any delay or failure by Bentham Science Publishers in enforcing your compliance with this License Agreement constitute a waiver of any of its rights.

3. You acknowledge that you have read this License Agreement, and agree to be bound by its terms and conditions. To the extent that any other terms and conditions presented on any website of Bentham Science Publishers conflict with, or are inconsistent with, the terms and conditions set out in this License Agreement, you acknowledge that the terms and conditions set out in this License Agreement shall prevail.

Bentham Science Publishers Pte. Ltd.
80 Robinson Road #02-00
Singapore 068898
Singapore
Email: subscriptions@benthamscience.net

**BENTHAM
SCIENCE**

CONTENTS

PREFACE

The book "Role of Microbes and Microbiomes in Ecosystem Restoration" focuses on basic to advanced techniques in various roles of microbes and microbiomes in the abatement and restoration of polluted ecosystems, climate change, production of renewable energy sources, and waste management. It covers ecosystem sustainability, the UN decade of ecosystem restoration, efficient utilization of microbes and microbiomes and their role in socio-economic development, and the current status of polluted and degraded ecosystems.

Stepping into an unusual era of concurrent buffer leads to a shifting global climate. At the beginning of the twenty-first century, one of the active concerns in the human ecological background is the destruction of ecology and ecosystems. Human actions have evolved a remarkable power to affect the ecosystem. To address this developing issue, the science of restoration ecology and its applied practices provide a potentially cost-effective, buoyant answer. The notion of restoration has emerged as the dominant subject in the global environmental context. One of the most important goals of the UN Convention on Biological Diversity from 2011 to 2020 is to restore at least 15% of the world's damaged ecosystems. World leaders adopted the "Bonn Challenge" in 2011, which is a global commitment to rehabilitate 150 million hectares of deforested and damaged land by 2020. Most significantly, in 2015, the UN formalized these worldwide pledges by endorsing the 2030 Sustainable Development Goals, one of which focuses on ecological restoration. Microbes are ubiquitous, providing many critical services to the ecosystem, such as sustainable plant productivity and a stable environment for human life. They help to keep atmospheric CO_2 and nitrogen levels stable, which are now reduced due to greenhouse gases and other hazardous pollutants. On a global scale, microbial organisms are extremely strong. Bacteria create approximately 50% of total oxygen, 75% of added nitrogen to the atmosphere, and 92% of nitrogen removal from the environment. As a result, this book covers the potential of bacteria and microbiomes in many ecosystems.

In Chapter 1, Prasad *et al.* provide an overview of the causes of ecosystem destruction, the need for ecosystem restoration, the significance of microbiome in biomining, restoration of farm and degraded land, control of heavy metals, production of renewable energy, crop growth, biofertilizer production, mitigation of greenhouse gases, and waste management. It also encompasses the role of molecular techniques in ecosystem restoration and the challenges involved in adopting microbiomes for ecosystem restoration.

Microbes are the crucial living elements of soils that contribute to the sustainability of ecosystems because of their capacity for stress tolerance, vast effective genetic pool, ability to survive in various conditions, and capacity for catabolism. However, various factors like soil conditions, geographical and climatic factors, and soil stressors (drought, submersion, pollutants, and salinity) may result in distinct microbial composition and characteristics, as well as its mechanism to support ecosystem restoration and defense against all of these stressors. Hence, Pooja *et al.*, in Chapter 2, deliver the vital edaphic (pH, temperature, oxygen, nutrients, and moisture), geographical, climatic (UV radiation, elevated CO_2, temperature, permafrost thaw), and abiotic factors (drought, submergence, salinity, pollutants) involved in the establishment of microbes and microbiome.

In Chapter 3, Sinduja *et al.* discuss the ecological role of microorganisms participating in biogeochemical cycles, hoping to delineate the role of microbes and microbiomes in biogeochemical cycles. Microorganisms play an essential role in moderating the Earth's biogeochemical cycles; nevertheless, despite our fast-increasing ability to investigate highly complex microbial communities and ecosystem processes, they remain unknown. Hence, this chapter covers the strategies for proper management of prevailing natural resources, considerations for management, its role in the biogeochemical cycle, and the influence of beneficial soil microbes, such as plant growth promoting rhizobacteria and cyanobacteria, on natural resource management, with special emphasis on the role of soil enzymes in nutrient cycling.

Bioleaching (microbial leaching) is being studied intensively for metal extraction since it is a cost-effective and environmentally benign technique. Bioleaching with acidophiles involves the production of ferric (Fe III) and sulfuric acid. Cyanogenic microorganisms, in particular, can extract metal(s) by creating hydrogen cyanide. Besides, bioremediation is one of the most effective approaches for reducing environmental contaminants since it restores the damaged site to its original state. Hence, Chapter 4 by Poornima *et al.* provides a baseline on bioleaching, its types, microbes involved in bioleaching, bioleaching pathways, and the role of microbes in the bioremediation of polluted habitats.

In recent years, microbial-assisted bioremediation has emerged as a promising and eco-friendly alternative for HM remediation. This approach utilizes microorganisms to transform, immobilize, or detoxify HMs, making them less harmful and more accessible for removal. Hence, Naik *et al.*, in Chapter 5, highlight the eco-friendly use of microorganisms, their mechanisms that contribute to the bioremediation of HMs, and their potential use in the future.

In Chapter 6, Sajish *et al.* present the basic principle of an MFC and the role of microbes in a microbial fuel cell, genetic engineering, biofilm engineering approaches, and electrode engineering approaches for increasing the overall efficiency of an MFC for its practical implementation. Microbial fuel cell, a type of BES, is a budding technology that exploits the potential of electroactive microorganisms for extracellular electron transfer to generate electricity. Hence, this chapter encompasses the history of MFC, bio-electrochemically active microorganisms, electroactive microbial genera in microbial fuel cells, factors affecting the development of anode biofilm, biofilm engineering, and the recent advances in strain improvement for improved MFC performance.

Energy crises resulting from the depletion of petroleum resources, hikes in the price of fossil fuel, and unpredictable climate change are some of the recent concerns that have provoked serious research on alternative energy sources that will be sustainable. In this regard, biofuels are a straightforward substitute for fossil fuels. Renewable feedstocks are suitable ingredients that sustainably produce biofuels using microbial-based bioconversion processes. Industrially important enzymes are capable of degrading long-chained biopolymers into short-chained monomeric sugars and fermenting them into energy-dense biomolecules. Hence, Chapter 7, authored by Oyelade *et al.*, comprehensively reviews how sustainable bioenergy production through microbes using feedstocks can provide clean and green energy that can consequently facilitate ecosystem restoration. Feedstocks are pivotal to this biotechnological process.

In recent decades, biofertilizers have gained popularity as a viable alternative to unsafe chemical fertilizers in pursuing sustainable agriculture. They have an essential role in enhancing crop output and preserving long-term soil fertility, both of which are critical for fulfilling global food demand. Therefore, Chavada *et al.*, in Chapter 8, deliver the various microbes involved in nitrogen fixing, phosphorus and potassium solubilizing and mobilizing, sulfur oxidizing, and zinc solubilizing. The role of arbuscular fungi and plant growth-promoting rhizobacteria in biofertilizer production is also discussed.

Knowingly or unknowingly, agricultural systems face stress and resource quality degradation and their depletion by the activities of humans. Abiotic stresses, such as nutrient deficiency, water logging, extreme cold, frost, heat, and drought, affect agricultural productivity. Similarly, biotic factors like insects, weeds, herbivores, pathogens, bacteria, viruses, fungi, parasites, algae, and other microbes also limit good-quality products. Thus, Vijayalakshmi *et al.* discuss the application of microbes and microbiomes in biotic and abiotic stress management in Chapter 9. This chapter especially discusses the adaptive mechanisms of salt tolerance in plants, tolerance to abiotic stress, the emerging microbiome in soil biota, and nanomaterials' efficacy on stress.

Microbes play a significant role as either generators or consumers of greenhouse gases such as carbon dioxide (CO_2), methane (CH_4), and nitrous oxide (N_2O) through various processes. Sethupathi *et al.*, in Chapter 10, discuss the role of microbes and microbiomes in the emission of major greenhouse gases like CO_2, CH_4, N_2O, and NH_3. The potential of the microbiome in mitigating these greenhouse gases is also delivered in this chapter.

Given that there is potential for warmth to boost the release of carbon dioxide from dirt to the atmosphere due to better microbial disintegration of dirt raw material, the impact of environmental change on the soil carbon sink remains uncertain. If forecasted climate modification situations are precise, this boost in soil carbon loss might significantly worsen the dirt carbon cycle responses. Therefore, Chapter 11 by Al-Jawhari introduces us to the soil CO2 balance, environmental effects, and the significance of the soil carbon cycle and microbial decomposers, carbon cycle in soil, ocean, and ecosystem restoration under climate change perspective.

The generation of wastewater increases multi-fold because of industries and the overexploitation of freshwater resources. Wastewater treatment is always linked with waste recovery and its optimum utilization, which broadens the amplitude of wastewater treatment, enhancing the quality of the byproducts and as an efficient alternative for non-potable purposes. Microbiomes are crucial in biological wastewater treatment methods such as activated sludge, anaerobic digestion, and bioelectrochemical systems. The microbial population's activity and resilience in the microbiome significantly impact the performance and stability of these activities. Suganthi *et al.* present the biological wastewater treatment, growth and kinetics, and different microbial community types, including bacteria and fungus, actinomycetes, algae, plants, and the range of microbial wastewater treatment in Chapter 12.

Solid waste disposal is a significant issue that worsens daily as more people move into cities. In Chapter 13, Velusamy *et al.* provide the status of solid waste management in India, sources and types of solid wastes, various conventional solid waste management techniques, and the role of microbes in solid waste management through composting and anaerobic digestion.

Microorganisms are pervasive and genuinely make up the "unseen majority" in the marine environment. Although marine isolates have been the subject of laboratory-based culture methods for more than ten years, we still do not completely understand the ecology of marine microorganisms. Thus, in Chapter 14, Poornachandhra *et al.* explore marine microbial diversity, its utilization in bioremediation, and understanding their role in ecosystem sustainability.

Mangroves and wetlands are critical intermediary ecosystems between terrestrial and marine environments. These ecosystems offer a wide range of invaluable ecological and economic services. However, under the influence of natural and anthropogenic threats, mangroves and wetlands face rapid degradation. Hence, Chapter 15 by Haghani *et al.* is dedicated to enlightening us regarding the most critical features of microbial groups, including archaea, bacteria, algae, and fungi in mangroves and wetlands. Moreover, the biochemical transformations brought about by wetlands' microbial groups and the degree of complexity in microbial interactions are explained.

Jerome *et al.*, in Chapter 16, articulate the significance of forest microbiomes in ecosystem restoration and sustainability. Generally, forest microorganisms are essential to how plants interact with the soil environment and are necessary to access critically limiting soil resources. This chapter focuses on the ecosystems below and above ground level of a forest microbiome, including the soil microorganisms, their importance, and the diverse interrelationships among soil microorganisms (parasitism, mutualism, commensalism).

Employing field-based monitoring and restoration assessment techniques, surveying microbes or microbial populations is challenging or impossible. In contrast, it is now possible to precisely and quickly describe and quantify these diverse and functional taxonomic groups by sequencing large quantities of environmental DNA or RNA utilizing genomic and, in particular, meta-omic technologies. Hence, Nagendran *et al.*, in Chapter 17, throw light on using meta-omics techniques to monitor and assess the outcomes of ecological restoration projects and to monitor and evaluate interactions between the various organisms that make up these networks, such as metabolic network mapping. An overview of functional gene editing with CRISPR/Cas technology to improve microbial bioremediation is also provided herewith.

Chapter 18 by Satpathy *et al.* provides details on metagenomic approaches like Multi-Locus Sequence Typing (MLST), MOTHUR, Quantitative Insight into Microbial Ecology (QIIME), and PHAge Communities From Contig Spectrum (PHACCS) in the restoration of the temperate and tropical ecosystem.

Soil microorganisms also play a fundamental role in ecosystem functioning and conserving plant diversity. Exploring voluminous beneficial microorganisms and promoting the reestablishing of those beneficial microbes in the soil will preserve Earth's diverse native plant populations. Hence, Prasad *et al.*, in Chapter 19, delve into fundamental and conventional techniques and approaches that can be employed to maintain soil microbial populations. Furthermore, the chapter investigates the possibility of creating protocols for regulatory or commercial objectives, emphasizing the significance of ecological restoration by using bioinoculants or microbial colonies in degraded sites.

In Chapter 20, Shivakumar *et al.* examine the application of molecular methods to ecosystem regeneration. The various available molecular methods and how they have been applied to monitor ecosystem health, identify microbial communities in ecosystems, and comprehend interactions between microbes and plants are discussed. The chapter also discusses the application of molecular methods to the restoration of ecosystems that have been damaged, including the use of plant-microbe interactions to promote plant development in contaminated soils.

The sustainable industrial revolution is the way forward to help humankind to prolong its existence on Earth. John *et al.* enlighten us with the role of the microbiome in a sustainable industrial production system. In Chapter 21, they disclose the energy sector's current status, microbes' role in organic and amino acid production, and the role of microalgae in sustainable agriculture.

The human microbiome plays a vital role in human development, immunity, and nutrition, where beneficial bacteria establish themselves as colonizers rather than destructive invaders. In Chapter 22, Pradyutha *et al.* introduce microbes' role in human and animal health security. The various human and animal diseases and the potential of microbiota, such as probiotics, in disease treatment are also discussed in this chapter.

Shiv Prasad

&

Govindaraj Kamalam Dinesh
Division of Environment Science
ICAR-Indian Agricultural Research Institute
New Delhi-110012, India

Murugaiyan Sinduja
National Agro Foundation, Taramani, Chennai
Tamil Nadu, India

Velusamy Sathya
Tamil Nadu Pollution Control Board, Chennai
Tamil Nadu, India

Ramesh Poornima

&

Sangilidurai Karthika
Department of Environmental Sciences
Tamil Nadu Agricultural University, Coimbatore
India*v*

List of Contributors

Anushka Satpathy	Department of Bioengineering & Biotechnology, Birla Institute of Technology, Mesra, Ranchi-835215, Jharkhand, India
B. Balaji	ICAR-National Institute for Plant Biotechnology, LBS Centre, Pusa Campus, New Delhi, India Post Graduate School, Indian Agricultural Research Institute, Pusa, New Delhi-110012, India
Christobel R. Gloria Jemmi	V.V. Vanniaperumal College for Women, Virudhunagar, Tamil Nadu, India
C. Avinash	Division of Environment Science, ICAR-Indian Agricultural Research Institute (IARI), New Delhi, India
Divya Pooja	Division of Environment Science, ICAR-Indian Agricultural Research Institute (IARI), New Delhi, India
G. Ramanathan	Sri Paramakalyani College, Alwarkurichi, Tirunelveli, Tamil Nadu, India
Govindaraj Kamalam Dinesh	Division of Environment Science, ICAR-Indian Agricultural Research Institute (IARI), New Delhi, India Division of Environment Sciences, Department of Soil Science and Agricultural Chemistry, Faculty of Agricultural Sciences, SRM College of Agricultural Sciences, SRM Institute of Science and Technology, Baburayanpettai - 603201, Chengalpattu, Tamil Nadu, India INTI International University, Persiaran Perdana BBN, Putra Nilai, Negeri Sembilan, Malaysia
Konderu Niteesh Varma	Division of Microbiology, ICAR-Indian Agricultural Research Institute, New Delhi, India
Koel Mukherjee	Department of Bioengineering & Biotechnology, Birla Institute of Technology, Mesra, Ranchi-835215, Jharkhand, India
K. Boomiraj	Department of Environmental Sciences, Tamil Nadu Agricultural University, Coimbatore, Tamil Nadu-641003, India
M. Prasanthrajan	Centre for Agricultural Nanotechnology, Tamil Nadu Agricultural University, Coimbatore, Tamil Nadu-641003, India
Murugaiyan Sinduja	National Agro Foundation, Taramani, Chennai, Tamil Nadu, India
M. Vijayalakshmi	V.V. Vanniaperumal College for Women, Virudhunagar, Tamil Nadu, India
M. Shankar	Bureau of Plant Genetic Resources, ICAR- Indian Agricultural Research Institute, New Delhi, India
Nikul B. Chavada	Om College of Science, Bheshan Highway, Junagadh- 362001, Gujarat, India
Pooja Mehta	SVKM`s Mithibai College of Arts, Chauhan Institute of Science & Amrutben Jivanlal College of Commerce and Economics Vile Parle (W), Mumbai, Maharashtra 400056, India
Rohit Das	Department of Microbiology, School of Life Sciences, Sikkim University, Gangtok - 737102, Sikkim, India

Ramesh Poornima	Department of Environmental Sciences, Tamil Nadu Agricultural University, Coimbatore, India
R. Vinothini	MIT College of Agricultural and Technology, Musiri, Tamil Nadu, India
R.V. Akil Prasath	Department of Environmental Science and Management, Bharathidasan University, Tiruchirappalli–620024, India
R. Raveena	Department of Environmental Sciences, Tamil Nadu Agricultural University, Coimbatore, Tamil Nadu-641003, India
R. Shivakumar	Department of Biotechnology, Centre for Plant Molecular Biology and Bioinformatics, Tamil Nadu Agricultural University, Coimbatore-641003, India
Ramesh Poornima	Department of Environmental Sciences, Tamil Nadu Agricultural University, Coimbatore, India
Shiv Prasad	Division of Environment Science, ICAR-Indian Agricultural Research Institute (IARI), New Delhi, India
Sangilidurai Karthika	Department of Environmental Sciences, Tamil Nadu Agricultural University, Coimbatore, India
S.T.M. Aravindharajan	Division of Microbiology, ICAR-Indian Agricultural Research Institute, New Delhi, India
S. Karthika	Tamil Nadu Agricultural University, , , Coimbatore, Tamil Nadu, India
Suganthi Rajendran	JSA College of Agriculture and Technology, Avatti, Cuddalore, Tamil Nadu, India
Saraswathy Nagendran	SVKM`s Mithibai College of Arts, Chauhan Institute of Science & Amrutben Jivanlal College of Commerce and Economics Vile Parle (W), Mumbai, Maharashtra 400056, India
S. Akila	National Agro Foundation, Research & Development Centre, Anna University Taramani Campus, Taramani, Chennai, Tamil Nadu-600113, India
Selvaraj Keerthana	Department of Environmental Sciences, Tamil Nadu Agricultural University, Coimbatore, Tamil Nadu-641003, India
Santosh Kumar	Department of Microbiology, School of Life Sciences, Sikkim University, Gangtok - 737102, Sikkim, India
Viabhav Kumar Upadhayay	Department of Microbiology, College of Basic Sciences & Humanities, Dr. Rajendra Prasad Central Agricultural University, Pusa, Samastipur, Bihar, India
Vinod Kumar Nigam	Department of Bioengineering & Biotechnology, Birla Institute of Technology, Mesra, Ranchi-835215, Jharkhand, India
Velusamy Sathya	Tamil Nadu Pollution Control Board, Chennai, Tamil Nadu, India
Yogesh Dashrath Naik	Department of Agricultural Biotechnology and Molecular Biology, Dr. Rajendra Prasad Central Agricultural University, Pusa, Samastipur, Bihar, India

Part 1: Introduction to Microbes and Microbiomes in the Ecosystem Restoration and Green Environment

Overview of Microbes and Microbiomes in the Restoration of Terrestrial, Aquatic, and Coastal Ecosystems

Shiv Prasad[1,*], **Sangilidurai Karthika**[2], **Murugaiyan Sinduja**[3], **Ramesh Poornima**[2], **Govindaraj Kamalam Dinesh**[1,4,5] and **Velusamy Sathya**[6]

[1] *Division of Environment Science, ICAR-Indian Agricultural Research Institute (IARI), New Delhi, India*

[2] *Department of Environmental Sciences, Tamil Nadu Agricultural University, Coimbatore, India*

[3] *National Agro Foundation, Taramani, Chennai, Tamil Nadu, India*

[4] *Division of Environment Sciences, Department of Soil Science and Agricultural Chemistry, Faculty of Agricultural Sciences, SRM College of Agricultural Sciences, SRM Institute of Science and Technology, Baburayanpettai - 603201, Chengalpattu, Tamil Nadu, India*

[5] *INTI International University, Persiaran Perdana BBN, Putra Nilai, Negeri Sembilan, Malaysia*

[6] *Tamil Nadu Pollution Control Board, Chennai, Tamil Nadu, India*

Abstract: Ecosystems consist of biotic and abiotic components, including flora and fauna, along with the conducive environmental factors of a particular place. These are imperative for maintaining the ecosystem's structure and energy flow between trophic levels and providing ecosystem services for the well-being of humans and other living organisms. However, ecosystems are being threatened by human activities, which disrupt the balance of nature. Thus, it impacts billions of people by causing economic loss and threats to the survival of terrestrial, aquatic, and other species. Climate change and increasing pollution also adversely affect the functions of the ecosystem. Microbes and microbiomes are reported to restore terrestrial, aquatic, and coastal ecosystems. The diverse microbes such as bacteria, archaea, algae, fungi, and protozoa help detoxify the polluted ecosystems through various physical, chemical, and biological mechanisms. They also help with the nutrient cycling and mineralization of nutrients from the soil to plants in their available forms. With the focus on ecorestoration, there is a need to take collective action to protect the environment and prevent ecosystem degradation worldwide.

Keywords: Clean environment, Ecosystems, Ecosystem restoration, Green environmenta, Microbes, Microbiomes.

---
[*] **Corresponding author Shiv Prasad:** Division of Environment Science, ICAR-Indian Agricultural Research Institute (IARI), New Delhi, India; E-mail: shiv_drprasad@yahoo.co.in

INTRODUCTION

The ecosystem is the complex of living organisms and non-living things interacting with each other in their physical environment called habitat, and all the interrelationships between organisms occur in a particular space unit. It is divided into (i) terrestrial, *e.g.*, desert, forest, grassland, taiga, and tundra, and (ii) non-terrestrial, *e.g.*, aquatic, marine, and wetlands. Ecosystems play a vital role in balancing the natural phenomenon, structural organization, energy flow, and nutrient cycling and provide various ecosystem services and benefits to human society. However, human activities negatively influence the well-being of 3.2 billion people. These activities cost more than 10% of the annual global gross product by losing biodiversity and vital ecosystem services. Human activities are reported to reduce productivity in 23% of global terrestrial areas [1]. Vegetation cover is decreasing, influencing grasslands, croplands, woodlands, and rangelands, particularly in vulnerable regions. Desertification has severe consequences for 38% of the global population. Wetlands have declined 70% over the previous century [2]. The global forest area has decreased by 100 million hectares since 2000 [3]. Reversing this fact can have substantial advantages. It can help improve the food and water supply, reduce GHG emissions, and mitigate adverse effects related to climate change.

The restoration of ecosystems is vital across global international conventions and agreements to achieve their goals and priorities regarding biodiversity, climate change, desertification, and a sustainable future. Global action is required to restore ecosystems and enhance positive global impact. Investment in ecosystem restoration projects can deliver many advantages to society, including biodiversity conservation [4]. At an international level, restoring the ecosystem degradation is essential to maintain temperatures below 2°C [5, 6]. IUCN calls for collective action to set the world on a transformational trajectory in the UN Decade on Ecosystem Restoration, enabling the implementation of the Post-2020 Framework [7]. The UN also calls for accelerated and scaled-up ecosystem restoration by 50% to reverse loss in the area by 2030. They focused on spatial integrated planning in all ecosystems to cover 50% of the land, freshwater, and ocean regions by 2030 to reduce pressure on ecosystems and maximize biodiversity and ecosystem services.

Microbes and microbiomes are vital in ecosystem restoration [6, 8]. They contribute to nutrient cycling, decomposition of organic materials, soil fertility maintenance, and crop productivity enhancement [6]. Bacteria, archaea, and fungi exhibit distinct assemblies along vertical and horizontal profiles in reforested ecosystems. The diversity of bacteria and fungi decreases with increasing soil depth while archaeal diversity increases. As reforestation progresses, bacterial communities' vertical and spatial variation declines while archaeal and fungal communities proliferate. The distribution patterns of soil microbes are linked to the soil's physical and biochemical properties and the existence of plant roots. Bacterial and archaeal communities play influential roles in deep and superficial soil layers in multi-nutrient cycling. Soil fungi comprise various dynamic kingdoms of eukaryotes and are vital for maintaining ecosystem processes and functions [6].

Understanding microbial community assembly processes and biogeochemical cycling during ecosystem restoration is critical for optimizing management strategies [9]. Factors such as variation in community assembly processes, measurable microbial community attributes, and linkages to ecosystem function must be considered. By examining microbial succession, insights can be gained into microbial community structures in various ecosystem recoveries. They can help determine the success of ecosystem restoration efforts and maintain ecosystem stability [10]. Native mycorrhizal fungal communities play a significant role in restoring native plants [11]. Understanding microbial community assembly processes, managing microbial contamination risks, and identifying effective bio-indicators of soil health are crucial for optimizing ecosystem restoration efforts and promoting ecosystem stability and resilience. This chapter discusses an overview of microbes and microbiomes' role in ecosystem restoration.

CAUSES OF ECOSYSTEM DESTRUCTION AND THEIR IMPACTS

Ecosystem degradation is defined as a long-term reduction in the structure and functionality of the ecosystem or loss in delivering services and proficiency to benefit people. The significant causes of ecosystem destruction and their impacts are overviewed in the following sub-headings. The essential causes of ecosystem destruction are shown in Fig. (**1**), which is happening at an alarming rate due to various anthropogenic activities such as changes in land use practices, unrestricted natural resources utilization, deforestation, habitat loss, climate change, warming ocean waters, ocean acidity, and pollution.

Resource Exploitation

Resource exploitation, particularly mining and extraction activities, can positively and negatively impact ecosystems, socio-economics, and the environment. Saputra and co-workers [12] investigated the effect of sand mining in Indonesia and reported its promising impact on the socio-economics and environment. They also stated that the river deepens after sand mining, enabling it to hold considerable water and control overflow during the rainy season. Similarly, exploiting mineral resources like coal can impact local communities and ecosystems. Aigbedion and Iyayi [13] highlighted the negative impacts of mineral exploitation and its processing, which disturbed environmental settings and ecosystems, devasted natural flora and fauna, polluted air, soil, and water, and caused many other hazards. These impacts can worsen poverty and income inequality in resource-rich regions.

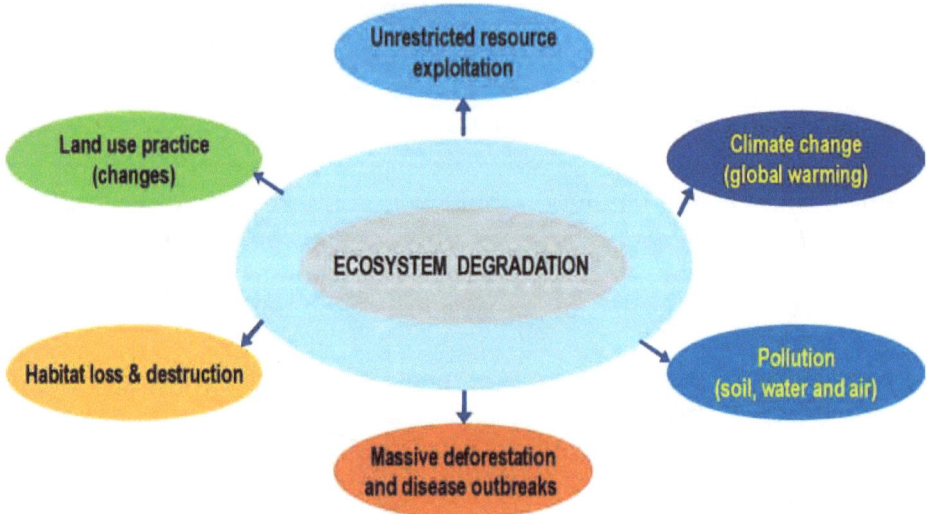

Fig. (1). Causes of ecosystem degradation.

Industrial enterprises associated with resource exploitation, particularly in Arctic regions, can also contribute to environmental pollution and safety concerns [14], including global warming and irreversible ecosystem damage. Expanding mining areas to exploit mineral and other natural resources causes ecological damage. In mining areas, authorities must focus on reducing environmental impact, health and risk assessment, and ecosystem services valuation for effective ecosystem restoration planning [15] to minimize implications in biodiversity conservation and the balance of native ecosystems [16]. Adopting these approaches, we can enhance the food resources base and promote ecosystem restoration by exploiting the potential of novel microbes and conserving threatened species [17].

In the Arctic tundra, the exploitation and depletion of scarce resources, including pasture grasses, can contribute to the irreversible degradation of grasslands. Animal husbandry, which heavily relies on pasture grasses, is a principal economic activity for indigenous peoples in the Arctic. However, the unique Arctic ecosystem requires special attention to prevent over-exploitation and degradation of these valuable resources [18]. The impact of resource exploitation on ecosystems is not limited to environmental factors but also extends to social and cultural aspects [19]. The effect of resource exploitation on ecosystems can also be influenced by factors such as tree diversity and the positive relationship between tree and fungal diversity in subtropical Chinese forests. Greater fungal diversity promotes better resource exploitation and confers higher resilience due to functional redundancy, which helps ecosystem service [20].

Climate Change

Climate change dramatically impacts ecosystem functions and their services [21]. Its impact assessment on ecosystem services has many challenges due to long time scales and high uncertainties [22]. However, climate change affects both abiotic and biotic elements of ecosystems, leading to changes in ecosystem functions and processes. Climate change can impact tropical and subtropical forests due to temperature increases, ecosystem structure, and function alterations. Its impacts on the hydrological cycle and the availability of water resources can reduce ecosystem services and freshwater supply [23]. Coastal habitat is highly vulnerable to climate change and is expected to be negatively impacted.

Climate change can alter biodiversity and species distribution, functions, and productivity [24] and negatively impact human well-being. Climate change affects water sources, leading to water quality and availability alterations, affecting human and ecosystem health. Temperature elevation and changes in precipitation resulted in extreme weather conditions, which directly and indirectly affect biodiversity and ecosystems. Studies show that the economic value created by ecosystem service functions of rivers has declined due to climate change and human activities [25, 26]. Climate change can disrupt biodiversity, change ecosystem structure and functions, alter water sources, and impact human well-being. Understanding climate change's effects on ecosystems and their services is crucial for effective conservation and adaptation strategies.

Increasing Environmental Pollution

Environmental pollution is a leading cause of many disturbances and destruction in the ecosystem and adversely affects its components. Significant air, water, and soil pollution sources are biomass burning, fossil fuel combustion, trash, toxic gaseous emissions, oil spills, agricultural chemicals, and pesticides. Pollution significantly impacts ecosystems, affecting the organisms and the environment's health. Various types of pollution, such as organic, chemical, heavy metal, and plastic, can harm ecosystem health. These pollutants can enter ecosystems through different pathways, including wastewater discharge, industrial activities, agricultural practices, and urbanization. The consequences of pollution on ecosystems can be wide-ranging, including changes in biodiversity, disruption of ecological processes, and degradation of water and soil quality [27, 28].

Polycyclic aromatic hydrocarbons (PAHs) are another pollutant type that can significantly impact marine ecosystems. PAHs are widespread in marine environments and can enter the ecosystem through chronic or acute pollution by oil spills [29]. Pollution derived from both point and non-point sources can transform aquatic ecosystems and impact the structure of local communities, including zooplankton communities. The interactions between multiple environmental factors and ecological processes determine different local communities' structures in contaminated river ecosystems. Understanding these interactions is crucial for assessing the impact of pollution on zooplankton communities and developing effective management strategies for polluted river ecosystems [30].

Microplastic pollution is a growing global problem that threatens marine ecosystems. Plastic production has increased significantly recently, accumulating plastic litter in marine environments. Plastic pollution and other artificial impacts, such as global heating, ocean acidification, eutrophication, and chemical pollution, can push marine ecosystems to the brink and harm ecosystem functioning and services. The persistence of plastic in the environment and its ability to accumulate in ecosystems make it a particularly concerning pollutant. Efforts to reduce and remove plastic litter from marine ecosystems are crucial for mitigating the impacts of plastic pollution on marine ecosystems [31].

Various strategies and technologies have been proposed to mitigate pollution's impacts on ecosystems. Bioremediation, which utilizes the capacity of microorganisms to degrade pollutants, is a promising approach for the recovery of contaminated environments. Microorganisms play a crucial role in the degradation of organic pollutants and can be harnessed to remove pharmaceutical compounds and other organic contaminants from impacted environments [29]. Rhodococcus spp., a microorganism, has shown potential for degrading pharmaceutical pollutants and producing valuable products. The use of Rhodococcus *spp.* and other microorganisms in biodegradation processes and biotechnology can contribute to the remediation of polluted environments and the production of targeted pharmaceutical products [14]. Coral reefs, for example, are highly vulnerable to climate change impacts and marine pollution. Evaluating the resilience of coral reefs and implementing management strategies to protect them is crucial for their conservation and the sustainable use of natural resources. Indicator frameworks incorporating various dimensions, such as coral diversity, biodiversity, and environmental factors, can provide suitable methods for decision-makers to make better management strategies to protect coral reefs and use natural resources strategies effectively [32].

Deforestation

Deforestation significantly impacts ecosystems, affecting various aspects such as biodiversity, carbon cycle, hydrological regimes, indigenous populations, and human health [33]. It can change atmospheric circulation patterns and rainfall distribution, affecting climate locally and in other regions [34]. Reducing forest cover can decrease evapotranspiration and moisture availability, reducing rainfall and longer dry seasons. Studies have shown that deforestation in the Amazon basin can reduce dry-season rainfall by up to 20% in regions far from the deforested area [35]. These climatological effects can have long-term consequences, potentially leading to large-scale forest loss and ecosystem degradation.

Loss of species due to deforestation can reduce functional redundancy, making ecosystems more vulnerable to further species loss. Functional redundancy refers to multiple species that perform similar ecological functions, providing a buffer against the loss of individual species [36]. The impacts of deforestation extend beyond terrestrial ecosystems to aquatic ecosystems as well. Deforestation in the catchment area can disrupt the structure and processes of riparian ecosystems, leading to changes in aquatic assemblages. These changes can occur through direct and indirect pathways, affecting the abundance and structural composition of benthic macro-invertebrate assemblages [37].

Studies have shown that deforestation increases the risk of vector-borne diseases, such as malaria, by enriching human exposure to mosquito vectors [38]. In addition to the direct impacts on ecosystems and human health, deforestation also affects the provision of ecosystem services. Ecosystem services are the benefits humans have by providing clean water, climate regulation, and nutrient cycling. Deforestation can disrupt the provision of these services, potentially leading to negative consequences for human well-being. For example, deforestation in Indonesia, one of the world's mega-biodiversity countries, has been severe and threatens the future provision of ecosystem services [39].

Habitat Loss and Destruction

Habitat loss and destruction significantly impact ecosystems, biodiversity, ecosystem services, species abundance reductions, species extinction, and disruptions to species interactions. When habitats are destroyed or fragmented, it can lead to a regime shift in the ecosystem, where reinforcing feedback mechanisms intensify and result in a new community configuration with ecological, social, and economic consequences. Habitat destruction can also lead to the erosion of environmental resilience, making ecosystems more vulnerable to disturbances [20]. One of the critical impacts of habitat loss and destruction is biodiversity loss. Biodiversity refers to an area's variety of species, genes, and ecosystems. Habitat destruction can result in the loss of species, as well as the loss of interactions between species. This biodiversity loss can have cascading effects on ecosystem functions and vital services. For example, losing keystone species, various ecosystem engineers species, and habitat-forming species can significantly damage functional redundancy, ecosystem resilience, and habitat complexity. Additionally, habitat loss can lead to the loss of carbon stored in vegetation biomass, contributing to climate change [20].

Urban green spaces (UGS) are crucial in promoting resilience and health in urban areas. UGS provides numerous benefits, including improved mental and physical health, increased social cohesion, and enhanced resilience to climate change and other stressors. Studies have shown that UGS positively impacts the promotion of resilience and health in urban citizens. However, the extent of the relationship between UGS and health and resilience is still being explored [40]. Habitat loss and destruction can also have specific impacts on marine ecosystems. For example, in temperate rocky reefs, habitat destruction can lead to the transition from kelp habitats to sea urchin barrens. This phase shift is commonly linked to destructive overgrazing behavior by sea urchins and developing urchin consumer fronts along the edges of kelp beds. This shift in ecosystem structure can have significant ecological consequences and reduce the resistance of the ecosystem to invasion, overgrazing, and the downfall of turf dominance [41].

The impacts of habitat loss and destruction are not limited to individual species or specific ecosystems. They can also have broader effects on ecological processes and dynamics. For example, habitat destruction can alter the coevolutionary trajectories of species and their response to habitat loss. Coevolution, the reciprocal evolutionary change between interacting species, can mitigate the adverse effects of habitat devastation in mutualistic networks. However, the impact of coevolution on antagonistic communities' persistence tends to be minor and less predictable. Additionally, habitat loss can simplify and destabilize soil microbial networks, affecting ecosystem functioning [42].

Humans' Activities

anthropogenic activities have a significant impact on ecosystems, including both marine and terrestrial environments [43]. In terrestrial ecosystems, For instance, human activities such as land use change and forage harvest in grassland ecosystems can lead to changes in net primary productivity (NPP) and grassland degradation [44]. Human activities can also, directly and indirectly, impact specific ecosystems, such as mangrove forests [45]. Activities adjacent to mangrove forests, such as land use changes and pollution, threaten the ecosystem.

The cumulative impact of these activities on mangrove forests can be quantified using models that overlay human activities onto maps of mangrove coverage [46].

The impacts of human activities on ecosystems are not limited to specific regions or ecosystems. They have global implications and can contribute to the emergence of zoonotic diseases. Human activities, such as deforestation, urbanization, and wildlife trade, disrupt the wildlife balance and can fuel zoonotic disease's emergence [47]. The COVID-19 pandemic exemplifies relationships between human activities and zoonotic epidemics. Understanding and mitigating the impacts of anthropogenic activities on ecosystems is crucial for preventing future outbreaks of zoonotic diseases. Understanding the cumulative effects of anthropogenic activities on ecosystems is essential for conservation and management strategies. Sustainable practices and conservation efforts are needed to mitigate the adverse effects of human activities and ensure ecosystems' long-term health and resilience.

Disease Outbreaks

Disease outbreaks can significantly impact ecosystems, affecting various ecological factors such as population size, population structure, species interactions, and ecosystem functioning [48]. It can also have long-lasting effects on ecosystems, *e.g.*, a decline of the coral cover due to bleaching events. These outbreaks are linked to significant ecological and structural changes in coral reef ecosystems [49]. Additionally, disease outbreaks can affect the dynamics of species interactions, altering predator-prey relationships and the composition and functioning of ecosystems [50]. Biodiversity plays a crucial role in shaping disease outbreaks. A meta-analysis of 61 parasites found broad evidence that host diversity inhibits parasite abundance, indicating that the dilution effect is robust across different ecological contexts; observational studies overwhelmingly documented dilution effects, including for zoonotic parasites of humans. Therefore, maintaining or restoring healthy ecosystems with high biodiversity can help prevent zoonotic diseases and mitigate their impact [51].

Climate change can also influence disease outbreaks and their impacts on ecosystems. Temperature and other climatic factors can alter pathogen evolution, host-pathogen interactions, and the range of pathogens, increasing plant disease's spread in new areas [52]. For example, warming temperatures have increased the prevalence of wasting disease in eelgrass meadows, which are essential coastal habitats. Changes in lake surface water temperatures can also impact the probability of cyanobacteria outbreaks, significantly affecting lake ecosystems [53]. Therefore, climate change mitigation and adaptation strategies are crucial for reducing the risks and impacts of disease outbreaks on ecosystems. It is important to note that disease outbreaks can also indirectly impact ecosystems through their effects on human activities. For example, the COVID-19 pandemic has led to social distancing measures and reduced economic activities, resulting in land use changes and reduced pollution levels. These changes may have positive and negative impacts on ecosystems. Reduced human activity can improve air and water quality, benefiting the urban ecosystem. On the other hand, changes in land use and human behaviour can also have negative consequences, such as increased deforestation or the disposal of sanitary consumables in natural environments.

Land-use Changes

Land use changes to satisfy the human population are increasing alarmingly—the development of more landscapes, clear land for housing, roads, and infrastructure projects enhancing ecosystem degradation. The conversion of natural habitats and land-use changes for agriculture, urbanization, and infrastructure development predominantly contribute to the destruction of ecosystems, shifts in species interactions resulting in habitat loss, and a decline in biodiversity, ecosystem services, and human well-being [44, 86]. Land-use changes have economic implications, as they can affect the availability of resources, such as water and timber, and impact local economies.

In order to address problems related to land-use changes, we need to promote sustainable development by coordinating resources, functional regions, and ecological development, including management of regional landfills and municipal waste management with concerned authorities [44, 86]. Some developing nations have successfully managed their land use change by executing farming intensification, land use zoning, forest site protection, and augmented reliance on imported products. Sound policies and innovations are crucial in achieving sustainable land use practices and mitigating ecosystem degradation [54].

ECOSYSTEM RESTORATION

Ecosystem restoration is crucial to managing biodiversity and ensuring ecosystem services are provided to humanity [57]. Restoration actions can be evaluated based on diverse metrics and indicators. In marine coastal ecosystems, short-term survival rates are typically used to assess restoration success. Diversity measurements, vegetation arrangement, and ecological processes are often used in terrestrial ecosystems. Pilot-scale restoration investigations can deliver valuable initial data to improve the efficacy and accomplishment of large-scale restoration projects. These investigations can assist in assessing habitat suitability and optimize site selection for restoration [55]. The cost of ecological restoration can differ depending on factors such as the ecosystem type, degradation level, and the restoration procedures employed [56]. However, a lack of knowledge of functional traits and their link to ecosystem services poses a barrier to operationalizing various ecosystem restoration approaches [58, 59].

Restoration Ecology

Restoration ecology is a field of study that focuses on assisting the recovery of ecosystems that have been damaged, degraded, or destroyed. It involves restoring ecosystems' ecological structure, function, and biodiversity to a more natural and sustainable state. The goal of restoration ecology is to reverse the adverse impacts of human activities and promote the recovery of ecosystems, thereby enhancing their resilience and ability to provide ecosystem services. One of the critical aspects of restoration ecology is understanding the factors that drive the success of restoration efforts. A global meta-analysis by Crouzeilles and co-workers [60] examined the ecological drivers of forest restoration success. The study found that forest restoration boosts biodiversity by 15-84% and vegetation cover structure by 36-77% compared to degraded or damaged ecosystems. The primary ecological drivers of restoration success were the time elapsed since restoration started, disturbance types, and various landscape contexts. The time that has elapsed since restoration significantly influenced ecosystem restoration achievement in secondary forests but not selectively in logged forest types. Landscape restoration was most fruitful after the earlier disturbance was less severe and reduced habitat fragmentation.

Microbe and Microbiomes in Terrestrial Ecosystem Restoration

Microbes and microbiomes play a crucial role in terrestrial ecosystem restoration. Several investigations have emphasized the extent of microbial diversity in keeping multi-functionality and delivering essential ecosystem services in terrestrial ecosystems. Microbial communities in the soil have been found to drive multi-nutrient cycling in reforested ecosystems, with different microbial groups showing distinct assemblies along vertical soil profiles. Artificial revegetation has been shown to promote soil microbial diversity and restoration community in degraded ecosystems, increasing the alpha diversity of soil microbes within restoration periods [61]. In ecological restoration, it is essential to conserve microbial biodiversity, especially in drylands, as it is vital to soil survival and ecosystem functioning. Incorporating the microbial component into ecosystem restoration planning is pivotal in rebuilding the disturbed ecosystem microbiome and enhancing land management practices. The response of soil microbial communities to grassland degradation has also been studied, with changes in microbial community structural compositions and diversity at diverse levels of ecosystem degradation [62].

Removing and storing topsoil can negatively affect soil microbial communities and nutrient cycling in mining and post-mining rehabilitation. Successful restoration of degraded terrestrial ecosystems needs effective soil microbial inoculum to restore their populations and promote plant community development. Additionally, the enhancement of soil phosphorus cycling has been observed following ecological restoration, with the relative abundances of critical genes and genomes leading to higher soil microbial phosphorus cycling at restored sites than at unrestored sites [63]. Global warming consequences in glacier-fed stream (GFS) ecosystems are irreversible, and the microbiome cannot be preserved, restored, or managed like the microbiome of terrestrial environments. However, microdiversity within microbial communities may mitigate the impacts of climate change on microbial life in GFSs [64]. Many studies highlight the importance of microbes and microbiomes in restoring terrestrial ecosystems and maintaining soil fertility, as shown in Table **1**.

Table 1. Microbes in maintaining soil fertility.

Microbes	N	P	K	Fe	OM	AF	AS	References
Rhizophagus irregularis	-	-	-	-	*	*	*	[65]
Agaricus lilaceps	-	-	-	-	-	*	*	[5]
Paraglomus occultum	-	-	-	-	*	*	*	[66]
Glomus mosseae	*	*	-	-	*	*	*	[67]
Paxillus involutus	-	-	-	-	-	*	*	[68]
Aspergillus spp.	-	-	-	*	-	-	-	[69]
Rhizobacteria	*	-	-	-	*	-	-	[70]
Pseudomonas aeruginosa	*	*	-	*	-	-	-	[71]
Bradyrhizobium diazoefficiens	*	-	-	-	*	-	-	[72]
Pseudomonas alcaligenes	-	*	-	-	-	-	-	[73]

Microbe and Microbiomes in Aquatic and Coastal Ecosystem Restoration

Microbes and microbiomes are crucial in restoring and managing aquatic and coastal ecosystems [74]. In freshwater ecosystems, microalgae biofilms are essential as they contribute significantly to the primary productivity of shallow waters and stabilize sediments, encouraging the establishment of microbial communities that operate bio-geochemical cycles [75]. Sedimentary microbes also play vital roles in sustaining the functional resilience of aquatic ecosystems. Still, due to the environmental complexity, their taxonomic composition and community processes must be better understood in estuarine-coastal margins. However, investigations have revealed that the abundance, diversity, and composition of microbes in sediments vary spatially and are influenced by various factors such as salinity gradients [76].

Bacteria-derived vesicles have been notified to be abundant in coastal and open-ocean seawater and are implicated in marine carbon flux. However, investigations on bacteria-derived vesicles in freshwater ecosystems are limited. Microplastics, which act as novel substrates for microbial colonization, are a growing problem due to their potential to multiply foreign or invasive species across various aquatic ecosystems. Microbes on microplastics build biofilms, and salinity gradients and ecological processes influence their dynamics and capacity to displace microorganisms across diverse marine ecosystems [77]. When studying microbial communities in aquatic and coastal ecosystems for restoration purposes, it is essential to consider the risks of microbial contamination and take measures to mitigate them. Microbes are ubiquitous, and uncontrolled contamination can compromise sample integrity and research quality. Therefore, researchers should carefully plan and execute field sampling, transport, and storage of microbial samples to ensure accurate and reliable results [5].

The relationship between microbes and aquatic plants has been observed in coastal ecosystems. Cable bacteria, for instance, are associated with aquatic plant roots in diverse ecosystems, including marine coastal habitats, estuaries, freshwater streams, isolated pristine lakes, and intensive farming systems. This plant-microbe association is widespread and not species-specific, and it has implications for vegetation vitality, primary productivity, coastal ecosystem restoration practices, and gaseous balance. Anthropogenic activities, such as fishing, shipping, and tourism, can harm coastal ecosystems, including deterioration of water quality, habitat destruction, and biodiversity loss. Microbes in these ecosystems are sensitive to environmental changes and can respond to anthropogenic stress by activating various adaptation strategies. Understanding anthropogenic activities' impact on microbial communities is crucial for effective ecosystem restoration and management [78].

Coastal wetlands, which act as transitional zones between terrestrial and aquatic ecosystems, undergo habitat transformations that can affect soil microbial community system and their functions. The decline and deterioration of aquatic macrophytes in these wetlands can simulate root-linked microbial communities and the larger micro-eukaryotes that rely on these interactions. Further, coastal wetlands are significant sinks for nitrogen reduction, and microbial-mediated functions such as denitrification and anaerobic ammonium oxidation are paramount in decreasing nitrogen overload in estuarine and coastal ecosystems [79]. In ecosystem restoration, exploiting microbiomes, macroalgae, and seagrasses has been proposed as a nature-based solution to fight the negative impacts of global climate change and human perturbation on coastal ecosystems. These approaches can improve water quality, enhance carbon sequestration, and contribute to the restoration and conservation of coastal ecosystems [80].

SIGNIFICANCE OF MICROBIOME IN NATURAL RESOURCE MANAGEMENT

The microbiome refers to the microorganism community, which includes bacteria, fungi, and viruses that significantly participate in natural resource management. Their significance is shown in Fig. (2). The microbiome directly impacts plant health, productivity, and nutrient cycling. The composition and function of the microbiome are crucial for sustainable intensification strategies in agriculture, particularly in organic systems where natural resources are the primary source of crop growth, yield, and environmental sustainability [81]. Microbes play a significant role in renewable energy by employing cellulolytic fungal strains and fermenting yeast and bacteria [82]. Various types of microbes are reported to remove hexavalent chromium [Cr(VI)] form in both terrestrial and aquatic ecosystems [83]. Microbes play an essential role in biofertilizer formulations, which contain live or latent cells of effective microbial strains cultured in the lab and packed in appropriate carriers [84]. Organic manure amended with effective microbial biofertilizers is found to reduce greenhouse gases and global warming potential farming systems [85]. Microbes, particularly siderophores, are an efficient biomining agent in recovering rare earth elements (REEs) [86].

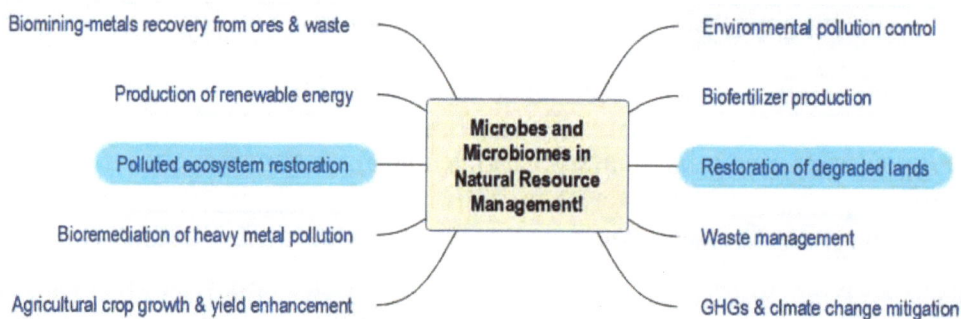

Biomining-metals recovery from ores & waste — Environmental pollution control

Production of renewable energy — Biofertilizer production

Polluted ecosystem restoration — **Microbes and Microbiomes in Natural Resource Management!** — Restoration of degraded lands

Bioremediation of heavy metal pollution — Waste management

Agricultural crop growth & yield enhancement — GHGs & climate change mitigation

Fig. (2). Microbes and microbiomes in natural resource management.

Role of Microbes and Microbiomes in Biomining

Biomining is a biotechnological process that employs microbes to retrieve metals and metalloids from ores and industrial and municipal waste materials. It has gained attention due to its prospect of addressing pressing issues linked to climate concerns, such as habitat devastation induced by mining effluent-related pollution, metal supply chains, and growing needs for cleantech-critical metals. However, biomining's drawbacks hinder its commercial applications, including prolonged processing periods, low recovery speeds, and narrow metal selectivity [87]. Microbes and microbiomes play a vital role in biomining operations. Specific microbial species, such as Geobacter and Shewanella, depend on extracellular electron transfer (EET) to decrease minerals and sustain growth, which has essential biotechnological applications in biomining. EET can also be harnessed to create biofuels and nano-materials [56]. These metal-reducing bacteria have extracellular electron transfer pathways that can route electrons across cell membranes to change the redox state of exogenous metals, allowing microbial-driven mineral conversions and the extraction of metals from ores, industrial and municipal waste [86].

Microbes employ microbial consortia rather than individual species in biomining, providing stable and efficient mineral degradation. These consortia consist of various classes of acidophilic prokaryotes, including Fe-oxidizing microbes, S-oxidizing microbes, and janitors that help to degrade organic carbon-containing compounds. The synergistic impact of these microbial consortia promotes sulfur-containing minerals' degradation and produces sulfuric acid, which sustains the optimum acidity levels for the consortium [88]. Advancements in synthetic biology present options to engineer iron/sulfur-oxidizing microbes to manage the limitations of biomining, such as low recovery rates and metal selectivity. Synthetic biology can improve the abilities of microbes applied in biomining operations, permitting the engineering of distinctive traits and functionalities. For instance, the engineering of polyhistidine tags on surface proteins of Acidithiobacillus ferrooxidans improved the metal binding ability of the microbes, extending further opportunities for metal bioseparation and their recovery [89]. The use of extremophiles, microbes that flourish in extreme conditions, is another area of interest in biomining and bio-leaching study [90]. Extremophiles have been studied for *in-situ* resource utilization in astrobiology, where biomining and bio-leaching processes extract resources from extra-terrestrial regolith and rocks. The science around space biomining is yet immature, and study is required to comprehend the consequences of space conditions on biomining methods and create specific strategies for space applications [91].

Microbiomes in Restoring Polluted Ecosystems and Environment

Microbes and microbiomes play an essential role in restoring polluted ecosystems and environments by adapting themselves to survive in perturbed ecosystems, resulting in environmental remediation. They are critical for removing heavy metals, pesticide microplastic, and PAH degradation [92]. Microbiomes play a crucial role in environmental pollution control. Various environmental factors, including air pollution, water pollution, soil contamination, and exposure to pollutants such as heavy metals and microplastics, can influence the composition and diversity of microbiomes [113]. Several studies have shown the potential of microbes to remediate organic and inorganic pollutants, as presented in Table **2**.

Table 2. Microbes involved in remediating organic and inorganic pollutants.

Microbes	Pollutant type	References
Pseudomonas, Acinetobacter Xanthomonas, Bacillus, Trichoderma and Penicillium spp.,	Petroleum hydrocarbon	[100, 101]
Kocuriarhizophila, Arthrobacter methylotrophs, Bacillus pp., Paenibacillus spp., Azospirillum brasilense, Aspergillus niger	Fertilizer	[102]
Rhodococcus erythropolis	Organics, nitrogen from landfill leachate	[103]
Streptomyces coelicolor	N – Hexadecane	[104]
Escherichia coli	Atrazine	[105]
Sphingomonas spp.	Hexachlorocyclohexane and methyl parathion	[106]
Acinetobacter sp., Sporosarcina saromensis, Bacillus cereeus, Bacillus circulans, Staphylococcus spp., Pseudomonas aeruginosa	Chromium	[107, 108]
Cellulosimicrobium spp., Methylobacterium organophilum, Bacillus firmus, Staphylococcus spp., Streptomyces spp.	Lead	[109]
Desulfovibrio desulfuricans, Flavobacterium sp., Bacillus firmus, Micrococcus spp., Pseudomoas spp., Acinetobacter spp.	Copper	[110, 111]
Enterobacter cloacae, Klebsiella pneunomiae, Pseudomonas aeruginosa, Vibrio parahaemolyticus, Bacillus licheniformis	Mercury	[112]

Phytoremediation, using plants and plant-associated microbes to remediate polluted sites, is a cost-effective and sustainable approach to ecological restoration [93]. Soil pollution significantly impacts biological diversity, ecosystem structure, and functioning [94]. Different bio-indicators assess contaminated ecosystems and habitat restoration, including microbial community analysis, microbial biomass, respiration, enzymatic activity, and molecular markers. Vergani and co-workers [95] employed DNA stable isotope probing to distinguish the effects of various biostimulant treatments on the structure of PCB-degrading bacteria populations in PCB-polluted soil. They found that plant biostimulants caused the enrichment of Actinobacteria and recognized the considerably abundant taxa involved in PCB degradation. Mine-site restoration constantly focuses on above-ground aspects of biodiversity and ecosystem functioning, ignoring the significance of the soil microbiome. D'Agui and co-workers [96] highlighted the necessity for a holistic strategy for mine-site restoration that considers the soil microbiome to achieve secure, sound, non-polluting terrains with comparable levels of biodiversity and ecosystem functioning to undisturbed sites.

Understanding the ecological roles of rare and abundant microbial taxa is crucial for restoring soil ecosystem functions in heavy metal-polluted soil. Chen and co-workers [97] assessed the efficiencies of different soil amendments on ecosystem recovery. They emphasized the contributions of rare and abundant microbial communities to ecosystem multi-functionality. Bacterial and fungal communities are essential elements of soil microbiomes and play paramount roles in ecosystem restoration. Addressing the challenges associated with bacterial and fungal communities in soil microbiomes is necessary for adequate ecosystem restoration. Shelyakin and co-workers [98] investigated the impact of kerosene leakage on the topsoil and found that the microbiome arrangement differed depending on the soil's physico-chemical properties.

Microbial analysis can serve as biomarkers and indicators for ecosystem restoration. Lian and co-workers [99] applied microbial and spatial analysis to identify wetland sediments affected by various fish farming and non-point source pollution. They demonstrated the potential of microbial information as biomarkers for environmental restoration. Water quality and microbial communities are closely related in aquatic ecosystems. The many studies provide valuable insights into the role of microbiomes in restoring polluted ecosystems. Manipulating plant microbiomes, phytoremediation, and biostimulation of microbial degraders are effective strategies for restoring contaminated sites. Understanding the responses of soil microbiomes to pollution and land-use changes is crucial for successful restoration. Research on the coral *Acropora tenuis* indicated that the microbiome associated with different coral genotypes was extremely host-genotype specific and upheld high composition stability irrespective of ecological fluctuations [94, 114].

Microbiomes can also be engineered for ecological use, such as in bioremediation, to degrade intractable contaminants and withstand engineered microbiomes' performance in fluctuating conditions [115]. The relationship between air pollution and microbiomes has been a topic of interest, particularly in respiratory health and disease. Exposure to air pollutants, like nitrogen oxides, carbon mono-oxides, ozone, and particulate matter, can alter the gut microbiota and boost obesity and type 2 diabetes *via* inflammatory pathways [116]. Air pollution can also influence the cutaneous microbiome, with pollutants like nitrogen dioxide negatively affecting the composition of skin microbiomes. Understanding the interactions between air pollution and microbiomes is vital for developing strategies to mitigate the health effects of pollution.

Microbes and Microbiomes in the Restoration of Farm and Degraded Lands

Microbes and microbiomes play a crucial role in restoring farms and degraded lands. Restoration efforts can provide a habitat for microbes, which can help lower extinction risks. Restoring high-diversity ecosystems on degraded and abandoned lands can increase carbon capture and sequestration, contributing to climate moderation [117]. One group of beneficial microbes that has gained attention in agricultural practices is cyanobacteria. Cyanobacteria have been found to enhance crop productivity and mitigate greenhouse gas emissions. They can fulfill their nitrogen requirements through nitrogen fixation, promoting crop growth and improving soil nutrient status [118]. Cyanobacteria have also been proposed as vital bio-agents in the ecological restoration of degraded lands. Soil microbial communities are critical players in connecting above-and-below-ground level terrestrial communities in ecosystems. Artificial revegetation, such as aerial seeding, has been shown to promote soil microbe restoration and maintain community diversity in degraded ecosystems. Restoring subsurface soil microbial communities is also vital to land restoration [87]. In soil microbial diversity and community structure, vegetation restoration in planted forests has been observed, highlighting the of soil microbial communities in ecosystem functioning.

Microbial inoculants, which contain beneficial soil microbes, have been used to increase plant growth and assist with the restoration of degraded dryland ecosystems. These inoculants allow land managers to enhance plant growth while avoiding the negative impacts of fertilizer application [119]. Using native grassland soils as inoculants has been shown to improve native plant species germination in highly disturbed soil, emphasizing the importance of re-establishing native soil microbiomes for successful restoration. Restoration strategies that revive biodiversity loss can ameliorate soil and balance ecosystems and human health in rural and urban areas. Degraded soil microbes have been seen to influence human health and rejuvenate soil biodiversity.

Microbes also can assist in improving soil fertility and ecosystem functioning. This approach is especially appropriate in arid, semiarid, and dryland areas that laboriously depend on microbial communities for maintaining high productivity *via* nutrient cycling [120]. The role in restoring degraded lands is also apparent in post-mining rehabilitation. Symbiotic microbes, like mycorrhiza, help to improve soil quality, seedling survival, and plant growth in coal-mined polluted areas. Typical phosphate (P) solubilizing bacteria have also been seen to enhance soil P-cycling after restoring laboriously degraded mined areas [121]. Comprehending the interplay between desert plants and soil microbes is essential for restoring shrub ecosystems in arid regions. The plant-soil microbes interaction can enhance degraded land restoration and boost the overall productivity of desert ecosystems. Soil fungi, in particular, play vital roles in maintaining ecosystem processes and functions.

Bioremediation Approach for Heavy Metal Pollution Control

Bioremediation is a highly effective approach for controlling environmental heavy metal pollution. It utilizes biological mechanisms, such as microorganisms and plants, to remove or recover heavy metals from polluted sites. Bioremediation is more environmentally friendly and cost-effective than conventional chemical and physical methods, especially for low-metal concentrations [122]. Depending on the application site, Different bioremediation techniques and approaches can be used. *in situ* bioremediation techniques, such as bioventing, biosparging, and phytoremediation, have successfully treated heavy metal-polluted areas.

These techniques rely on the natural abilities of microorganisms and plants to degrade or accumulate heavy metals in the environment. Microorganisms play a crucial role in bioremediation by utilizing various mechanisms to adapt to the toxicity of heavy metals. They can degrade heavy metals through enzymatic reactions or bind them to their cell surfaces through biosorption. The effectiveness of microbial bioremediation is influenced by factors, such as the concentration of toxic compounds, pH, contact time, and temperature. Optimization of these conditions can enhance the biosorption and bioremediation of heavy metals [123, 124].

In recent years, there has been increasing research on the application of synthetic biology in bioremediation. Synthetic biology allows for the design and construction of organisms with enhanced capabilities for heavy metal removal [125]. This approach involves using microbial biosensors for detection, phytoremediation for metal removal, and heavy metal recovery. Synthetic biology offers new possibilities for improving the efficiency and effectiveness of bioremediation strategies [126]. However, it is crucial to consider factors such as the suitability of environmental conditions, adherence to biosafety regulations, and the selection of appropriate microorganisms for practical bioremediation. Continued research and development in bioremediation techniques, including synthetic biology, will further enhance its potential for heavy metal pollution control.

Microbes in the Production of Renewable Energy

Microbes play an essential role in producing renewable energy for microbial fuel cells (MFCs), where bacteria are used as catalysts to oxidize organic and inorganic matter and generate current. The bacteria consume organic matter as their food source, breaking it into simpler compounds such as carbon dioxide, protons, and electrons. The electrons are then transferred to an electrode, where they can be used to generate electricity [127]. Microbes play an essential role in sustainable bioenergy and biofuel production. Microbes can also be used to produce biofuels through microbial fermentation. Microbial fermentation is how microorganisms convert organic compounds into usable energy sources such as ethanol or methane. The process produces biofuels from various feedstocks, including corn, sugarcane, switchgrass, algae, and other biomass sources.

One example is the engineering of *Pichia pastoris*, an industrial yeast, to produce the biofuel isobutanol from C-sources like pentose, hexose, and glycerol [128]. Researchers increased the production of isobutanol by using the yeast's amino acid biosynthetic pathway and diverting the amino acid intermediates to the 2-keto acid degradation pathway. Anaerobic digestion is another microbial process that helps to produce renewable energy. It relies on the interactions between fermentative and syntrophic bacteria and methanogens to decompose organic compounds into CO_2 and CH_4, which can be used as an energy carrier [46, 82]. Metagenomics is a powerful tool in the study of microbial biofuel production. It allows for the analysis of functional genes and the discovery of new enzymes essential for the large-scale production of biofuels from plant biomass [129].

Microbiomes in Agriculture Crop Growth and Development

Microbiomes are crucial in agriculture, impacting plant growth, health, and productivity. Microorganisms, including bacteria, fungi, archaea, and viruses, reside in and on plants and the surrounding soil [6, 84, 130]. These microorganisms interact with plants and each other, influencing nutrient cycling, disease resistance, stress tolerance, and overall ecosystem functioning. An example of microbiomes that play an essential role in alleviating abiotic stress (cold, heat, drought, and salinity) on crops is shown in Table **3**. Comprehending the ecological aspects of the rumen microbes could lead to improved food resources and eco-friendly livestock-based agriculture [131].

The root-associated microbes greatly influenced plant health, growth, and yield [6]. Various agriculture systems, *i.e.*, conventional, no-till, and organic farming, can affect the structure and complexity of root microbes. Investigations have shown that organic agriculture harbors a better complex of fungi with higher connectivity than conventional and no-till agriculture systems [62]. The abundance of keystone taxa is essential for ecosystem resilience and functioning. They are also found higher under organic farming, where farm intensification is lower. Soil properties, bulk density, pH, and mycorrhizal colonization can impact the occurrence of keystone taxa. These findings highlight the importance of considering farming practices and their impact on the root microbiome for sustainable agriculture [132].

Table 3. Microbiome in alleviating abiotic stress on crops.

Microbes	Crop	References
Cold and Freezing stress	-	-
Pseudomonas fluorescens A506	Pear and Apple	[137]
S. marcescens SRM, Pseudomonas spp. and Pantoeadispersa 1A	Wheat	[138]
Paraburkholderia phytofirmans	Grapevine	[103]
Pseudomonas chlororaphis, Pseudomonas fluorescens, Pseudomonas fragi, Brevibacterium frigoritolerans, and Pseudomonas proteolytica	Bean	[139]
Heat stress	-	-
Pseudomonas putida AKMP7 and Pseudomonas sp. AKMP6	Wheat and Sorghum	[112]
Azospirillum brasilense and Bacillus amyloliquefaciens	Wheat	[140]
Bacillus aryabhatthai SRBO2	Soybean	[110]
Glomus sp	Tomato	[141]
Paraburkholderia phytofirmans	Potato	[142]
Drought	-	-
Bacillus thuringiensis AZP2	Wheat	[143]
Funneliformis mosseae	Orange	[144]
Staphylococcus sp.	Mung bean	[145]
Pseudomonas libanensis EU-LWNA-33	Wheat	[146]
Penicillium sp. strain EU-DSF-10 and Streptomyces laurentii EU-LWT3-69	Millet	[146]
Salinity	-	-
Leclercia adecarboxylata MO1	Tomato	[147]
Paenibacillus sp. and Aneurinibacillus aneurinilyticus	French bean	[148]
Bacillus spp.	Potato	[149]

The plant genotype and soil type are important factors that shape the assembly of the plant rhizosphere microbiome. The rhizosphere microbiome refers to the microorganisms that colonize the root surface and the surrounding soil. Studies have shown that both soil type and plant genotype can cooperatively modulate microbiome assembly, with soil type predominantly shaping rhizosphere microbiome assembly. Different plant genotypes can have varying levels of rhizosphere diversity, with undomesticated progenitor species often having higher diversity than domesticated genotypes. Specific microbial taxa, such as Rhizobium, Novosphingobium, Phenylobacterium, Streptomyces, and Nocardioides, are associated with the soybean rhizosphere microbiome. These findings highlight the complex interactions between plants, soil, and microorganisms in shaping the rhizosphere microbiome [133].

Plant secondary metabolites (PSMs) and their interactions with the plant microbiome are another vital aspect of agriculture. PSMs are chemical compounds produced by plants that play various roles, including defense against pathogens, pests, and herbivores and response to environmental stresses. The plant microbiome can influence the production and composition of PSMs, and in turn, PSMs can shape the composition and function of crop-associated microbiomes. Studies have shown that plants can secrete metabolites that influence the composition of their microbiome, and the microbiome can also impact the metabolome of the host plant. Understanding the interactions between PSMs and the plant microbiome can have implications for sustainable crop production and the development of methods to manipulate these interactions for desired outcomes [134].

Fungi are crucial in the soil's cycling of matter and energy and are involved in plant health and nutrient uptake. Certain fungal species, such as mycorrhizal fungi, can form symbiotic associations with plants and enhance nutrient uptake [135]. The plant microbiome can help suppress soil-borne pathogens and improve plant disease resistance. For example, Fusarium wilt disease has been found to substantially impact the microbiome of vegetative organs (roots and stems) more than reproductive organs (fruits). Fungal communities in the microbiome are often more sensitive to disease-induced changes than bacterial communities. Manipulating the plant microbiome can enhance plant growth-promoting mechanisms and improve plant tolerance to harsh environments. Understanding the interactions between plants and their associated microbial communities can inform strategies for enhancing microbial functions that support plant growth and production [136].

Microbes and Microbiomes for Biofertilizer Production

Biofertilizers, based on specific strains of plant growth-promoting (PGP) bacteria, have emerged as an alternative to chemical fertilizers in recent decades [6, 84]. These biofertilizers interact with the root microbiome and root-associated microbes to colonize roots and benefit plants. The rhizosphere's bacterial and fungal communities composition can be manipulated by biofertilizer application, which may contribute to disease suppression, nitrogen fixation, phosphate solubilization, produce phytohormone [150], increase crop yield, and maintain long-term soil fertility. Certain microorganisms in biofertilizers solubilize insoluble or complex forms of essential nutrients, making them available to plants [102]. Efficient, beneficial microbial strains for use as biofertilizers are often screened from the rhizosphere, and they possess plant growth-promoting (PGP) traits and practical colonization ability.

PGP microbes decompose organic matter, enhance nutrient availability, produce phytohormones, and mitigate abiotic and biotic stresses [151]. The presence of phosphate-solubilizing microbial populations in soils is a positive indicator for utilizing these microbes as biofertilizers for sustainable agriculture [152]. However, various factors influence microbe-based biofertilizers, including the targeted crop, soil conditions, competition with indigenous strains, microbial parasites and predators, and climatic factors. Biofertilizers have been shown to increase cucumber growth, decrease soil-borne pathogens, and alter microbial diversity and taxonomic structure in the rhizosphere. These findings suggest that biofertilizers can produce high-quality crops cost-effectively while controlling soil-borne diseases [74].

Microbiome in Mitigating Ghg Emission and Climate Change

Microorganisms mitigate greenhouse gas (GHG) emissions and climate change [79]. They provide the opportunity for climate change mitigation by sequestering atmospheric carbon into stable reservoirs such as soil. The host genome also influences the microbiome's function, including its role in methane emissions, highlighting the potential for genetic manipulation to mitigate GHG emissions [153]. Methanotrophic bacteria are reported to cause methane consumption or oxidation under both anoxic and oxic conditions, acting as a sink for atmospheric methane [64]. The rhizosphere microbiome, influenced by root exudation patterns, can contribute to plant health and resilience to climate change. The integrated application of extreme environments-selected prebiotics and probiotics in the rhizosphere has the potential to generate a resilient holobiont that can mitigate the adverse effects of climate change on crops.

Microbial inoculants have the potential to reshape soil microbiomes and influence agroecosystem functions relevant to climate change adaptation and mitigation, such as soil organic matter turnover, nutrient cycling, and GHG regulation [154]. Soil microbiome can curb soil GHG emissions under global change conditions [155]. The microbiome is central to mitigating GHG emissions and climate change. Understanding how microorganisms affect climate change and how they will be affected by climate change and human activities is crucial for achieving an environmentally sustainable future. Manipulating the microbiome through environmental microbiome engineering, microbial inoculants, and plant-microbe interactions offers promising strategies for climate change mitigation. However, further research is needed to fully understand the complex interactions between microorganisms, climate change, and GHG emissions and to develop effective strategies for harnessing the potential of the microbiome in mitigating climate change.

Microbes and Microbiomes in Waste Management

Microbes and microbiomes play a crucial role in waste management, particularly in bioremediation, composting, and the degradation of organic waste. Municipal solid waste (MSW) landfills are the most prevalent waste disposal method. Harbour diverse microbial communities that contribute to the decomposition of waste and the production of greenhouse gases. The microbial diversity in landfills can vary with depth, with different phyla and genera dominating different layers of the landfill. For example, Firmicutes, Proteobacteria, and Bacteroidetes are commonly found in landfill microbial communities.

These microbial communities are responsible for organic matter degradation. Microbes such as bacteria, fungi, protozoa, and actinomycetes are also involved in the biodegradation of agricultural waste, including cellulose, lignin, chitin, keratin, and pectin. These microbes are significant in breaking down farm waste and preventing environmental pollution. Microbial cultures can be added to agricultural waste to facilitate the breakdown process and create compost, which can then be used as a bio-organic fertilizer. This approach helps manage waste and promotes an eco-friendly environment [119].

Microplastic pollution poses significant hazards to the environment and organisms. The ingestion of microplastics by organisms, transfer of toxic chemicals, and disruption of ecological processes. Mitigation efforts are necessary to minimize these hazardous pollutants to protect ecosystems and living beings [124]. Many microbes in plastic waste degradation produce vital enzymes in plastic degradation, as shown in Table **4**. The application of microbial species could lead to more efficient and effective plastic waste management strategies.

Composting is another essential process facilitated by microbial activities under aerobic or anaerobic conditions for efficient waste management. Various microbes and their metabolites contribute to the composting process, *e.g.*, β-glucosidas--produced by microbial communities during cattle manure-rice straw composting plays a crucial role in cellulose degradation, a vital part of the global carbon cycle. The dynamics of microbial communities during the co-composting of different types of waste, such as swine and poultry manure with spent mushroom substrates, have been studied by Lin and co-workers [156]. This study provides insights into the microbial processes of composting and can help optimize waste management practices. Microbial communities contribute to the breakdown and transformation of organic matter, the release of nutrients, and the production of stable compost. Understanding these microbial communities' dynamics and their metabolic activities can help optimize composting processes and improve the quality of the resulting compost.

Table 4. Microbes and their enzymes involved in plastic degradation.

Microbes	Enzymes	References
Alcanivorax spp. 24	Alkane hydroxylase and Alkane monooxygenases	[158]
Desulfobulbaceae, Desulfobacteraceae and Desulfovibrio	Adenyl sulfate reductase, depolymerase, and sulfite reductase	[159]
Thermobifidafusca	Hydrolases	[160]
Moritella spp.	Lipases	[161]
P. chlororaphis	Polyurethenase	[70]
Comamonas testoteroni, Alcaligenes faecalis, Pseudozyma antarctica	Depolymerase	[125, 159, 162]

In the context of waste management, consider the potential risks associated with microbial activities. For example, landfill sites can release bioaerosols containing pathogenic microbes, posing health risks to employees and nearby communities. The microbial diversity in landfill air samples has been studied, and various pathogenic microbes have been identified, including *Enterobacteriaceae, Staphylococcus aureus, Clostridium perfringens, Acinetobacter calcoaceticus*, and *Aspergillus fumigatus*. These bioaerosols can penetrate the respiratory system and cause respiratory symptoms and chronic pulmonary diseases [145]. In waste management, ongoing research is also on using genetically modified organisms (GMOs) for enhanced degradation strategies. GMOs can potentially improve waste management practices, but there are concerns about the unintended effects of gene transfer to other microbes. In order to address these concerns, the development of bioluminescent suicidal GMOs has been proposed [157]. Further research is needed to explore the use of GMOs in waste management and address the associated challenges.

MOLECULAR TECHNIQUES IN ECOSYSTEM RESTORATION

Molecular techniques play a crucial role in ecosystem restoration by providing valuable insights into the processes and mechanisms involved in restoring degraded ecosystems. These techniques enable scientists and practitioners to assess the effectiveness of restoration methods, monitor the recovery of ecosystems, and make informed decisions for successful restoration outcomes. High-throughput amplicon sequencing, a culture-independent molecular technique, has been developed to determine the composition of microbial communities in various environments, including soil ecosystems [144]. This technique identifies dominant microbial groups and their dynamics during natural succession, providing valuable information for developing strategies to restore degraded soil ecosystems.

Furthermore, molecular techniques have been used during ecological restoration to study the interactions between plants and beneficial microorganisms, such as arbuscular mycorrhizal (AM) fungi. AM fungi form mutualistic associations with plant roots, enhancing nutrient uptake and promoting plant growth and resilience. Advances in molecular techniques have allowed for a deeper understanding of these plant-microbe interactions and their potential benefits for ecosystem recovery. By considering the potential benefits of mutualistic interactions during ecological restoration, such as associations between plants and AM fungi, restoration practitioners can improve ecosystem recovery and enhance restoration success [19].

Molecular techniques are also valuable in assessing restoration efforts' success and monitoring ecosystem recovery. For example, DNA sequencing technologies have enabled the analysis of soil microbial community (SMC) data, which can be used to monitor ecological restoration trajectories. By examining the composition of SMCs, scientists can evaluate the effectiveness of restoration practices and guide restoration efforts. However, it is essential to note that interpreting SMC data and its application in guiding restoration practice remains an ongoing research and debate. Complementing molecular approaches with non-molecular approaches can enhance the utility of species inventory data for ecological restoration.

In addition to assessing microbial communities, molecular techniques have been used to study the genetic diversity of plant and animal populations in the context of conservation translocations, which are widely used management tools for species and ecosystem restoration. Genetic and genomic approaches can inform conservation management by revealing the resistance and resilience of species to environmental change and guiding seed-sourcing strategies for restoration. These techniques can also monitor restored ecosystems for invasive species and reconstruct historical baselines for reference points in restoration efforts [9].

Overall, molecular techniques have revolutionized the field of ecosystem restoration by providing valuable insights into the processes and mechanisms involved in restoring degraded ecosystems. These techniques enable scientists and practitioners to assess the effectiveness of restoration methods, monitor ecosystem recovery, and make informed decisions for successful restoration outcomes. By leveraging the power of molecular techniques, ecosystem restoration efforts can be more targeted, efficient, and effective in achieving their goals.

THE PROSPECT OF MICROBE AND MICROBIOMES IN ECOSYSTEM RESTORATION

Microbes and microbiomes play a crucial role in ecosystem restoration. They contribute to nutrient cycling, decomposition of organic matter, and the overall functioning of terrestrial and aquatic ecosystems. Studies have shown that the composition and diversity of soil microbiomes change with soil depth and restoration progress. Bacterial and fungal diversity in reforestation sites decreases with increasing soil depth, while archaeal diversity increases. As reforestation proceeds, the vertical variation in bacterial communities decreases while that in archaeal and fungal communities increases. Soil properties and plant roots influence these changes in microbiome composition. Bacterial and archaeal communities play significant roles in nutrient cycling in deep and superficial soil layers.

The interplay between plants and soil microbiomes is crucial for ecosystem restoration. Desert plants and soil microbiomes have complex interactions that must be fully understood. However, an improved understanding of these interactions can enhance degraded land restoration programs and increase the productivity of desert ecosystems. Soil fungi contribute to nutrient cycling, organic matter decomposition, and plant growth promotion [99]. Microbial contamination is a potential risk in microbiota restoration studies. Microbes are ubiquitous, and uncontrolled contamination can compromise the integrity of samples and the validity of research findings. Restoration ecologists must be aware of and manage these risks to ensure the quality of microbial samples and the accuracy of study results. Proper study design, field sampling techniques, and sample transport and storage protocols can help mitigate microbial contamination risks [5].

Microbial rewilding has been proposed to restore and enhance microbial communities in various ecosystems, including the gut microbiome. Rewilding involves promoting diverse environmental microbiomes to improve host-microbe symbiosis and adaptability to new environments. This approach can benefit captive animals transitioning between settings or ecosystems and promote resilient natural ecosystems [103]. Incorporating soil microbes into restoration projects has become increasingly important. Soil microbes shape plant health and ecosystem function, and their inclusion in restoration efforts can improve the success of revegetation and native plant establishment. Commercial microbial products are available for agricultural applications, and their use in restoration projects is being explored. These products can enhance plant-microbe interactions and support ecosystem functions and services [112].

Overall, the prospects of microbes and microbiomes in ecosystem restoration are promising. Understanding the dynamics of soil microbiomes during restoration, managing microbial contamination risks, and promoting microbial rewilding can contribute to successful ecosystem recovery and enhance the health and functioning of restored ecosystems. Incorporating soil microbes into restoration projects and utilizing microbial products can improve the establishment of native plants and support ecosystem functions and services. Further research is needed to explore the specific roles of different microbial taxa and their interactions in ecosystem restoration.

CHALLENGES WITH MICROBES AND MICROBIOMES IN ECOSYSTEM RESTORATION

Microbial ecosystem restoration is critical to ecological restoration efforts to mitigate global biodiversity loss and restore degraded ecosystems. The assembly of below-ground microbial communities is crucial in the restoration process. Understanding the challenges associated with microbial ecosystem restoration is essential for optimizing restoration strategies and achieving successful outcomes. One of the significant challenges in microbial ecosystem restoration is the narrow insight into processes of distinct soil microbes during the natural restoration of ex-arable ecosystems. Another challenge is understanding vital microbial taxa and functional activities in natural and restored ecosystems. Lambin and co-workers [54] examined the effect of 30 years of carefully managed restoration on soil microbial communities in tallgrass prairies. They found that older restorations had communities statistically distinct from newer restorations, and these communities converged toward those in local prairie remnants. The recovery of microbial clades within the Verrucomicrobia and Acidobacteria was a vital characteristic of this convergence, indicating that plant-focused restoration has yielded soil bacterial communities reflective of a successful restoration. Study findings underline the significance of considering microbial communities as indicators of restoration success and targeting specific microbial groups for soil-focused restoration measures [2].

Soil degradation induced by anthropogenic disruptions and climate change poses another challenge to microbial ecosystem restoration. Numerous investigations have demonstrated the efficacy of introducing microbial communities in degraded soil ecosystems to restore N-cycling. However, these investigations also reveal the complexity of microbial community dynamics and the challenges associated with restoring ecosystem functions. These results indicate that exploiting microbial communities for soil restoration needs a complete understanding of the underlying processes and the precise conditions of the degraded ecosystem [130]. The persistence of soil legacies and their impact on restoration projects is another challenge in microbial ecosystem restoration. Soil legacies, which include chemical composition and microbial community abundance and composition, are difficult to predict and assess. The persistence of soil legacies depends on the characteristics of the invaded ecosystem, further complicating restoration efforts. Considering soil legacies in restoration projects is crucial for understanding the dynamics of microbial communities and their role in ecosystem recovery [163].

Vegetation restoration plays a significant role in shaping rhizosphere microbial communities. Kong and co-workers [163] investigated the rhizosphere microbial diversity and community composition in coniferous forests across a chronosequence of restored lands. They found significant differences in soil bacterial and fungal communities among stand ages, with bacterial diversity decreasing and fungal diversity increasing with stand development. The specific bacterial taxa, such as Acidobacteria relative abundance, varied with stand age, highlighting the importance of vegetation restoration in shaping microbial communities. These findings emphasize the need to consider the dynamics of rhizosphere microbial communities in restoration efforts.

The linkages between plant and soil microbial diversity pose another challenge in restoring microbial ecosystems. Xu and co-workers [164] conducted a study investigating soil microbial diversity during 30 years of grassland restoration. They found that below-ground microbial communities profoundly impact ecological restoration by improving restoration outcomes and accelerating ecological succession. Soil microbial communities can respond rapidly to alterations in local environmental conditions induced by ecological restoration, making them valuable indicators for monitoring the restoration process. These findings highlight the importance of considering plant-microbe interactions and the role of microbial communities in driving ecosystem recovery.

Fire is an ecological disturbance that alters soil microbiomes and the functions they mediate in terrestrial ecosystems. Méndez-Albores and co-workers [165] studied the recovery of microbial processes after a fire. They found that microbial functions related to organic matter decomposition and nutrient cycling showed recovery patterns over time. Abiotic and biotic drivers such as soil nitrogen concentration and microbial diversity influenced microbial function recovery. These findings suggest that the long-term recovery of soil biodiversity creates resilience to restore essential ecosystem functions after fire.

Controlling complex microbial communities is another challenge in microbial ecosystem restoration. Kumar and co-workers [166] developed a control framework based on structural accessibility to identify minimum sets of driver species that can control the whole microbial community. Their framework provides a systematic pipeline to drive complex microbial communities toward desired states efficiently. Understanding the mechanisms underlying soil microbial community dynamics and the interactions between microbial communities is essential for effectively controlling and manipulating microbial communities during restoration efforts.

As stated above, microbial ecosystem restoration faces several challenges, including a limited understanding of small-scale spatial assembly processes, key microbial taxa and functional activities, soil degradation, soil legacies, biogeographic patterns, plant-microbe interactions, fire disturbance, microbial community assembly processes, and microbial manipulation strategies. Overcoming these challenges requires a thorough understanding of microbial community dynamics, ecosystem functioning, and the interactions between microbial communities and their environment. Integrating microbial ecology into restoration ecology can enhance restoration outcomes and contribute to the conservation and restoration of biodiversity and ecosystem services.

CONCLUDING REMARKS

In conclusion, microbes and microbiomes are crucial in restoring terrestrial, aquatic, and coastal ecosystems. Understanding the assembly processes, dynamics, and interactions of soil and marine microbiomes during restoration is essential for successful ecological restoration. Incorporating microbial components and considering below-ground microbial communities can improve restoration and accelerate ecological succession. However, contamination risks should be adequately managed to ensure the reliability of microbiome restoration studies. Understanding the variations in microbial diversity and community composition within degraded landscapes is crucial for practical restoration efforts. Incorporating microbial community assembly processes into restoration planning and management strategies can optimize restoration outcomes. Managing microbial contamination risks is essential to ensure the integrity of microbiome restoration studies. Soil microbial communities, including fungi and bacteria, serve as indicators of ecosystem health and can be used to evaluate the success of restoration efforts. Restoring the native microbiome, particularly AM fungi, can enhance plant establishment and growth in restoration projects. Harnessing the potential of microbes and microbiomes can contribute to sustainable and successful ecosystem restoration. Overall, restoring and rehabilitating microbiomes are essential tools to mitigate ecosystem decline and support organismal and ecosystem resilience in the face of environmental challenges. Overall, the prospects of microbe and microbiome research are vast and hold great potential for advancing our understanding of microbial communities and their interactions with the environment and host organisms. Continued research in these areas can lead to the development of innovative strategies for disease prevention, sustainable agriculture, environmental conservation, and biotechnological applications.

ACKNOWLEDGEMENTS

The Authors thank the Division of Environmental Science, ICAR-*Indian Agricultural Research Institute, New Delhi, India*, for their continual encouragement and unwavering support.

REFERENCES

[1]   Jiao S, Chen W, Wang J, Du N, Li Q, Wei G. Soil microbiomes with distinct assemblies through vertical soil profiles drive the cycling of multiple nutrients in reforested ecosystems. Microbiome 2018; 6(1): 146.
[http://dx.doi.org/10.1186/s40168-018-0526-0] [PMID: 30131068]

[2]   Kiyasudeen SK, Ibrahim MH, Quaik S, Ahmed Ismail S, Ibrahim MH, Quaik S, *et al.* An introduction to anaerobic digestion of organic wastes. Prospect Org Waste Manag Significance Earthworms 2016; pp. 23-44.

[3]   Köhl M, Ehrhart HP, Knauf M, Neupane PR. A viable indicator approach for assessing sustainable forest management in terms of carbon emissions and removals. Ecol Indic 2020; 111: 106057.
[http://dx.doi.org/10.1016/j.ecolind.2019.106057]

[4]   Marasco R, Ramond JB, Van Goethem MW, Rossi F, Daffonchio D. Diamonds in the rough: Dryland microorganisms are ecological engineers to restore degraded land and mitigate desertification. Microb Biotechnol 2023; 16(8): 1603-10.
[http://dx.doi.org/10.1111/1751-7915.14216] [PMID: 36641786]

[5]   Cando-Dumancela C, Davies T, Hodgson RJ, *et al.* A practical guide for restoration ecologists to manage microbial contamination risks before laboratory processes during microbiota restoration studies. Restor Ecol 2023; 31(1): e13687.
[http://dx.doi.org/10.1111/rec.13687]

[6]   Prasad Shiv, Malav Lal Chand, Choudhary Jairam, *et al.* Soil microbiomes for healthy nutrient recycling. Current trends in microbial biotechnology for sustainable agriculture, 2021; 1-21.
[http://dx.doi.org/10.1007/978-981-15-6949-4_1]

[7]   Graham EB, Knelman JE. Implications of soil microbial community assembly for ecosystem restoration: patterns, process, and potential. Microb Ecol 2023; 85(3): 809-19.
[http://dx.doi.org/10.1007/s00248-022-02155-w] [PMID: 36735065]

[8]   Rebollido R, Martinez J, Aguilera Y, Melchor K, Koerner I, Stegmann R. Microbial populations during composting process of organic fraction of municipal solid waste. Appl Ecol Environ Res 2008;

6(3): 61-7.
[http://dx.doi.org/10.15666/aeer/0603_061067]

[9]   Huet S, Romdhane S, Breuil MC, *et al.* Experimental community coalescence sheds light on microbial interactions in soil and restores impaired functions. Microbiome 2023; 11(1): 42.
[http://dx.doi.org/10.1186/s40168-023-01480-7] [PMID: 36871037]

[10]   Bhaduri D, Sihi D, Bhowmik A, Verma BC, Munda S, Dari B. A review on effective soil health bio-indicators for ecosystem restoration and sustainability. Front Microbiol 2022; 13: 938481.
[http://dx.doi.org/10.3389/fmicb.2022.938481] [PMID: 36060788]

[11]   Koziol L, Bauer JT, Duell EB, *et al.* Manipulating plant microbiomes in the field: Native mycorrhizae advance plant succession and improve native plant restoration. J Appl Ecol 2022; 59(8): 1976-85.
[http://dx.doi.org/10.1111/1365-2664.14036]

[12]   Saputra I, Ferry , Rahmawati D, Aulia SS, Sidik H. The Impact of Sand Mining on Socio-Economic and Environmental Sectors: A Case Study on Sedau Village, Narmada District, West Lombok Regency, Indonesia. IOP Conf Ser Earth Environ Sci 2023; 1175(1): 012022.
[http://dx.doi.org/10.1088/1755-1315/1175/1/012022]

[13]   Aigbedion I, Iyayi SE. Environmental effect of mineral exploitation in Nigeria. International journal of physical sciences. 2007, 1;2(2):33-8.

[14]   Tsukerman VA, Ivanov SV. Problems of Reducing Air Pollution from Industrial Enterprises in the Arctic Regions. IOP Conf Ser Earth Environ Sci. IOP Publishing. 2022; p. 988: 32006.
[http://dx.doi.org/10.1088/1755-1315/988/3/032006]

[15]   Dong J, Meng L, Bian Z, Fang A. Investigating the characteristics, evolution and restoration modes of mining area ecosystems. Pol J Environ Stud 2019; 28(5): 3539-49.
[http://dx.doi.org/10.15244/pjoes/97390]

[16]   Hempel CA, Buchner D, Mack L, Brasseur MV, Tulpan D, Leese F, *et al.* Predicting environmental stressor levels with machine learning: a comparison between amplicon sequencing, metagenomics, and total RNA sequencing based on taxonomically assigned data. BioRxiv 2022.
[http://dx.doi.org/10.1101/2022.11.18.517107]

[17]   Blanco G, Romero-Vidal P, Tella JL, Hiraldo F. Novel food resources and conservation of ecological interactions between the Andean Araucaria and the Austral parakeet. Ecol Evol 2022; 12(10): e9455.
[http://dx.doi.org/10.1002/ece3.9455] [PMID: 36311393]

[18]   Amosova AV, Zoshchuk SA, Rodionov AV, *et al.* Molecular cytogenetics of valuable Arctic and sub-Arctic pasture grass species from the Aveneae/Poeae tribe complex (Poaceae). BMC Genet 2019; 20(1): 92.
[http://dx.doi.org/10.1186/s12863-019-0792-2] [PMID: 31801460]

[19]   Curley A. *T'áá hwó ají t'éego* and the Moral Economy of Navajo Coal Workers. Ann Am Assoc Geogr 2019; 109(1): 71-86.
[http://dx.doi.org/10.1080/24694452.2018.1488576]

[20]   Ke X, van Vliet J, Zhou T, Verburg PH, Zheng W, Liu X. Direct and indirect loss of natural habitat due to built-up area expansion: A model-based analysis for the city of Wuhan, China. Land Use Policy 2018; 74: 231-9.
[http://dx.doi.org/10.1016/j.landusepol.2017.12.048]

[21]   Trisos CH, Merow C, Pigot AL. The projected timing of abrupt ecological disruption from climate change. Nature 2020; 580(7804): 496-501.
[http://dx.doi.org/10.1038/s41586-020-2189-9] [PMID: 32322063]

[22]   Runting RK, Bryan BA, Dee LE, *et al.* Incorporating climate change into ecosystem service assessments and decisions: a review. Glob Change Biol 2017; 23(1): 28-41.
[http://dx.doi.org/10.1111/gcb.13457] [PMID: 27507077]

[23]   Pang Z, Chen J, Wang T, *et al.* Linking plant secondary metabolites and plant microbiomes: a review.

Front Plant Sci 2021; 12: 621276-9.
[http://dx.doi.org/10.3389/fpls.2021.621276] [PMID: 33737943]

[24]   Çamur D, Köker L, Ayça O, Akçaalan R, Albay M. The Effects of Climate Change on Aquatic Ecosystems in Relation to Human Health. Aquat Sci Eng 2022; 37: 123-8.

[25]   Cheng B, Li H. Impact of climate change and human activities on economic values produced by ecosystem service functions of rivers in water shortage area of Northwest China. Environ Sci Pollut Res Int 2020; 27(21): 26570-8.
[http://dx.doi.org/10.1007/s11356-020-08963-2] [PMID: 32372355]

[26]   TSOMB EIBT, Atangana HO. Do Multilateral Environmental Agreements Explain Vulnerability to Climate Change in Developing Countries? Research Square; 2022.
[http://dx.doi.org/10.21203/rs.3.rs-2317636/v1.]

[27]   Sures B, Nachev M, Selbach C, Marcogliese DJ. Parasite responses to pollution: what we know and where we go in 'Environmental Parasitology'. Parasit Vectors 2017; 10(1): 65.
[http://dx.doi.org/10.1186/s13071-017-2001-3] [PMID: 28166838]

[28]   Wen Y, Schoups G, van de Giesen N. Organic pollution of rivers: Combined threats of urbanization, livestock farming and global climate change. Sci Rep 2017; 7(1): 43289.
[http://dx.doi.org/10.1038/srep43289] [PMID: 28230079]

[29]   Duran R, Cravo-Laureau C. Role of environmental factors and microorganisms in determining the fate of polycyclic aromatic hydrocarbons in the marine environment. FEMS Microbiol Rev 2016; 40(6): 814-30.
[http://dx.doi.org/10.1093/femsre/fuw031] [PMID: 28201512]

[30]   Xiong W, Ni P, Chen Y, Gao Y, Shan B, Zhan A. Zooplankton community structure along a pollution gradient at fine geographical scales in river ecosystems: The importance of species sorting over dispersal. Mol Ecol 2017; 26(16): 4351-60.
[http://dx.doi.org/10.1111/mec.14199] [PMID: 28599072]

[31]   Onyena A, Aniche D, Ogbolu B, Rakib M, Uddin J, Walker T. Governance strategies for mitigating microplastic pollution in the marine environment: a review. Microplastics 2021; 1(1): 15-46.
[http://dx.doi.org/10.3390/microplastics1010003]

[32]   Linh NT, Tue NT, Nhuan MT. Assessing coral reef resilience for sustainable resource management (case study in Hon La island, Quang Binh province, Vietnam). Tạp chí Khoa học và Công nghệ biển 2019; 19(3): 385-94.
[http://dx.doi.org/10.15625/1859-3097/19/3/13516]

[33]   Vancutsem C, Achard F, Pekel JF, et al. Long-term (1990–2019) monitoring of forest cover changes in the humid tropics. Sci Adv 7(10): eabe1603.

[34]   Division USEPAeg, Monitoring E. Division USEPAEG, Monitoring E, (Cincinnati SL. Sampling and analysis procedures for screening of industrial effluents for priority pollutants. US Environmental Protection Agency, Environmental Monitoring and Support. 1977.

[35]   Abatenh E, Gizaw B, Tsegaye Z, Wassie M. The role of microorganisms in bioremediation-A review. Open Journal of Environmental Biology 2017; 2(1): 038-46.
[http://dx.doi.org/10.17352/ojeb.000007]

[36]   Guevara EA, Dehling DM, Bello C, et al. Asymmetric effect of deforestation on the functional roles of interacting plants and hummingbirds, 31 January 2023, PREPRINT (Version 1) available at Research Square
[http://dx.doi.org/10.21203/rs.3.rs-2527763/v1]

[37]   Linares MS, Macedo DR, Hughes RM, Castro DMP, Callisto M. The past is never dead: legacy effects alter the structure of benthic macroinvertebrate assemblages. Limnetica 2023; 42(1): 1.
[http://dx.doi.org/10.23818/limn.42.05]

[38]   Estifanos TK, Fisher B, Ricketts TH. Impacts of deforestation on childhood malaria depends on wealth

and vector biology, 08 July 2022, PREPRINT (Version 1) available at Research Square [http://dx.doi.org/10.21203/rs.3.rs-1809058/v1]

[39]   Firdaus N. SUPRIATNA S, Supriatna J. Ecosystem services research trends in Indonesia: a bibliometric analysis. Biodiversitas (Surak) 2022; 23(2): 1105-7.0.

[40]   Huma Z, Lin G, Hyder SL. Promoting Resilience and Health of Urban Citizen through Urban Green Space. Water and Environmental Sustainability 2021; 1(1): 37-43. [http://dx.doi.org/10.52293/WES.1.1.3743]

[41]   Reeves SE, Kriegisch N, Johnson CR, Ling SD. Kelp habitat fragmentation reduces resistance to overgrazing, invasion and collapse to turf dominance. J Appl Ecol 2022; 59(6): 1619-31. [http://dx.doi.org/10.1111/1365-2664.14171]

[42]   Gawecka KA, Pedraza F, Bascompte J. Effects of habitat destruction on coevolving metacommunities. Ecol Lett 2022; 25(12): 2597-610. [http://dx.doi.org/10.1111/ele.14118] [PMID: 36223432]

[43]   Halpern BS, Frazier M, Afflerbach J, *et al.* Recent pace of change in human impact on the world's ocean. Sci Rep 2019; 9(1): 11609. [http://dx.doi.org/10.1038/s41598-019-47201-9] [PMID: 31406130]

[44]   Foley JA, DeFries R, Asner GP, *et al.* Global consequences of land use. Science 2005; 309(5734): 570-4. [http://dx.doi.org/10.1126/science.1111772] [PMID: 16040698]

[45]   Yadav KK, Gupta N, Prasad S, *et al.* An eco-sustainable approach towards heavy metals remediation by mangroves from the coastal environment: A critical review. Marine Pollution Bulletin 2023; 1;188:114569. [http://dx.doi.org/10.1016/j.marpolbul.2022.114569]

[46]   Saravanakumar K, SivaSantosh S, Sathiyaseelan A, *et al.* Unraveling the hazardous impact of diverse contaminants in the marine environment: Detection and remedial approach through nanomaterials and nano-biosensors. J Hazard Mater 2022; 433: 128720. [http://dx.doi.org/10.1016/j.jhazmat.2022.128720] [PMID: 35366447]

[47]   Tounta DD, Nastos PT, Tesseromatis C. Human activities and zoonotic epidemics: a two-way relationship. The case of the COVID-19 pandemic. Global Sustainability 2022; 5: e19. [http://dx.doi.org/10.1017/sus.2022.18]

[48]   Cheval S, Mihai Adamescu C, Georgiadis T, Herrnegger M, Piticar A, Legates DR. Observed and potential impacts of the COVID-19 pandemic on the environment. Int J Environ Res Public Health 2020; 17(11): 4140. [http://dx.doi.org/10.3390/ijerph17114140] [PMID: 32532012]

[49]   González-Barrios FJ, Estrada-Saldívar N, Pérez-Cervantes E, Secaira-Fajardo F, Álvarez-Filip L. Legacy effects of anthropogenic disturbances modulate dynamics in the world's coral reefs. Glob Change Biol 2023; 29(12): 3285-303. [http://dx.doi.org/10.1111/gcb.16686] [PMID: 36932916]

[50]   Ridha SN. Investigating the genetic basis of disease resistance in animal populations. World Journal of Advanced Research and Reviews 2023; 18(1): 073-9. [http://dx.doi.org/10.30574/wjarr.2023.18.1.0443]

[51]   Buij R, Bugter RJF, Henkens R, Moonen S, Jones-Walters LM, van der Grift EA. Review: the impact of coronavirus disease (COVID-19) on wildlife: with a focus on Europe. Wageningen: Wageningen Environmental Research; 2021; p. 59.

[52]   Gautam K, Sharma P, Dwivedi S, *et al.* A review on control and abatement of soil pollution by heavy metals: Emphasis on artificial intelligence in recovery of contaminated soil. Environ Res 2023; 225: 115592. [http://dx.doi.org/10.1016/j.envres.2023.115592] [PMID: 36863654]

[53]   Wang L, Chen M, Li J, Jin Y, Zhang Y, Wang Y. A novel substitution-based method for effective leaching of chromium (III) from chromium-tanned leather waste: The thermodynamics, kinetics and mechanism studies. Waste Manag 2020; 103: 276-84.
[http://dx.doi.org/10.1016/j.wasman.2019.12.039] [PMID: 31911374]

[54]   Lambin EF, Meyfroidt P. Global land use change, economic globalization, and the looming land scarcity. Proc Natl Acad Sci USA 2011; 108(9): 3465-72.
[http://dx.doi.org/10.1073/pnas.1100480108] [PMID: 21321211]

[55]   Benjamin ED, Handley SJ, Jeffs A, Olsen L, Toone TA, Hillman JR. Testing habitat suitability for shellfish restoration with small-scale pilot experiments. Conserv Sci Pract 2023; 5(2): e12878.
[http://dx.doi.org/10.1111/csp2.12878]

[56]   Gu L, Xiao X, Zhao G, *et al.* Rewiring the respiratory pathway of *Lactococcus lactis* to enhance extracellular electron transfer. Microb Biotechnol 2023; 16(6): 1277-92.
[http://dx.doi.org/10.1111/1751-7915.14229] [PMID: 36860178]

[57]   Carlucci MB, Brancalion PHS, Rodrigues RR, Loyola R, Cianciaruso MV. Functional traits and ecosystem services in ecological restoration. Restor Ecol 2020; 28(6): 1372-83.
[http://dx.doi.org/10.1111/rec.13279]

[58]   Su J, Friess DA, Gasparatos A. A meta-analysis of the ecological and economic outcomes of mangrove restoration. Nat Commun 2021; 12(1): 5050.
[http://dx.doi.org/10.1038/s41467-021-25349-1] [PMID: 34413296]

[59]   Derhé MA, Murphy H, Monteith G, Menéndez R. Measuring the success of reforestation for restoring biodiversity and ecosystem functioning. J Appl Ecol 2016; 53(6): 1714-24.
[http://dx.doi.org/10.1111/1365-2664.12728]

[60]   Crouzeilles R, Curran M, Ferreira MS, Lindenmayer DB, Grelle CEV, Rey Benayas JM. A global meta-analysis on the ecological drivers of forest restoration success. Nat Commun 2016; 7(1): 11666.
[http://dx.doi.org/10.1038/ncomms11666] [PMID: 27193756]

[61]   Chao L, Ma X, Tsetsegmaa M, *et al.* Response of Soil Microbial Community Composition and Diversity at Different Gradients of Grassland Degradation in Central Mongolia. Agriculture 2022; 12(9): 1430.
[http://dx.doi.org/10.3390/agriculture12091430]

[62]   Xiong C, Lu Y. Microbiomes in agroecosystem: Diversity, function and assembly mechanisms. Environ Microbiol Rep 2022; 14(6): 833-49.
[http://dx.doi.org/10.1111/1758-2229.13126] [PMID: 36184075]

[63]   Liang JL, Liu J, Jia P, *et al.* Novel phosphate-solubilizing bacteria enhance soil phosphorus cycling following ecological restoration of land degraded by mining. ISME J 2020; 14(6): 1600-13.
[http://dx.doi.org/10.1038/s41396-020-0632-4] [PMID: 32203124]

[64]   Battin TJ, Ezzat L, Peter H, *et al.* Towards a global biogeography of the glacier-fed stream benthic microbiome 2023.
[http://dx.doi.org/10.21203/rs.3.rs-2697617/v1]

[65]   Leifheit EF, Verbruggen E, Rillig MC. Arbuscular mycorrhizal fungi reduce decomposition of woody plant litter while increasing soil aggregation. Soil Biol Biochem 2015; 81: 323-8.
[http://dx.doi.org/10.1016/j.soilbio.2014.12.003]

[66]   Wu QS, Cao MQ, Zou YN, He X. Direct and indirect effects of glomalin, mycorrhizal hyphae and roots on aggregate stability in rhizosphere of trifoliate orange. Sci Rep 2014; 4(1): 5823.
[http://dx.doi.org/10.1038/srep05823] [PMID: 25059396]

[67]   Xu P, Liang LZ, Dong XY, Shen RF. Effect of arbuscular mycorrhizal fungi on aggregate stability of a clay soil inoculating with two different host plants. Acta Agric Scand Sect B—Soil. Plant Sci 2015; 65: 23-9.

[68]   Liu Q, Zhang Q, Jarvie S, *et al.* Ecosystem restoration through aerial seeding: Interacting plant–soil microbiome effects on soil multifunctionality. Land Degrad Dev 2021; 32(18): 5334-47.
[http://dx.doi.org/10.1002/ldr.4112]

[69]   Prajapati KK, Yadav M, Singh RM, Parikh P, Pareek N, Vivekanand V. An overview of municipal solid waste management in Jaipur city, India - Current status, challenges and recommendations. Renew Sustain Energy Rev 2021; 152: 111703.
[http://dx.doi.org/10.1016/j.rser.2021.111703]

[70]   Song X, Liu M, Wu D, *et al.* Interaction matters: Synergy between vermicompost and PGPR agents improves soil quality, crop quality and crop yield in the field. Appl Soil Ecol 2015; 89: 25-34.
[http://dx.doi.org/10.1016/j.apsoil.2015.01.005]

[71]   Goswami D, Patel K, Parmar S, *et al.* Elucidating multifaceted urease producing marine Pseudomonas aeruginosa BG as a cogent PGPR and bio-control agent. Plant Growth Regul 2015; 75(1): 253-63.
[http://dx.doi.org/10.1007/s10725-014-9949-1]

[72]   Vigneshvar S, Sudhakumari CC, Senthilkumaran B, Prakash H. Recent advances in biosensor technology for potential applications–an overview. Front Bioeng Biotechnol 2016; 4: 11.
[http://dx.doi.org/10.3389/fbioe.2016.00011] [PMID: 26909346]

[73]   Cui H, Zhou Y, Gu Z, Zhu H, Fu S, Yao Q. The combined effects of cover crops and symbiotic microbes on phosphatase gene and organic phosphorus hydrolysis in subtropical orchard soils. Soil Biol Biochem 2015; 82: 119-26.
[http://dx.doi.org/10.1016/j.soilbio.2015.01.003]

[74]   Wu L, An Z, Zhou J, *et al.* Effects of aquatic acidification on microbially mediated nitrogen removal in estuarine and coastal environments. Environ Sci Technol 2022; 56(9): 5939-49.
[http://dx.doi.org/10.1021/acs.est.2c00692] [PMID: 35465670]

[75]   Fanesi A, Martin T, Breton C, Bernard O, Briandet R, Lopes F. The architecture and metabolic traits of monospecific photosynthetic biofilms studied in a custom flow-through system. Biotechnol Bioeng 2022; 119(9): 2459-70.
[http://dx.doi.org/10.1002/bit.28147] [PMID: 35643824]

[76]   Yue Y, Tang Y, Cai L, *et al.* Co-occurrence relationship and stochastic processes affect sedimentary archaeal and bacterial community assembly in estuarine–coastal margins. Microorganisms 2022; 10(7): 1339.
[http://dx.doi.org/10.3390/microorganisms10071339] [PMID: 35889058]

[77]   Wang S, Xu C, Song L, Zhang J. Anaerobic digestion of food waste and its microbial consortia: A historical review and future perspectives. Int J Environ Res Public Health 2022; 19(15): 9519.
[http://dx.doi.org/10.3390/ijerph19159519] [PMID: 35954875]

[78]   Caruso G, Giacobbe MG, Azzaro F, *et al.* All-In-One: Microbial Response to Natural and Anthropogenic Forcings in a Coastal Mediterranean Ecosystem, the Syracuse Bay (Ionian Sea, Italy). J Mar Sci Eng 2021; 10(1): 19.
[http://dx.doi.org/10.3390/jmse10010019]

[79]   Wu J, Shi Z, Zhu J, *et al.* Taxonomic response of bacterial and fungal populations to biofertilizers applied to soil or substrate in greenhouse-grown cucumber. Sci Rep 2022; 12(1): 18522.
[http://dx.doi.org/10.1038/s41598-022-22673-4] [PMID: 36323754]

[80]   Reusch TBH, Schubert PR, Marten SM, *et al.* Lower Vibrio spp. abundances in Zostera marina leaf canopies suggest a novel ecosystem function for temperate seagrass beds. Mar Biol 2021; 168(10): 149.
[http://dx.doi.org/10.1007/s00227-021-03963-3]

[81]   Schmidt JE, Kent AD, Brisson VL, Gaudin ACM. Agricultural management and plant selection interactively affect rhizosphere microbial community structure and nitrogen cycling. Microbiome 2019; 7(1): 146.

[http://dx.doi.org/10.1186/s40168-019-0756-9] [PMID: 31699148]

[82]   Prasad S, Kumar S, Yadav KK, *et al.*   Screening and evaluation of cellulytic fungal strains for saccharification and bioethanol production from rice residue. Energy. 2020; 1: 190: 116422.
[http://dx.doi.org/10.1016/j.energy.2019.116422]

[83]   Prasad S, Yadav KK, Kumar S, *et al.* Chromium contamination and effect on environmental health and its remediation: A sustainable approaches. Journal of Environmental Management. 2021, 1;285:112174.
[http://dx.doi.org/10.1016/j.jenvman.2021.112174]

[84]   Renjith PS, Sheetal KR, Kumar S, Choudhary J, Prasad S. Microbial and biotechnological approaches in the production of biofertilizer. Environmental Microbiology and Biotechnology: Volume 1: Biovalorization of Solid Wastes and Wastewater Treatment. 2020:201-19.
[http://dx.doi.org/10.1007/978-981-15-6021-7_10]

[85]   Malav MK, Prasad S, Jain N, Kumar D, Kanojiya S. Effect of organic rice (Oryza sativa) cultivation on greenhouse gas emission. Indian J Agric Sci 2020; 90(9): 1769-75.
[http://dx.doi.org/10.56093/ijas.v90i9.106625]

[86]   Ambaye TG, Vaccari M, Castro FD, Prasad S, Rtimi S. Emerging technologies for the recovery of rare earth elements (REEs) from the end-of-life electronic wastes: a review on progress, challenges, and perspectives. Environ Sci Pollut Res Int 2020; 27(29): 36052-74.
[http://dx.doi.org/10.1007/s11356-020-09630-2] [PMID: 32617815]

[87]   Ma Y, Zu L, Long F, *et al.* Promotion of Soil Microbial Community Restoration in the Mu Us Desert (China) by Aerial Seeding. Sustainability (Basel) 2022; 14(22): 15241.
[http://dx.doi.org/10.3390/su142215241]

[88]   Opara CB, Kamariah N, Spooren J, Pollmann K, Kutschke S. Interesting halophilic sulphur-oxidising bacteria with bioleaching potential: implications for pollutant mobilisation from mine waste. Microorganisms 2023; 11(1): 222.
[http://dx.doi.org/10.3390/microorganisms11010222] [PMID: 36677514]

[89]   Jung H, Inaba Y, Jiang V, West AC, Banta S. Engineering polyhistidine tags on surface proteins of Acidithiobacillus ferrooxidans: impact of localization on the binding and recovery of divalent metal cations. ACS Appl Mater Interfaces 2022; 14(8): 10125-33.
[http://dx.doi.org/10.1021/acsami.1c23682] [PMID: 35170950]

[90]   Tonietti L, Barosa B, Pioltelli E, *et al.* Exploring the Development of Astrobiology Scientific Research through Bibliometric Network Analysis: A Focus on Biomining and Bioleaching. Minerals (Basel) 2023; 13(6): 797.
[http://dx.doi.org/10.3390/min13060797]

[91]   Santomartino R, Zea L, Cockell CS. The smallest space miners: principles of space biomining. Extremophiles 2022; 26(1): 7.
[http://dx.doi.org/10.1007/s00792-021-01253-w] [PMID: 34993644]

[92]   Ambaye TG, Formicola F, Sbaffoni S, *et al.* Treatment of petroleum hydrocarbon contaminated soil by combination of electro-Fenton and biosurfactant-assisted bioslurry process. Chemosphere 2023; 319: 138013.
[http://dx.doi.org/10.1016/j.chemosphere.2023.138013] [PMID: 36731662]

[93]   Papik J, Strejcek M, Musilova L, *et al.* Legacy Effects of Phytoremediation on Plant-Associated Prokaryotic Communities in Remediated Subarctic Soil Historically Contaminated with Petroleum Hydrocarbons. Microbiol Spectr 2023; 11(2): e04448-22.
[http://dx.doi.org/10.1128/spectrum.04448-22] [PMID: 36975310]

[94]   Du S, Li XQ, Hao X, *et al.* Stronger responses of soil protistan communities to legacy mercury pollution than bacterial and fungal communities in agricultural systems. ISME Communications 2022; 2(1): 69.
[http://dx.doi.org/10.1038/s43705-022-00156-x] [PMID: 37938257]

[95] Vergani L, Mapelli F, Folkmanova M, *et al.* DNA stable isotope probing on soil treated by plant biostimulation and flooding revealed the bacterial communities involved in PCB degradation. Sci Rep 2022; 12(1): 19232.
[http://dx.doi.org/10.1038/s41598-022-23728-2] [PMID: 36357494]

[96] D'Agui HM, van der Heyde ME, Nevill PG, *et al.* Evaluating biological properties of topsoil for post-mining ecological restoration: different assessment methods give different results. Restor Ecol 2022; 30(S1): e13738.
[http://dx.doi.org/10.1111/rec.13738]

[97] Chen X, Zhang Z, Han X, *et al.* Impacts of land-use changes on the variability of microbiomes in soil profiles. J Sci Food Agric 2021; 101(12): 5056-66.
[http://dx.doi.org/10.1002/jsfa.11150] [PMID: 33570760]

[98] Shelyakin PV, Semenkov IN, Tutukina MN, *et al.* The influence of kerosene on microbiomes of diverse soils. Life (Basel) 2022; 12(2): 221.
[http://dx.doi.org/10.3390/life12020221] [PMID: 35207510]

[99] Lian Y, Zhen L, Fang Y, *et al.* Microbial biomarkers to identify areas of wetland sediments affected by massive fish farming. Front Environ Sci 2022; 10: 1000437.
[http://dx.doi.org/10.3389/fenvs.2022.1000437]

[100] Das N, Chandran P. Microbial degradation of petroleum hydrocarbon contaminants: an overview. Biotechnol Res Int 2011;2011.
[http://dx.doi.org/10.4061/2011/941810]

[101] Ite AE, Ibok UJ. Role of plants and microbes in bioremediation of petroleum hydrocarbons contaminated soils. Int J Environ Bioremediat Biodegrad 2019; 7: 1-19.

[102] Nosheen S, Ajmal I, Song Y. Microbes as biofertilizers, a potential approach for sustainable crop production. Sustainability (Basel) 2021; 13(4): 1868.
[http://dx.doi.org/10.3390/su13041868]

[103] Theocharis A, Bordiec S, Fernandez O, *et al.* Burkholderia phytofirmans PsJN primes Vitis vinifera L. and confers a better tolerance to low nonfreezing temperatures. Mol Plant Microbe Interact 2012; 25(2): 241-9.
[http://dx.doi.org/10.1094/MPMI-05-11-0124] [PMID: 21942451]

[104] Gallo G, Lo Piccolo L, Renzone G, *et al.* Differential proteomic analysis of an engineered Streptomyces coelicolor strain reveals metabolic pathways supporting growth on n-hexadecane. Appl Microbiol Biotechnol 2012; 94(5): 1289-301.
[http://dx.doi.org/10.1007/s00253-012-4046-8] [PMID: 22526801]

[105] Neumann G, Teras R, Monson L, Kivisaar M, Schauer F, Heipieper HJ. Simultaneous degradation of atrazine and phenol by Pseudomonas sp. strain ADP: effects of toxicity and adaptation. Appl Environ Microbiol 2004; 70(4): 1907-12.
[http://dx.doi.org/10.1128/AEM.70.4.1907-1912.2004] [PMID: 15066779]

[106] Lu J, Guo C, Li J, *et al.* A fusant of Sphingomonas sp. GY2B and Pseudomonas sp. GP3A with high capacity of degrading phenanthrene. World J Microbiol Biotechnol 2013; 29(9): 1685-94.
[http://dx.doi.org/10.1007/s11274-013-1331-3] [PMID: 23529357]

[107] Nayak AK, Panda SS, Basu A, Dhal NK. Enhancement of toxic Cr (VI), Fe, and other heavy metals phytoremediation by the synergistic combination of native *Bacillus cereus* strain and *Vetiveria zizanioides* L. Int J Phytoremediation 2018; 20(7): 682-91.
[http://dx.doi.org/10.1080/15226514.2017.1413332] [PMID: 29723050]

[108] Chaturvedi MK. Studies on chromate removal by chromium-resistant Bacillus sp. isolated from tannery effluent. J Environ Prot (Irvine Calif) 2011; 2(1): 76-82.
[http://dx.doi.org/10.4236/jep.2011.21008]

[109] Kumar R, Bhatia D, Singh R, Rani S, Bishnoi NR. Sorption of heavy metals from electroplating

effluent using immobilized biomass Trichoderma viride in a continuous packed-bed column. Int Biodeterior Biodegradation 2011; 65(8): 1133-9.
[http://dx.doi.org/10.1016/j.ibiod.2011.09.003]

[110] Congeevaram S, Dhanarani S, Park J, Dexilin M, Thamaraiselvi K. Biosorption of chromium and nickel by heavy metal resistant fungal and bacterial isolates. J Hazard Mater 2007; 146(1-2): 270-7.
[http://dx.doi.org/10.1016/j.jhazmat.2006.12.017] [PMID: 17218056]

[111] Kim IH, Choi JH, Joo JO, Kim YK, Choi JW, Oh BK. Development of a microbe-zeolite carrier for the effective elimination of heavy metals from seawater. J Microbiol Biotechnol 2015; 25(9): 1542-6.
[http://dx.doi.org/10.4014/jmb.1504.04067] [PMID: 26032363]

[112] Jafari SA, Cheraghi S, Mirbakhsh M, Mirza R, Maryamabadi A. Employing response surface methodology for optimization of mercury bioremediation by Vibrio parahaemolyticus PG02 in coastal sediments of Bushehr, Iran. Clean (Weinh) 2015; 43(1): 118-26.
[http://dx.doi.org/10.1002/clen.201300616]

[113] Sbihi H, Boutin RCT, Cutler C, Suen M, Finlay BB, Turvey SE. Thinking bigger: How early-life environmental exposures shape the gut microbiome and influence the development of asthma and allergic disease. Allergy 2019; 74(11): 2103-15.
[http://dx.doi.org/10.1111/all.13812] [PMID: 30964945]

[114] Glasl B, Smith CE, Bourne DG, Webster NS. Disentangling the effect of host-genotype and environment on the microbiome of the coral *Acropora tenuis*. PeerJ 2019; 7: e6377.
[http://dx.doi.org/10.7717/peerj.6377] [PMID: 30740275]

[115] Hu H, Wang M, Huang Y, *et al.* Guided by the principles of microbiome engineering: Accomplishments and perspectives for environmental use. mLife 2022; 1(4): 382-98.
[http://dx.doi.org/10.1002/mlf2.12043] [PMID: 38818482]

[116] Bailey MJ, Naik NN, Wild LE, Patterson WB, Alderete TL. Exposure to air pollutants and the gut microbiota: a potential link between exposure, obesity, and type 2 diabetes. Gut Microbes 2020; 11(5): 1188-202.
[http://dx.doi.org/10.1080/19490976.2020.1749754] [PMID: 32347153]

[117] Yang Y, Tilman D, Furey G, Lehman C. Soil carbon sequestration accelerated by restoration of grassland biodiversity. Nat Commun 2019; 10(1): 718.
[http://dx.doi.org/10.1038/s41467-019-08636-w] [PMID: 30755614]

[118] Múnera-Porras LM, García-Londoño S, Ríos-Osorio LA. Action mechanisms of plant growth promoting cyanobacteria in crops *in situ*: a systematic review of the literature. Int J Agron 2020; 2020: 1-9.
[http://dx.doi.org/10.1155/2020/2690410]

[119] Dadzie F, Munoz-Rojas M, Slavich E, Pottier P, Zeng K, Moles A. Native and commercial microbial inoculants show equal effects on plant growth in dryland ecosystems. Authorea 2023.
[http://dx.doi.org/10.22541/au.167785097.78920686/v1]

[120] Aronson J, Goodwin N, Orlando L, Eisenberg C, Cross AT. A world of possibilities: six restoration strategies to support the United Nation's Decade on Ecosystem Restoration. Restor Ecol 2020; 28(4): 730-6.
[http://dx.doi.org/10.1111/rec.13170]

[121] Chadha U, Bhardwaj P, Agarwal R, *et al.* Recent progress and growth in biosensors technology: A critical review. J Ind Eng Chem 2022; 109: 21-51.
[http://dx.doi.org/10.1016/j.jiec.2022.02.010]

[122] Gümral R, Özhak-Baysan B, Tümgör A, *et al.* Dishwashers provide a selective extreme environment for human-opportunistic yeast-like fungi. Fungal Divers 2016; 76(1): 1-9.
[http://dx.doi.org/10.1007/s13225-015-0327-8]

[123] Taran M, Fateh R, Rezaei S, Khalifeh Gholi M. Isolation of arsenic accumulating bacteria from

garbage leachates for possible application in bioremediation. Iran J Microbiol 2019; 11(1): 60-6.
[http://dx.doi.org/10.18502/ijm.v11i1.707] [PMID: 30996833]

[124] Priya AK, Muruganandam M, Ali SS, Kornaros M. Clean-Up of Heavy Metals from Contaminated Soil by Phytoremediation: A Multidisciplinary and Eco-Friendly Approach. Toxics 2023; 11(5): 422.
[http://dx.doi.org/10.3390/toxics11050422] [PMID: 37235237]

[125] Yan C, Wang F, Geng H, *et al.* Integrating high-throughput sequencing and metagenome analysis to reveal the characteristic and resistance mechanism of microbial community in metal contaminated sediments. Sci Total Environ 2020; 707: 136116.
[http://dx.doi.org/10.1016/j.scitotenv.2019.136116] [PMID: 31874394]

[126] Bai S, Han X, Feng D. Shoot-root signal circuit: Phytoremediation of heavy metal contaminated soil. Front Plant Sci 2023; 14: 1139744.
[http://dx.doi.org/10.3389/fpls.2023.1139744] [PMID: 36890896]

[127] Nocera DG. Proton-coupled electron transfer: the engine of energy conversion and storage. J Am Chem Soc 2022; 144(3): 1069-81.
[http://dx.doi.org/10.1021/jacs.1c10444] [PMID: 35023740]

[128] Siripong W, Wolf P, Kusumoputri TP, *et al.* Metabolic engineering of *Pichia pastoris* for production of isobutanol and isobutyl acetate. Biotechnol Biofuels 2018; 11(1): 1-16.
[http://dx.doi.org/10.1186/s13068-017-1003-x] [PMID: 29321810]

[129] Pérgola M, Sacco NJ, Bonetto MC, Galagovsky L, Cortón E. A laboratory experiment for science courses: Sedimentary microbial fuel cells. Biochem Mol Biol Educ 2023; 51(2): 221-9.
[http://dx.doi.org/10.1002/bmb.21702] [PMID: 36495269]

[130] Guan X, Gao X, Avellan A, *et al.* CuO nanoparticles alter the rhizospheric bacterial community and local nitrogen cycling for wheat grown in a calcareous soil. Environ Sci Technol 2020; 54(14): 8699-709.
[http://dx.doi.org/10.1021/acs.est.0c00036] [PMID: 32579348]

[131] Shabat SKB, Sasson G, Doron-Faigenboim A, *et al.* Specific microbiome-dependent mechanisms underlie the energy harvest efficiency of ruminants. ISME J 2016; 10(12): 2958-72.
[http://dx.doi.org/10.1038/ismej.2016.62] [PMID: 27152936]

[132] Banerjee S, Walder F, Büchi L, *et al.* Agricultural intensification reduces microbial network complexity and the abundance of keystone taxa in roots. ISME J 2019; 13(7): 1722-36.
[http://dx.doi.org/10.1038/s41396-019-0383-2] [PMID: 30850707]

[133] Wu K, Su D, Liu J, Saha R, Wang JP. Magnetic nanoparticles in nanomedicine: a review of recent advances. Nanotechnology 2019; 30(50): 502003.
[http://dx.doi.org/10.1088/1361-6528/ab4241] [PMID: 31491782]

[134] Lawson CE, Harcombe WR, Hatzenpichler R, *et al.* Common principles and best practices for engineering microbiomes. Nat Rev Microbiol 2019; 17(12): 725-41.
[http://dx.doi.org/10.1038/s41579-019-0255-9] [PMID: 31548653]

[135] Frąc M, Hannula ES, Bełka M, Salles JF, Jedryczka M. Soil mycobiome in sustainable agriculture. Front Microbiol 2022; 13: 1033824.
[http://dx.doi.org/10.3389/fmicb.2022.1033824] [PMID: 36519160]

[136] Afridi MS, Javed MA, Ali S, *et al.* New opportunities in plant microbiome engineering for increasing agricultural sustainability under stressful conditions. Front Plant Sci 2022; 13: 899464.
[http://dx.doi.org/10.3389/fpls.2022.899464] [PMID: 36186071]

[137] Lindow SE, Brandl MT. Microbiology of the Phyllosphere. Appl Environ Microbiol 2003; 69(4): 1875-83.
[http://dx.doi.org/10.1128/AEM.69.4.1875-1883.2003] [PMID: 12676659]

[138] Mishra PK, Mishra S, Selvakumar G, *et al.* Characterisation of a psychrotolerant plant growth promotingPseudomonas sp. strain PGERs17 (MTCC 9000) isolated from North Western Indian

Himalayas. Ann Microbiol 2008; 58(4): 561-8.
[http://dx.doi.org/10.1007/BF03175558]

[139]   Tiryaki D, Aydın İ, Atıcı Ö. Psychrotolerant bacteria isolated from the leaf apoplast of cold-adapted wild plants improve the cold resistance of bean (*Phaseolus vulgaris* L.) under low temperature. Cryobiology 2019; 86: 111-9.
[http://dx.doi.org/10.1016/j.cryobiol.2018.11.001] [PMID: 30419217]

[140]   Abdel-Shafy HI, Mansour MSM. Solid waste issue: Sources, composition, disposal, recycling, and valorization. Egyptian Journal of Petroleum 2018; 27(4): 1275-90.
[http://dx.doi.org/10.1016/j.ejpe.2018.07.003]

[141]   Duc NH, Csintalan Z, Posta K. Arbuscular mycorrhizal fungi mitigate negative effects of combined drought and heat stress on tomato plants. Plant Physiol Biochem 2018; 132: 297-307.
[http://dx.doi.org/10.1016/j.plaphy.2018.09.011] [PMID: 30245343]

[142]   Sangiorgio D, Cellini A, Donati I, Pastore C, Onofrietti C, Spinelli F. Facing climate change: application of microbial biostimulants to mitigate stress in horticultural crops. Agronomy (Basel) 2020; 10(6): 794.
[http://dx.doi.org/10.3390/agronomy10060794]

[143]   Ullah N, Mansha M, Khan I, Qurashi A. Nanomaterial-based optical chemical sensors for the detection of heavy metals in water: Recent advances and challenges. Trends Analyt Chem 2018; 100: 155-66.
[http://dx.doi.org/10.1016/j.trac.2018.01.002]

[144]   Liu CY, Zhang F, Zhang DJ, Srivastava AK, Wu QS, Zou YN. Mycorrhiza stimulates root-hair growth and IAA synthesis and transport in trifoliate orange under drought stress. Sci Rep 2018; 8(1): 1978.
[http://dx.doi.org/10.1038/s41598-018-20456-4] [PMID: 29386587]

[145]   Jayakumar A, Padmakumar P, Nair IC, Radhakrishnan EK. Drought tolerant bacterial endophytes with potential plant probiotic effects from Ananas comosus. Biologia (Bratisl) 2020; 75(10): 1769-78.
[http://dx.doi.org/10.2478/s11756-020-00483-1]

[146]   Kour D, Kaur T, Devi R, Rana KL, Yadav N, Rastegari AA, *et al.* Biotechnological applications of beneficial microbiomes for evergreen agriculture and human health New Futur Dev Microb Biotechnol Bioeng. Elsevier 2020; pp. 255-79.

[147]   Khan MA, Asaf S, Khan AL, *et al.* Alleviation of salt stress response in soybean plants with the endophytic bacterial isolate Curtobacterium sp. SAK1. Ann Microbiol 2019; 69(8): 797-808.
[http://dx.doi.org/10.1007/s13213-019-01470-x]

[148]   Gupta S, Pandey S. ACC deaminase producing bacteria with multifarious plant growth promoting traits alleviates salinity stress in French bean (*Phaseolus vulgaris*) plants. Front Microbiol 2019; 10: 1506.
[http://dx.doi.org/10.3389/fmicb.2019.01506] [PMID: 31338077]

[149]   Tahir M, Ahmad I, Shahid M, *et al.* Regulation of antioxidant production, ion uptake and productivity in potato (Solanum tuberosum L.) plant inoculated with growth promoting salt tolerant Bacillus strains. Ecotoxicol Environ Saf 2019; 178: 33-42.
[http://dx.doi.org/10.1016/j.ecoenv.2019.04.027] [PMID: 30991245]

[150]   Sasmita KD, Rokmah DN, Hafif B, Putra S. The Effect of Biofertilizer from Waste Bioconversion on The Growth of Cocoa Seedlings. IOP Conf Ser Earth Environ Sci. IOP Publishing. 2022; p. 1038: 12008.
[http://dx.doi.org/10.1088/1755-1315/1038/1/012008]

[151]   Kumar S, Diksha , Sindhu SS, Kumar R. Biofertilizers: An ecofriendly technology for nutrient recycling and environmental sustainability. Current Research in Microbial Sciences 2022; 3: 100094.
[http://dx.doi.org/10.1016/j.crmicr.2021.100094] [PMID: 35024641]

[152]   Zahid M, Abbasi MK, Hameed S, Rahim N. Isolation and identification of indigenous plant growth promoting rhizobacteria from Himalayan region of Kashmir and their effect on improving growth and

nutrient contents of maize (*Zea mays* L.). Front Microbiol 2015; 6: 207.
[http://dx.doi.org/10.3389/fmicb.2015.00207] [PMID: 25852667]

[153] Alonso-Lomillo MA, Domínguez-Renedo O, Matos P, Arcos-Martínez MJ. Electrochemical determination of levetiracetam by screen-printed based biosensors. Bioelectrochemistry 2009; 74(2): 306-9.
[http://dx.doi.org/10.1016/j.bioelechem.2008.11.003] [PMID: 19059814]

[154] Dixit A, Singh D, Shukla SK. Changing scenario of municipal solid waste management in Kanpur city, India. J Mater Cycles Waste Manag 2022; 24(5): 1648-62.
[http://dx.doi.org/10.1007/s10163-022-01427-4]

[155] Selvi A, AlSalhi MS, Devanesan S, Maruthamuthu MK, Mani P, Rajasekar A. Characterization of biospheric bacterial community on reduction and removal of chromium from tannery contaminated soil using an integrated approach of bio-enhanced electrokinetic remediation. J Environ Chem Eng 2021; 9(6): 106602.
[http://dx.doi.org/10.1016/j.jece.2021.106602]

[156] Lin Q, Song Y, Zhang Y, Hao JL, Wu Z. Strategies for Restoring and Managing Ecological Corridors of Freshwater Ecosystem. Int J Environ Res Public Health 2022; 19(23): 15921.
[http://dx.doi.org/10.3390/ijerph192315921] [PMID: 36497995]

[157] Salehizadeh H, Shojaosadati SA. Removal of metal ions from aqueous solution by polysaccharide produced from Bacillus firmus. Water Res 2003; 37(17): 4231-5.
[http://dx.doi.org/10.1016/S0043-1354(03)00418-4] [PMID: 12946905]

[158] Zadjelovic V, Gibson MI, Dorador C, Christie-Oleza JA. Genome of *Alcanivorax* sp. 24: A hydrocarbon degrading bacterium isolated from marine plastic debris. Mar Genomics 2020; 49: 100686.
[http://dx.doi.org/10.1016/j.margen.2019.05.001]

[159] Yoshida S, Hiraga K, Takehana T, *et al.* s A bacterium that degrades and assimilates poly (ethylene terephthalate). Science (80-) 2016;351:1196–9.

[160] Jabloune R, Khalil M, Ben Moussa ie, *et al.* Enzymatic degradation of p-nitrophenyl esters, polyethylene terephthalate, cutin, and suberin by Sub1, a suberinase encoded by the plant pathogen Streptomyces scabies. Microbes Environ 2020; 35(1): n/a.
[http://dx.doi.org/10.1264/jsme2.ME19086] [PMID: 32101840]

[161] Puglisi E, Romaniello F, Galletti S, Boccaleri E, Frache A, Cocconcelli PS. Selective bacterial colonization processes on polyethylene waste samples in an abandoned landfill site. Sci Rep 2019; 9(1): 14138.
[http://dx.doi.org/10.1038/s41598-019-50740-w] [PMID: 31578444]

[162] Sameshima-Yamashita Y, Ueda H, Koitabashi M, Kitamoto H. Pretreatment with an esterase from the yeast Pseudozyma antarctica accelerates biodegradation of plastic mulch film in soil under laboratory conditions. J Biosci Bioeng 2019; 127(1): 93-8.
[http://dx.doi.org/10.1016/j.jbiosc.2018.06.011] [PMID: 30054060]

[163] Kong R, Li J, Orban C, *et al.* Spatial topography of individual-specific cortical networks predicts human cognition, personality, and emotion. Cereb Cortex 2019; 29(6): 2533-51.
[http://dx.doi.org/10.1093/cercor/bhy123] [PMID: 29878084]

[164] Xu P, Guo Y, Fu B. Regional Impacts of climate and land cover on ecosystem water retention services in the Upper Yangtze River Basin. Sustainability (Basel) 2019; 11(19): 5300.
[http://dx.doi.org/10.3390/su11195300]

[165] Méndez-Albores A, Tarín C, Rebollar-Pérez G, Dominguez-Ramirez L, Torres E. Biocatalytic spectrophotometric method to detect paracetamol in water samples. J Environ Sci Health Part A Tox Hazard Subst Environ Eng 2015; 50(10): 1046-56.
[http://dx.doi.org/10.1080/10934529.2015.1038179] [PMID: 26121020]

[166] Kumar R, Qureshi M, Vishwakarma DK, *et al.* A review on emerging water contaminants and the application of sustainable removal technologies. Case Studies in Chemical and Environmental Engineering 2022; 6: 100219.
[http://dx.doi.org/10.1016/j.cscee.2022.100219]

CHAPTER 2

Role of Environmental Factors Influencing Microbes and Microbiomes for Ecosystem Restoration

Divya Pooja[1,*], Shiv Prasad[1], Govindaraj Kamalam Dinesh[1,2,3] and **C. Avinash[1]**

[1] Division of Environment Science, ICAR-Indian Agricultural Research Institute (IARI), New Delhi, India

[2] Division of Environment Sciences, Department of Soil Science and Agricultural Chemistry, Faculty of Agricultural Sciences, SRM College of Agricultural Sciences, SRM Institute of Science and Technology, Baburayanpettai - 603201, Chengalpattu, Tamil Nadu, India

[3] INTI International University, Persiaran Perdana BBN, Putra Nilai, 71800 Negeri Sembilan, Malaysia

Abstract: Ecosystem degradation poses a significant and growing environmental threat. Restoring degraded ecosystems is vital to restoring their ability to provide essential services and benefits. In 2021, the United Nations declared the Decade of Ecosystem Restoration to emphasize the importance of coordinated efforts in this area. Microbes, with their stress tolerance, genetic diversity, adaptation to various conditions, and capacity to break down substances, are crucial for ecosystem sustainability. Their critical functions are vital in restoring ecosystem function and biodiversity. This chapter describes the role of microbes in a microbiome and their interactions, instilling optimism about their potential. It also covers how various factors shape the soil microbiome spatially and temporally. Soil microorganisms such as bacteria, archaea, and fungi are found around, on, and in plant roots, and they play an essential role in responding to abiotic stressors. Factors like soil conditions, geographical and climatic factors, and stressors like drought, pollutants, and salinity can result in distinct microbial compositions and characteristics. This chapter provides an in-depth overview of how these factors can impact soil microbial communities and their role in ecological restoration. This chapter also covers beneficial microbiome-based strategies, including microbial engineering for ecosystem restoration. These strategies are essential and a source of hope for the future.

Keywords: Abiotic stressors, Ecosystem restoration, Environmental factors, Microbes, Microbiome.

* **Corresponding author Divya Pooja:** Division of Environmental Science, ICAR-Indian Agricultural Research Institute, New Delhi, India; email: poojadivya75@gmail.com

Shiv Prasad, Govindaraj Kamalam Dinesh, Murugaiyan Sinduja, Velusamy Sathya, Ramesh Poornima & Sangilidurai Karthika (Eds.)

INTRODUCTION

Population growth, industrialization, and urbanization have greatly influenced the environment. As a result of developmental activities, the environment has degraded. The significant impacts of developmental activities include habitat degradation, biodiversity loss, environmental pollution, and global climate change [1]. One of the leading causes of the decline in biodiversity and the integrity of ecosystems is anthropogenic pressure. The ecosystems need to be more resilient to cope with the degradation activities. This necessitates the urgency for ecological restoration. According to the Society for Ecological Restoration Science and Policy Working Group, the process of assisting in recovering an ecosystem that has been degraded, damaged, or destroyed is known as ecological restoration [2]. It entails various tasks like eliminating invasive species, reintroducing native species, and enhancing soil health. In 2021, the United Nations (UN) declared the Decade of Ecosystem Restoration, a significant global initiative in response to the destruction and degradation of ecosystems. This proclamation recognizes the need to rapidly restore damaged ecosystems worldwide to restore their ability to offer necessary services and benefits [3]. The majority of ecological restoration projects aim to stop further changes and restore ecosystems and ecosystem processes to their pre-injured state. The goal of restoration is to bring an ecosystem back to a functional state by reinstating its biodiversity, enhancing nutrient cycling and energy flow, improving the fertility of soil, water regulation, protecting against soil erosion, pest control, maintenance of a variety of pollinator species services, and reduction in CO_2 emissions and its buildup, among other factors. Microbes play an essential role in ecosystem restoration. Table **1** provides an overview of the ecological functions of microorganisms.

Microbiome refers to distinctive microbial communities, including viruses, fungi, bacteria, archaea, protozoa, and other micro-eukaryotes occupying well-defined habitats [11]. Also, these are the rich reservoirs of biodiversity that possess the ability to conserve and restore the functioning of various ecosystems. Microbiomes exist in habitats such as humans, plants, soils, sediments, and livestock animals. Many scientific disciplines depend heavily on microbiomes, including the medical, veterinary, and environmental sciences. A schematic representation of the application of microbiomes in environmental sciences is given in Fig. (**1**).

Microbiomes are not just crucial for developing novel, long-lasting bioeconomy applications, such as industrial biotechnology and waste recycling. They also hold the potential to address urgent social concerns. Applications of the microbiome as pre and probiotic food supplements, biofertilizers, and biocontrol agents are anticipated to significantly aid in eradicating hunger, stopping biodiversity loss, and slowing climate change. This potential gives us hope and inspires us to continue our research and application of microbiomes in various sectors.

Table 1. The major groups of microbes and their ecological roles.

Type of Microbes	Beneficial Roles	Ecological Function	References
Archaea	Nitrification, methanogenesis, *etc*	Regulation of climatic and atmospheric conditions, nutrient cycling, *etc*	[4]
Chemoautotrophic bacteria	Sulfate reduction, iron oxidation, nitrogen fixation, *etc*	Purification of water, cycling of nutrients, and climate control	[5]
Heterotrophic bacteria	Mineralization, oxidation of organic matter, synthesis of polymers, *etc*	Decomposition, water filtration, nutrient cycling, carbon sequestration, and climate control	[6]
Photoautotrophic bacteria	Photosynthesis	Chlorophyll production	[7]
Fungi	Organic matter consumption and mineralization	Soil formation, nutrient cycling, *etc*	[8]
Arbuscular mycorrhizal fungi	Nutrient cycling	Primary product (indirect)	[9]
Protozoans	Mineralization and consumption of other microbes, wastewater management	Mineralization, nutrition cycling, organic matter decomposition, *etc*	[10]
Virus	Cell lysis	Biogeochemical cycling, microbial evolution	[10, 11]

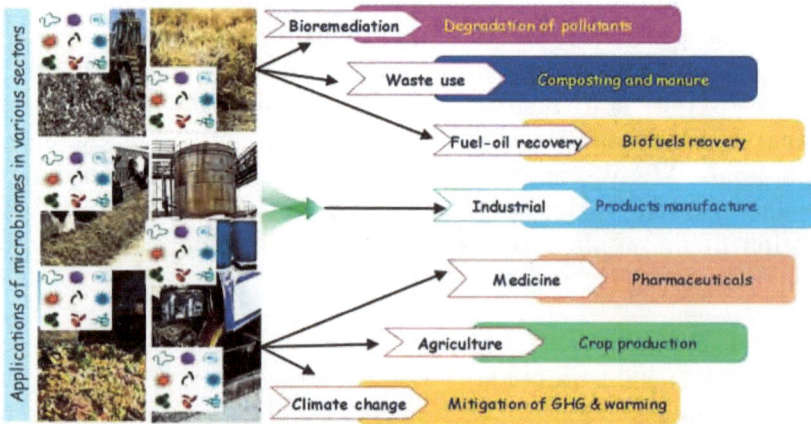

Fig. (1). Application of microbiomes in various sectors.

SIGNIFICANCE OF MICROBES AND MICROBIOMES IN ECOLOGICAL RESTORATION

Ecological restoration is being viewed more and more as being driven by the significant roles that microbes play in building soil structure, the establishment of biodiversity, the cycling of nutrients, the growth and development of plants, and the operation of ecosystems. Soil is essential for protecting terrestrial ecosystems and its plant community [6, 12]. Restoration experts integrate their understanding of particular microbial communities with fundamental information about soil quality, such as soil pH, bulk density, soil stability, organic carbon, and infiltration capacity, to guarantee that plants survive and grow into healthy communities. By preserving ecosystem services and controlling biogeochemical processes, microbes aid in restoration. Microbes aid in nitrogen fixation, phosphate solubilization, and plant growth hormone synthesis for better colonization of habitats. The availability of phosphorus and nitrogen is crucial for the effectiveness of ecosystem restoration. Particularly in semiarid and arid degraded environments, where nitrogen is the limiting resource, the role of denitrifiers in promoting nitrogen loss needs to be clarified. After repairing a severely degraded mined area, some phosphate-solubilizing bacteria improve the soil phosphorus cycle [9]. It has also been demonstrated that the active activity of microorganisms in a contaminated ecosystem drives the degradation of different contaminants. The secret to successful restoration is understanding the significance of soil microbes for the growth of higher plants and the health of the entire ecosystem.

Microorganisms are primarily responsible for controlling biogeochemical systems in nearly all of the ecosystems on our planet, even though the entire complement of living organisms and abiotic components that make up each ecosystem work together to coordinate biogeochemical cycles in that system. This includes living systems like the human gut and severe settings like acid lakes and hydrothermal vents. Nitrogen fixation, carbon fixation, methane metabolism, and sulfur metabolism are the four major collective metabolic activities that efficiently regulate global biogeochemical cycling in bacteria.

Agriculture also uses soil microbiomes to increase plant resistance to abiotic stress. Plants and soil microbes interact to promote plant growth and boost tolerance to abiotic stressors. However, their application for the restoration of degraded ecosystems is still unexplored. The soil microbial community (SMC), essential for carbon cycling, has been identified as the primary cause of variations in soils' capacity to store carbon. Since the structure and function of the SMC also control nutrient delivery and turnover, as well as the rate at which soil organic matter (SOM) decomposes, the composition of the SMC is essential for the sustainability of soil ecosystem services [4]. A modest amount of soil from unaltered habitats can be used to re-establish Soil Microbial Communities (SMC) in polluted sites, changing the course of floral succession. Ecosystems that are naturally occurring and those that are managed demonstrate the advantages of stress-conditioned soil microbial inoculations for re-establishing native plant communities in a changing environment.

Restoration of soil ecosystems using effective microorganisms and microbial fertilizer has gained momentum. To significantly increase microbial biodiversity, which promotes the sustainability of agroecosystems and the environment, direct Introduction of biofilm biofertilizers (BFBFs) communities can be a practical and effective strategy [13]. By coexisting in the biofilm structure and adapting to their surroundings, microorganisms display a high level of microbial function. When biofilms were discovered to be connected to plants, it was found that some bacteria, such as rhizobacteria, that encourage plant growth not only boost plant growth but also participate in a process known as biocontrol of plants to defend against soilborne infections. In addition, when employed at high cell densities, beneficial biofilms can be used as biofertilizers (also known as biofilm biofertilizers, or BBs), bioremediation, and nutrient mobilization [13, 14]. Restoration of degraded tropical agricultural land has been achieved in a few months by applying BFBFs developed from efficient microorganisms.

Rhizobacteria, also referred to as "plant growth promoting rhizobacteria" (PGPR), are microbial species capable of colonizing plant roots and influencing plant growth in a range of direct and direct ways to promote it and protect it from pest attack [15, 16]. A novel strategy for repairing damaged soils may be found in the symbiotic relationship between agricultural plants and potential PGPR. To accomplish a sustainable ecological restoration, the wise application of microbial fertilizers must be investigated. However, care must be taken when employing inoculants not native to the region because they may negatively impact the local soil bacteria, which could impede revegetation. Pilot studies must be carried out to determine whether SMC inoculants can adapt to the local environmental circumstances before applying microbial inoculants because commercial inoculants might not be as effective as native SMCs [4, 15]. Additionally, a long-term study of the impact of inoculants on the native SMCs is required.

Microorganisms have been found to play an active role in restoring heavy metal-contaminated habitats by cleaning up the pollutants in an environmentally safe manner and producing environmentally friendly by-products [17, 18]. Bioremediation is used to recover the contaminated soils, which uses microbial, botanical, microbial-plant related, and other novel approaches. Adding pollutant-degrading microbes (bioaugmentation) can decontaminate sites in contaminated land or water. Additionally, the contaminant levels may be brought down to acceptable levels by biostimulation of natural pollutant-degrading bacteria. *Bacillus cereus, Chlorella pyrendoidosa, Pseudomonas veronii 2E, P. aeruginosa, Serratia marcescens, Saccharomyces cerevisiae, Spirogyra* sp., *Spirulina* sp., and *Cladophora* sp. are examples of naturally occurring and genetically engineered microorganisms that are used to remediate heavy metals such as Cd, Pb, As, Cr, Mn, Cu, U, Se, and Zn from contaminated land and water [17, 19].

BENEFICIAL MICROBES AND THEIR ROLE IN ECOLOGICAL RESTORATION

Bacteria

The diverse bacterial species acting as microbial inoculants are crucial to the beneficial soil process [9, 20]. They are ubiquitous, single-celled organisms that can live a long time because they become dormant for a long time in soil or the environment. Various bacteria can be both helpful and harmful. Plant Growth Promoting Rhizobacteria (PGPR): Naturally occurring organisms colonize the plant roots and boost plant immunity and productivity. PGPR synthesizes low molecular weight siderophores, secondary metabolites that can solubilize and bind iron from the soil to make it available to plants [15]. These compounds can bind metal ions, and Fe (III) has a far stronger affinity for these ions than Fe (II) does. In addition to facilitating iron uptake into the rhizosphere, siderophores may also prevent the formation of diseases that could harm the plant. In rhizospheres with minimal iron, PGPR can create siderophores, which bind available iron and reduce its availability to pathogens, thereby encouraging plant growth. *Ochrobactrum anthropi* TRS-2, isolated from the tea rhizosphere, could solubilize phosphate and generate siderophore and IAA *in vitro* [21]. PGPR produces a large number of bioactive chemicals; it will boost soil diversity and enhance the ability of plants to withstand stress [22]. It is also reported that PGPR induces tolerance to abiotic stressors such as drought, salinity, nutritional deficiencies, *etc*. For example, *Paenibacillus polymyxa* is a PGPR with the same capacity for drought tolerance as *Arabidopsis thaliana*.

Endophytes and diazotrophic bacteria (DB): Endophytes are among the microorganisms with the greatest taxonomic and functional diversity. They are isolated from asymptomatic plant tissues. Since they can produce growth-promoting hormones, they actively enhance plant health by fixing nitrogen, solubilizing phosphate, producing siderophores, and lowering ethylene [23]. Endophytes also increase the ability of the host to withstand biotic, abiotic, and contaminant stressors [24]. *Methylobacterium populi* BJ001T may play a significant role in the metabolism of explosives in poplar plants [25]. Increased biomass production on marginal, nutrient-poor land and improved phytoremediation of volatile organic pollutants may be made possible by the endophyte *P. putida* W619 [79]. Without inducers like toluene or phenol, the poplar endophyte Enterobacter sp. PDN3 may dechlorinate trichloroethylene (TCE) well. This strain appears to be a potential, economical cleanup strategy for TCE-contaminated environments [26].

Diazotrophic bacteria, common in the rhizosphere and found in and around the roots of many different tree crops and herbaceous plants, are a significant source of biologically accessible nitrogen [27]. It may be possible to clarify how plants can develop under low nitrogen environments by enhancing growth through DB application [28]. *Actinobacteria*, or nitrogen-fixing bacteria that produce root nodules, form symbiotic relationships with Cucurbitales, Fagales, and Rosales orders, often known as actinorhizal plants [29]. As a result of symbiotic relationships, shrubs and trees can flourish in arid, moist, polluted, and saltwater environments. Because of this unique trait, Casuarinaceae actinorhizal plants were used to restore desert landscapes [27].

Fungi

These are plant-like microorganisms that lack the pigment chlorophyll (achlorophyllous). While some fungi are minute, others grow to produce much larger structures, such as mushrooms and bracket fungi that develop on moist logs or in soil. Fungi, unlike algae, lack chlorophyll and cannot perform photosynthesis. According to Žifčáková and co-workers [30], they produce a variety of extracellular enzymes that break down soil components to maintain nutritional balance while converting organic matter to carbon dioxide, biomass, and organic acid. The fungus can also take up heavy metals such as Cu, Pb, Hg, and Cd [31]. Several isolates might be exploited in upcoming mycoremediation research as good metal biosorbents. *Absidia cylindrospora* and *Chaetomium atrobrunneum*, two of the 28 studied fungi, absorbed more than 45% of Cd and Pb, whereas *Coprinellus micaceus* absorbed 100% of Pb [32]. Mainly, mycorrhizal fungi play a significant role in restoring degraded ecosystems by developing a long-lasting symbiotic connection with the roots of higher plants [33, 35],

Mycorrhizal fungi can colonize plant roots extracellularly (ectomycorrhizal fungi; EMF) or intracellularly (arbuscular mycorrhizal fungi; AMF) in a mycorrhizal association [33]. MF actively contributes to improved yields by promoting plant growth, nutrient absorption, reproductivity, and a greater ability to withstand environmental stress [34]. Arbuscular mycorrhizal fungi directly involve metal adsorption on the fungal surface and immobilization in the soil by glomalin to detoxify heavy metal contamination. The hyphae of AMF participate in the distribution of heavy metals by chelating and sequestering HMs in their fungal structure, which is one of the tolerance mechanisms that involve the direct involvement of fungi to create a physical barrier to HMs entering plant [36]. According to certain studies, euphoric acid, or fungal melanin, acts protectively when interacting with fungus and abiotic environmental stressors. The advantages of mycorrhizas, diazotrophs, and endophytes in interactions between plants and microbes are shown in Fig. (**2**).

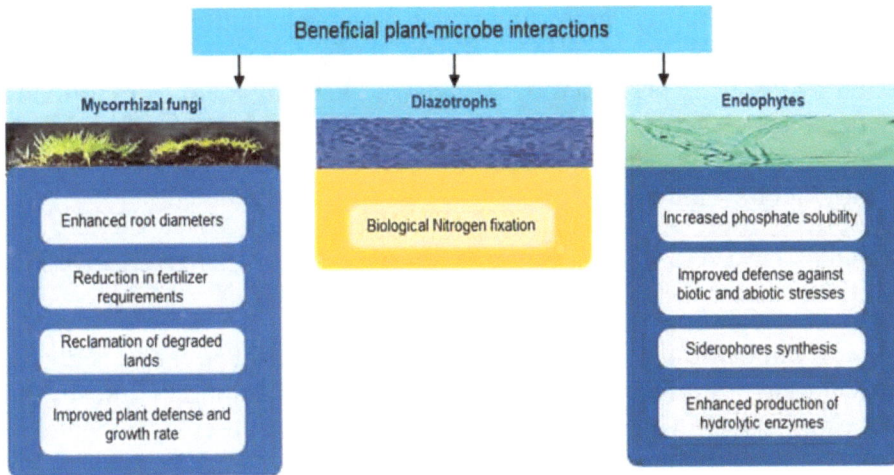

Fig. (2). Beneficial effects of endophytes, diazotrophs, and mycorrhizas on plant-microbe interactions.

Protists

Protists are unicellular eukaryotes that can function in the environment as either producers or consumers [10]. They can be photoautotrophic or chemoheterotrophic. Protists are essential components of food chains, especially in marine habitats. Their capacity to use extracellular cellulase to break down resistant organic substrates is critical to their ecological significance in the microbial food chain. They contribute to the restoration of the original environment by accelerating hydrocarbon decomposition. Sediments from mangrove forests and seagrass beds have been shown to contain exceptionally high concentrations of throustochytrid protists [37]. Their widespread distribution and capacity for degradation demonstrate their ecological importance as decomposers. On benthic ecosystems, they may have substantial effects.

Virus

Viruses are considered the most prevalent living entity in almost every ecosystem on earth, contributing significantly to biodiversity. Viruses are obligate intracellular parasites embedded in the food web's structure and are likely responsible for many ecosystem interactions. Viruses actively participate in the microbial loop and the population ecology of both prokaryotic and eukaryotic microorganisms. The abundance of viruses is influenced by host ecology. Viruses contribute significantly to biodiversity and are intricately linked to ecosystems. A more diverse ecosystem is more adaptable to change and has more functional redundancy. According to Short [38], a crucial mechanism for releasing cellular components necessary for bacterial production and the aquatic food web is thought to be the cell lysis of single-cellular algae by viruses. Viruses may maintain food webs by controlling the host population and limiting the environment's overexploitation by creatures. They can assist in preventing local extinction by lowering predator populations and keeping them from utilizing their prey excessively. Viruses are now recognized as ecological agents that influence the geochemical cycles of the planet. Consequently, viruses are not necessarily dangerous, and the natural virome may benefit ecosystem function [39].

ENVIRONMENTAL FACTORS INFLUENCING MICROBES AND MICROBIOMES FOR ECOSYSTEM RESTORATION

A complex population of microbes, including bacteria, archaea, fungi, algae, protists, and viruses, makes up the microbiome and is among the most important and intricate parts of all ecosystems. These microbes and environmental factors shape the microbiome spatially and temporally [40]. A wide range of environmental factors influence the performance and outcome of ecological interactions. The impact of several edaphic, geographical, and climatic factors on the microbiome is described in more detail below.

Edaphic Factors

Soil pH

The hydronium ion concentration $[H_3O]^+$ in soil (or other systems) is measured by the pH scale, which establishes the acidity or alkalinity of soils. According to Thomas and co-workers [41], soil pH is a common element that determines the organization of microbiome communities. Since soil microorganisms have a wide range of optimal pH tolerance, significant pH fluctuations upset microbial communities and soil microorganisms. Soil acidity or alkalinity variations are frequently followed by changes in the makeup and activity of the microbial community [42]. The ideal pH range needed for microbial growth is between 6-8. Bacterial species would not persist at pH levels above or below their optimum value. However, the optimal growth for bacteria occurs at about neutral pH levels or 6.5 to 7.5. In general, soil pH may have an impact on fungal communities. However, additional environmental and edaphic factors, such as soil physicochemical characteristics, significantly impact the structure and dynamics of soil fungal communities. The tolerance of various bacterial species to soil acidity and basicity varies. For example, the pH of the soil has little effect on the *azospirillum* population. B*radyrhizobium* communities, on the other hand, can flourish in acidic soils, but *Mesorhizobium* communities will struggle in low-pH soils [43].

Soil Temperature

Variations in soil temperature will impact the activity of soil microbes and enzymes. Microorganisms can be divided into mesophiles, psychrophiles, and thermophiles. Mesophiles have optimal growth temperatures between 20 and 45 degrees Celsius, psychrophiles live in cold environments and have ideal growth temperatures between 15 and 20 degrees Celsius, and thermophiles have higher optimum growth temperatures between 50 and higher degrees Celsius [44]. Varying temperatures will impact key enzymes present in N_2-fixing bacteria. Vanadium nitrogenase, for instance, works best for the N_2-fixation process when temperatures are around 5 degrees Celsius, whereas, in warmer temperatures, around 30 degrees Celsius, molybdenum nitrogenase has a stronger affinity for N_2 than vanadium nitrogenase.

Soil Aeration

According to Loreti and Perata [45], waterlogging causes soil hypoxia, a lack of oxygen with limited solubility. Closure of stomata in plant tissues due to hypoxia reduces their ability to assimilate water and inorganic nutrients, which results in energy shortages [46]. Oxygen irreversibly inhibits nitrogenase, even in aerobic species. To safeguard nitrogenase and carry out the process of N_2 fixation, diazotrophs have some physiological modifications that serve as defense mechanisms. This entails limiting oxygen through the growth strategy, geographically and temporally separating nitrogenase from oxygen, and employing biofilms to obstruct oxygen diffusion.

Soil Moisture

Soil moisture directly impacts the physiological state of microorganisms and may reduce their ability to break down specific substances, such as organic substrates [47]. Although microbes in moist soils are diverse, too much moisture in the soil can lead to oxygen levels that are unfavorable to aerobic bacteria like gram-negative, gram-positive, and mycorrhizal fungi, leading to decreased microbial biomass. A soil environment with too much water is particularly hazardous to aerobic microorganisms since water has a significantly lower O_2 availability than air [48].

Soil Nutrients

According to Miransari [49], soil, climate, and plant characteristics all affect nutrient availability in the soil. Soil nutrient content influences the abundance of microbes. Microorganisms are essential for the decomposition of organic materials and the cycling of nutrients for plants. PGPR produces a broad range of chemicals, including enzymes like phosphatases and plant hormones such as auxins, which can alter the soil's availability of nutrients like N by fixing them. Soil bacteria like Pseudomonas spp. might influence the availability of soil micronutrients. Pseudomonas may have accelerated the uptake of minerals into the host plants, particularly phosphate formation of IAA, cytokinin production, and regulation of ethylene synthesis in roots, as well as by solubilizing nutrients like phosphorus [50] and also by chelating micronutrients through the production of carboxylates and humic compounds, which can dissolve mineral oxides [51].

Geographical Factors

Due to environmental heterogeneity, the composition of soil microbial communities varies over vast spatial scales (such as landscapes, regions, and continents). Greater research is needed on microbial communities since geographic distance and factors like latitude, altitude, and the rate and degree of disturbance are crucial elements of the microorganism communities' organization and diversity. Microbe biogeography is still poorly understood. Altitude is the main element affecting both biodiversity and the physio-chemical characteristics of soil in a complex way [52, 53]. Low temperatures, variable precipitation, soil nutrient stress, and decreased air pressure are often attributes of high-altitude habitats, significantly impacting biodiversity. Cold-adapted microbes have successfully colonized high-altitude, cold habitats, which comprise the vast bulk of the planet's biosphere.

These organisms can live and grow in these conditions and function metabolically even when temperatures drop below zero. Generally, the biogeographical phenomenon is governed by various abiotic components rather than only by geographic considerations. The major abiotic factors are temperature, light, water, wind, pressure, and radiation [52], which have documented a decreasing diversity of soil bacteria in the Colorado Rocky Mountains (2400–3400 m) with elevation. In contrast, there was no proof of an elevational gradient in the colonies of soil bacteria in a Peruvian montane region (200-3550 m) [53]. Fierer and Jackson [54] noted a tendency whereby bacterial diversity changed with latitude. In addition, according to a study [55], the variety of bacteria in Arctic soil was comparable to that of other biomes.

Climatic Factors

Soil microbial taxa are imperative with global climate changes as they play vital and undisputable roles in the mineralization of soil organic carbon, stabilization of carbon inputs into organic form, and plant growth [56]. Climatic factors such as temperature, precipitation, ultraviolet radiation, atmospheric CO_2, and unpredictable rainfall impact ecological mechanisms and microbial colonies in several regions across the globe. In this section, we examine the responses soil microorganisms employ to adjust to the shifting environmental conditions brought on by climate change. The critical climatic factors are depicted in Fig. (2).

UV Radiation

Ultra-violet (UV) radiation increases as stratospheric O_3 decreases due to ozone-depleting compounds [57]. Due to the ozone layer's exceptional thinness in the Polar regions, these consequences are extreme, and ecosystems have become vulnerable. Increased UV-B radiation can affect plants in positive and negative ways. While diminishing crop yields, shoot and root biomass, crop respiration efficiency, and water usage efficiency, increased UV-B promotes increased hormones, flavonoids, and amino acids. According to Zeeshan and Prasad [58], UV-B radiation directly impacts soil microorganisms by changing their pigment content, growth, and ability to induce carbon assimilation. UV-B radiation has a detrimental effect on the soil's macro and mesofauna.

Additionally, it shifts the fungal community by strengthening the competitive abilities of fungi with dark pigments. UV-B tolerance exists in *Tolypocladium sp*, an entomopathogenic fungus. UV-B is known to hasten the decomposition of various contaminants, including phenyl urea herbicides, 2,4-dichloro phenoxy acetic acid, biphenol, Z, and PAHs [59].

Elevated CO_2

Atmospheric CO_2 concentration has increased due to the Industrial Revolution and the emission of fossil fuels, which has resulted in global warming. Increased CO_2 (e CO_2) causes altered physiology and composition of the microbial colonies in the soil, favoring bacteria [60]. This is due to decreased soil nitrogen inputs, a decrease in the population of taxonomic units within the Firmicutes, and an abundance of Gram-positive bacteria in rhizosphere soils. Microbial responses to elevated CO_2 have been studied with free-air CO_2 enrichment experiments to assess how long-term exposure to high and ambient CO_2 levels affects different ecosystems and observed shifts in microbiomes with elevated CO_2 [61]. According to a study that combined modeling and meta-analysis, eCO_2 initially promotes photosynthetic and carbon inputs into soil. On a decadal time scale, however, eCO_2 accelerated SOM microbial degradation [62]. Although the total fungal population was higher in the studies conducted on Australian grassland [34], there was a decline when eCO_2 was combined with heat. Increased soil enzymatic activities of protease, invertase, phenol oxidase, and aryl-sulphatase in the central root zone are brought on by elevated CO_2 rather than changes in microbial abundance [63].

Temperature

Increases in temperature typically result in positive responses from microbial biomass, albeit this also depends on other environmental factors, including the availability of nitrogen and the amount of water in the soil [64]. The capacity of microbes to spread, live, and occupy soil cavities is constrained by rising soil temperatures, which also decrease the amount of water in the soil [65]. The structure of the rhizosphere's microbiome is also changed by soil heating brought on by rising ambient temperatures. Modifications to the lipid content of cell membranes that reduce membrane fluidity and heat expression are among the physiological reactions of bacteria to higher temperatures [61]. According to [66], there were significant differences between temperature treatments (5–25°C) in the reservoirs of substrates for microbial respiration and the prevalence of Gram-positive and Gram-negative microorganisms used as biomarkers. That proves that microbial communities have changed in function and composition in response to soil warming.

Permafrost Thaw

Most of the earth's terrestrial cryosphere comprises permafrost, which provides a particular ecological niche for microorganisms that can withstand freezing temperatures. Although there is some variation in community composition between distinct permafrost features and between sites, permafrost has a relatively high level of microbial diversity. Certain bacteria are even active in permafrost at subzero temperatures [67]. The Arctic permafrost soil thawing is a severe effect of global warming. Liquid water becomes more accessible, and microbial activity rises as the permafrost thaws. As a result, there may be an increase in the decomposition of soil organic carbon and the formation of carbon dioxide and methane. Most studies show increased Actinobacteria with permafrost depth, but the species can differ depending on location. Edaphic and climatic factors involved in the microbiomes are demonstrated in Fig. (**3**).

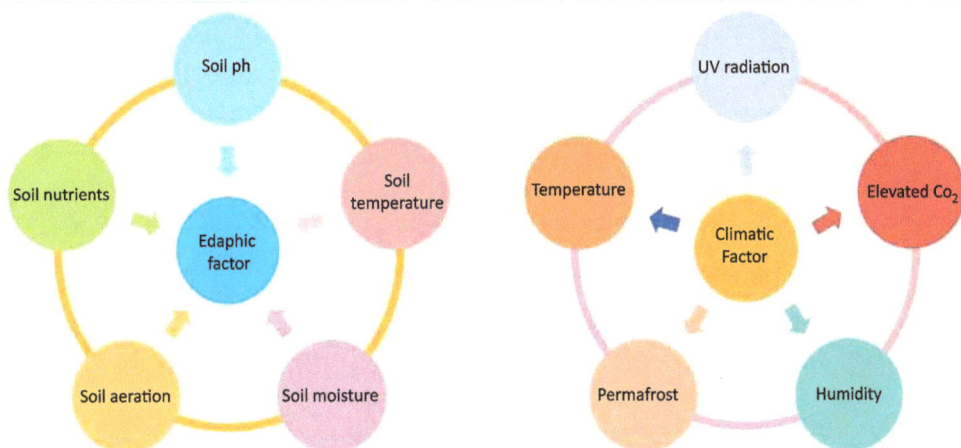

Fig. (3). Edaphic and climatic factors involved in the microbiomes.

ABIOTIC STRESSORS AFFECTING SOIL MICROBIOME

Drought

One of the significant barriers to plant growth, particularly in dry and semiarid regions, is water availability. It is anticipated that increasing dryness will lead to decreases in microbial services crucial for the sustainability of ecosystems [68]. A reduction in soil water potential is followed by a decline in nutrient mineralization and respiration, resulting in a fall in the potential metabolic microbial activity [69]. Numerous soil microorganisms, such as free-living and symbiotic bacteria and mycorrhizal fungi, can boost plant drought tolerance by altering biochemical and physiological crop characters and gene expression. There has not been a lot of research on how drought-tolerant or drought-affected soil microbes affect plants in natural environments, but if helpful soil microbes that promote tolerance to drought are found, they will make good targets for plant inoculation. Lau and Lennon [70] discovered that changes in the soil microbial colonies brought about by long-term adjustments to the water supply were more significant in determining the fitness of native plants than plants' adaptive responses. They discovered that drought-prone plants, when cultivated on soils with a history of drought, performed better than those grown in soils that had previously received adequate water, and this increased fitness was linked to rapid responses in the microbial colonies of soil. In multi-year field trials and network analyses in mesocosms, it was discovered that grassland bacteria are more vulnerable to dryness than fungi [71].

Submergence

Flooding causes the soil to become compacted with water, preventing the atmosphere, soil, and microorganisms from exchanging gases freely. As a result, there is a significant decrease in the amount of oxygen in the soil. Additionally, flooding is known to alter soil pH and nutrient status, which might impact the dispersion of the soil microbial population [72]. The soil's microbial population prefers anaerobic microbes during submergence, while obligate aerobic organisms eventually decline. Additionally, it results in the transfer of waterborne microorganisms to the soil. According to Furtak and co-workers [73], obligate aerobic bacteria like Xanthomonadaceae entirely vanished due to flooding, whereas anaerobic bacteria like Malikia and Anaeromyxobacter may only be found after flooding events. According to Rodriguez-Heredia and co-workers [74], AMF density rose in saline flood situations. Numerous investigations revealed that decreased precipitation boosted AMF community density by reducing soil humidity and elevating oxygen content. Higher precipitation, according to other studies, may promote AMF colonization. These discrepancies in the results could be brought about by variations in the host plant, soil temperature, soil texture, and soil pH [72].

Salinity

Especially in dryland ecosystems, improper irrigation, drainage, and land clearance techniques lead to soil salinization, resulting in land degradation becoming a more widespread issue. Salinity impairs plant growth and decreases soil microbes' activity due to high osmotic pressures, toxic ions, and nutritional imbalances [72]. Soil microorganisms can significantly improve plant growth and encourage plant salt tolerance. Arbuscular mycorrhizal fungi, for instance, can increase water use efficiency, osmo-protection, and the expression of genes for sodium and potassium ion transporters and channels. On the other hand, it has been demonstrated that specific ectomycorrhizal fungi strains improve plants' ability to omit sodium ions from the soil [75]. High salinity-conditioned soil microbial communities may efficiently raise the salinity tolerance of the plant. According to several studies, plants inoculated with salt-conditioned AMF grew more quickly and had better nutrient contents (P, K, Zn) and root proline than plants infected with unconditioned fungi [76]. Zhang and co-workers [77] found that AMF on saline-alkaline soils with patches of plants accelerated the re-establishment of native perennial grass in grassland damaged by salinity and a low pH condition.

Pollutants

Pollution is one of the main factors driving better stress tolerance in plants, supported by soil microbial conditioning. Heavy metal, hydrocarbon, and other toxic compound-contaminated soils constitute a substantial problem for revegetation since they can significantly restrict plant growth [78]. Belowground organisms may be crucial in restoring and phytoremediation places affected by hazardous waste, mining, military activities, and industrial pollution, even if these pollutants may significantly negatively influence the soil biota [79]. Phytoremediation involves several processes where the pollutants are extracted, degraded, immobilized, and volatilized. These processes may also involve plant endophytes, mycorrhizal fungi, and tolerant soil microorganisms [80].

In Silicon Valley, California, researchers discovered a naturally occurring bacterial endophyte in poplar trees that was isolated from a trichloroethylene (TCE) contaminated location. This bacterial endophyte quickly metabolized TCE, increased plant growth, and reduced the phytotoxic effects on plants [81]. This is just one example of the potential of microbially mediated phytoremediation. Through plant-microbial interactions, certain bacteria from the rhizosphere have been found to digest persistent organic pollutants [82] effectively. This evidence suggests that microbially mediated phytoremediation could be a powerful tool in restoring and detoxifying areas polluted by combinations of organic and inorganic pollutants and successful applications for single soil contaminants. Fig. (**4**) below illustrates the influence of abiotic stressors on the microbial population in the soil.

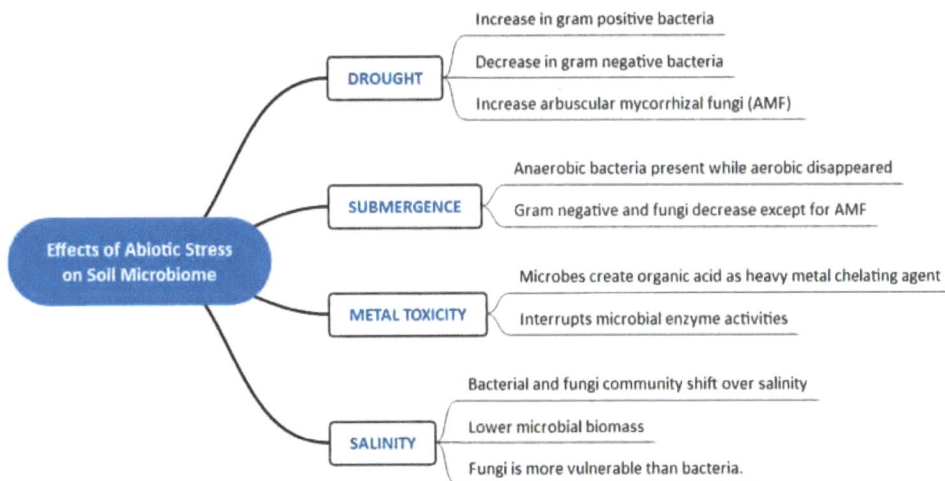

Fig. (4). Effect of changes in abiotic stressors on the soil microbial composition.

MICROBIOME ENGINEERING FOR ECOLOGICAL RESTORATION

Microbiomes, composed of various microbial populations, exist in every ecosystem. When these microbiomes are disturbed, the hosts experience unfavorable phenotypes, leading to chaos and disrupting the balance of surrounding ecosystems. However, microbiome engineering offers a promising solution. By altering the structural composition of the microbiota, it can help re-establish ecological equilibrium [83]. One particularly intriguing aspect of this research is the potential of microbiome transfer to reduce drought stress and increase biomass under drought conditions, using Arabidopsis thaliana as a specimen plant [84]. The microbial communities found in plant roots and the rhizosphere are the key components of the plant microbiome. The community structure and diversity of the plant microbiome are influenced by several factors relating to the host, microorganisms, and environment. The potential of microbiome engineering in improving plant growth, fitness, and health is a beacon of hope for the future of agriculture and environmental science. The microbiome connected to roots can be engineered to improve these characteristics to change its networking and composition. Techniques to create the root-associated microbiome include microbiome transfer, synthetic microbiomes, and host-mediated artificial selection [83].

Another method of microbiota engineering is the creation of 'synthetic microbiomes'. These are artificially assembled microbial communities with specific desired functionalities. For instance, rhizobacteria, both singular and in combinations, have been extensively used to enhance plant growth and yield [85]. The advantage of synthetic microbiomes is that they can be precisely controlled in terms of their microbial composition. However, it is important to note that only culturable bacteria can be incorporated into synthetic microbiomes, and our understanding of the core microbiome limits the creation of synthetic microbiomes. The ecology can change, and plant microbiomes can evolve through host-mediated artificial selection. For example, the improved tolerance of *Brassica rapa* to environmental stressors due to repeated replication of its best-performing plants is an example of how a plant's microbiome can develop to adapt to its growing conditions for increased fitness [86]. Chemical compounds, sometimes known as root exudates, have demonstrated the potential for altering the plant microbiome.

Lebeis [87] found that the assembly of a typical root microbiome in *A. thaliana* requires the manufacture of salicylic acid. This study is the first to demonstrate how a key plant defense system regulator quickly affects the makeup of the root-based microbiome. The primary regulators are used to modify the composition of plant microorganisms for greater production and resilience to environmental stressors. The engineering of soil microbiomes can be employed through organic farming, changes in land utilization, *etc*. A significant difference in the composition and functionality of the soil microbiome can result from a change in vegetation, as demonstrated by the Southeast Asian oil palm plantations that replaced tropical forests [88]. However, the field of microbiome engineering is relatively young and has a lot of untapped potential. Strategies used for engineering microbiomes for ecological restoration [83] are mentioned in Table **2**.

Table 2. Strategies used in microbiome engineering for ecosystem restoration.

Engineering methods	Microbiomes and their examples	Purpose	Remarks	References
Microbiome transfer	*Plants* a) Transplanting the local microflora from the roots	Combat environmental challenges, prevent plant infections, and encourage plant growth and development.	Transfer of advantageous soil microbes to less favorable ones, simple to manipulate, limited by the lack of a local microbiome that is capable of functioning.	[89]
	b) Synthetic transplant of the microbiota associated with roots	Protect against plant infections, withstand environmental challenges, and encourage growth	Only suitable to culturable microorganisms, customizable microbial make-up for positive effects, knowledge of the core microbiome limits application.	[90]
Signaling molecules	*Plants* Using root exudates, such as salicylic acid, as a treatment	Withstand environmental pressures while encouraging growth.	Encourage the development of a healthy microbiome, restricted by the lack of signaling molecules.	[91]
Agricultural management	*Soil* Altering land use, cropping, tillage, logging, and organic farming	Increasing fertility of the soil	Widespread implementation, Modify the structure and function of the soil microbiome, increase soil microbial diversity, biological cycles, and activity	[92]

BIOAUGMENTATION FOR RE-ESTABLISHMENT OF A HEALTHY MICROBIOME

Bioaugmentation is the process of adding functionally essential species to a complex system to enhance its bioremediation ability [93, 94]. In wastewater-activated sludge treatment plants, 'sludge seeding' is a common bioaugmentation method to establish a new bioreactor. When 'sludge seeding,' an entire microbial community is transplanted; the community must come from a system with a similar operating principle to the augmented one. Isolated microbial strain sourcing through selective enrichment culture is frequently utilized to improve bioremediation when a wastewater-activated sludge treatment system cannot break down the waste effectively. These chosen strains serve as bioaugmentation agents with clearly defined and essential functions. Many attempts to improve the bioremediation capacity of wastewater-activated sludge treatment have repeatedly failed. 'Sludge seeding' is the most effective way to tackle this issue if chosen functional strains or other methods [95], such as altering the essential physicochemical characteristics of wastewater (*e.g.,* adding a carbon source) and switching the operation mode, are unsuccessful. The introduction of microbes to re-establish a healthy microbiome for successful ecosystem restoration [95, 96] is shown in Fig. (**5**).

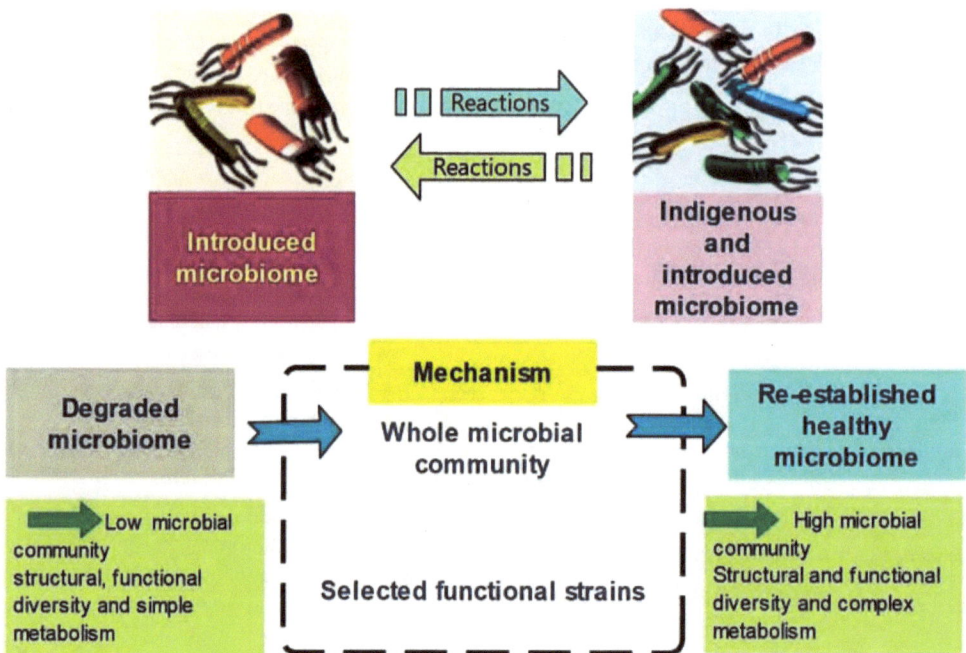

Fig. (5). Introduction of microbes for re-establishment of a healthy microbiome.

CONCLUDING REMARKS

The UN's designation of 2021–2030 as the Decade on Ecosystem Restoration emphasizes the severity of the ecosystem degradation issue and focuses on speeding up ecosystem restoration. The use of microbes and microbiomes can significantly improve attempts to restore ecosystems. As one of the most significant and intricate parts of all terrestrial ecosystems, the soil microbiome is home to hundreds of millions of microorganisms. Through the breakdown of organic matter and nutrient cycling, these microbes play an essential part in the soil ecosystem. These organisms and the environment impact how the soil microbiome develops across time and space. To improve climate impact projections and ultimately create microbial strategies, a greater understanding of how climate change affects soil microorganism functions is necessary to counteract further climate warming and soil degradation. Future studies must evaluate how changes to one ecosystem's microbiome affect the functioning of connected ecosystems' microbiomes and ecosystems. Successful applications of microbiome engineering to change the microbiota have shown promising outcomes for ecological restoration. Complete ecological restoration will speed up the achievement of objectives and aims specified in the ecosystem restoration programs and Sustainable Development Goals (SDGs) with the careful integration of microbe-assisted practices, their analysis, and evaluation. A brighter future for everyone can be ensured by carefully integrating microbe-assisted practices, analysis, and assessment to enable holistic ecological restoration.

ACKNOWLEDGEMENTS

The authors would like to thank the Division of Environmental Science, ICAR-Indian Agricultural Research Institute, New Delhi, India, for their continual encouragement and unflinching support.

REFERENCES

[1]   Valliere JM, Wong WS, Nevill PG, Zhong H, Dixon KW. Preparing for the worst: Utilizing stress☐tolerant soil microbial communities to aid ecological restoration in the Anthropocene. Ecol Solut Evid 2020; 1(2): e12027.
[http://dx.doi.org/10.1002/2688-8319.12027]

[2]   Clewell A, Aronson J, Winterhalder K. The SER international primer on ecological restoration https://cdn.ymaws.com/www.ser.org/resource/resmgr/custompages/publications/ser_publications/ser_primer.pdf

[3]   Moss ED, Evans DM, Atkins JP. Investigating the impacts of climate change on ecosystem services in UK agro-ecosystems: An application of the DPSIR framework. Land Use Policy 2021; 105: 105394.
[http://dx.doi.org/10.1016/j.landusepol.2021.105394]

[4]   Prasad S, Malav LC, Choudhary J, *et al.* Soil microbiomes for healthy nutrient recycling. Current trends in microbial biotechnology for sustainable agriculture. 2021:1-21.
[http://dx.doi.org/10.1007/978-981-15-6949-4_1]

[5]   Fan K, Wang W, Xu X, *et al.* Recent advances in biotechnologies for the treatment of environmental pollutants based on reactive Sulfur species. Antioxidants 2023; 12(3): 767.
[http://dx.doi.org/10.3390/antiox12030767] [PMID: 36979016]

[6]   Rao DL, Aparna K, Mohanty SR. Microbiology and biochemistry of soil organic matter, carbon sequestration, and soil health. Indian J Fert 2019; 15(2): 124-38.

[7]   Tarhriz V, Hirose S, Fukushima SI, *et al.* Emended description of the genus Tabrizicola and the species Tabrizicola aquatica as aerobic anoxygenic phototrophic bacteria. Antonie van Leeuwenhoek. 2019 Aug 15;112:1169-75.

[http://dx.doi.org/10.1007/s10482-019-01249-9]

[8]   Koshila Ravi R, Anusuya S, Balachandar M, Muthukumar T. Microbial interactions in soil formation and nutrient cycling. Mycorrhizosphere and pedogenesis. 2019:363-82.

[9]   Saia S, Tamayo E, Schillaci C, De Vita P. Arbuscular mycorrhizal fungi and nutrient cycling in cropping systems. Carbon and nitrogen cycling in soil. 2020:87-115.
       [http://dx.doi.org/10.1007/978-981-13-7264-3_4]

[10]   Singh AK, Sisodia A, Sisodia V, Padhi M. Role of microbes in restoration ecology and ecosystem services. New and future developments in microbial biotechnology and bioengineering 2019 Jan 1 (pp. 57-68). Elsevier.
       [http://dx.doi.org/10.1016/B978-0-444-64191-5.00004-3]

[11]   Berg G, Rybakova D, Fischer D, *et al.* Microbiome definition re-visited: old concepts and new challenges. Microbiome 2020; 8: 1-22.

[12]   Meisner A, Wepner B, Kostic T, *et al.* Calling for a systems approach in microbiome research and innovation. Curr Opin Biotechnol 2022; 73: 171-8.
       [http://dx.doi.org/10.1016/j.copbio.2021.08.003] [PMID: 34479027]

[13]   Seneviratne G, Kulasooriya SA. Reinstating soil microbial diversity in agroecosystems: The need of the hour for sustainability and health. Agric Ecosyst Environ 2013; 164: 181-2.
       [http://dx.doi.org/10.1016/j.agee.2012.10.002]

[14]   Seneviratne G, Thilakaratne RM, Jayasekara AP, Seneviratne KA, Padmathilake KR, De Silva MS. Developing beneficial microbial biofilms on roots of non legumes: a novel biofertilizing technique. Microbial strategies for crop improvement. 2009:51-62.
       [http://dx.doi.org/10.1007/978-3-642-01979-1_3]

[15]   Gupta G, Parihar SS, Ahirwar NK, Snehi SK, Singh V. Plant growth promoting rhizobacteria (PGPR): current and future prospects for development of sustainable agriculture. J Microb Biochem Technol. 2015 Mar;7(2):096-102.

[16]   Beneduzi A, Ambrosini A, Passaglia LMP. Plant growth-promoting rhizobacteria (PGPR): their potential as antagonists and biocontrol agents. Genet Mol Biol 2012; 35(4 suppl 1): 1044-51.
       [http://dx.doi.org/10.1590/S1415-47572012000600020] [PMID: 23411488]

[17]   Priya AK, Gnanasekaran L, Dutta K, Rajendran S, Balakrishnan D, Soto-Moscoso M. Biosorption of heavy metals by microorganisms: Evaluation of different underlying mechanisms. Chemosphere 2022; 307(Pt 4): 135957.
       [http://dx.doi.org/10.1016/j.chemosphere.2022.135957] [PMID: 35985378]

[18]   Gebregiorgis Ambaye T, Vaccari M, Franzetti A, *et al.* Microbial electrochemical bioremediation of petroleum hydrocarbons (PHCs) pollution: Recent advances and outlook. Chem Eng J 2023; 452: 139372.
       [http://dx.doi.org/10.1016/j.cej.2022.139372]

[19]   Prasad S, Yadav KK, Kumar S, *et al.* Chromium contamination and effect on environmental health and its remediation: A sustainable approaches. J Environ Manage 2021; 285: 112174.
       [http://dx.doi.org/10.1016/j.jenvman.2021.112174] [PMID: 33607566]

[20]   Singh RK, Chang HW, Yan D, *et al.* Influence of diet on the gut microbiome and implications for human health. J Transl Med 2017; 15(1): 73.
       [http://dx.doi.org/10.1186/s12967-017-1175-y] [PMID: 28388917]

[21]   Chakraborty U, Chakraborty BN, Basnet M, Chakraborty AP. Evaluation of *Ochrobactrum anthropi* TRS-2 and its talc based formulation for enhancement of growth of tea plants and management of brown root rot disease. J Appl Microbiol 2009; 107(2): 625-34.
       [http://dx.doi.org/10.1111/j.1365-2672.2009.04242.x] [PMID: 19426277]

[22]   Egamberdieva D, Wirth S, Bellingrath-Kimura SD, Mishra J, Arora NK. Salt-tolerant plant growth promoting rhizobacteria for enhancing crop productivity of saline soils. Front Microbiol 2019; 10:

2791.
[http://dx.doi.org/10.3389/fmicb.2019.02791] [PMID: 31921005]

[23]　Santoyo G, Moreno-Hagelsieb G, del Carmen Orozco-Mosqueda M, Glick BR. Plant growth-promoting bacterial endophytes. Microbiological research. 2016 Feb 1;183:92-9. Recent Pat Biotechnol 2016; 183: 92-.
[PMID: 20201804]

[24]　Mei C, Flinn BS. The use of beneficial microbial endophytes for plant biomass and stress tolerance improvement. Recent patents on biotechnology. 2010 Jan 1;4(1):81-95.

[25]　Van Aken B, Agathos SN. Biodegradation of nitro-substituted explosives by white-rot fungi: A mechanistic approach. Adv Appl Microbiol 2001; 48: 1-77.
[http://dx.doi.org/10.1016/S0065-2164(01)48000-2] [PMID: 11677677]

[26]　Kang JW, Khan Z, Doty SL. Biodegradation of trichloroethylene by an endophyte of hybrid poplar. Appl Environ Microbiol 2012; 78(9): 3504-7.
[http://dx.doi.org/10.1128/AEM.06852-11] [PMID: 22367087]

[27]　Peters JW, Boyd ES. Exploring alternative paths for the evolution of biological nitrogen fixation. Biological Nitrogen Fixation. 2015 Aug 3:167-76.
[http://dx.doi.org/10.1002/9781119053095.ch16]

[28]　Knoth JL, Kim SH, Ettl GJ, Doty SL. Biological nitrogen fixation and biomass accumulation within poplar clones as a result of inoculations with diazotrophic endophyte consortia. New Phytol 2014; 201(2): 599-609.
[http://dx.doi.org/10.1111/nph.12536] [PMID: 24117518]

[29]　Reed J, van Vianen J, Foli S, *et al.* Trees for life: The ecosystem service contribution of trees to food production and livelihoods in the tropics. For Policy Econ 2017; 84: 62-71.
[http://dx.doi.org/10.1016/j.forpol.2017.01.012]

[30]　Žifčáková L, Větrovský T, Howe A, Baldrian P. Microbial activity in forest soil reflects the changes in ecosystem properties between summer and winter. Environ Microbiol 2016; 18(1): 288-301.
[http://dx.doi.org/10.1111/1462-2920.13026] [PMID: 26286355]

[31]　Ali H, Khan E, Sajad MA. Phytoremediation of heavy metals—Concepts and applications. Chemosphere 2013; 91(7): 869-81.
[http://dx.doi.org/10.1016/j.chemosphere.2013.01.075] [PMID: 23466085]

[32]　Albert Q, Baraud F, Leleyter L, *et al.* Use of soil fungi in the biosorption of three trace metals (Cd, Cu, Pb): promising candidates for treatment technology? Environ Technol 2020; 41(24): 3166-77.
[http://dx.doi.org/10.1080/09593330.2019.1602170] [PMID: 30924724]

[33]　Norby RJ, De Kauwe MG, Walker AP, Werner C, Zaehle S, Zak DR. Comment on "Mycorrhizal association as a primary control of the CO_2 fertilization effect". Science 2017; 355(6323): 358.
[http://dx.doi.org/10.1126/science.aai7976] [PMID: 28126781]

[34]　Averill C, Turner BL, Finzi AC. Mycorrhiza-mediated competition between plants and decomposers drives soil carbon storage. Nature 2014; 505(7484): 543-5.
[http://dx.doi.org/10.1038/nature12901] [PMID: 24402225]

[35]　Smith SE, Read DJ. Mycorrhizal Symbiosis, 3rd ed. Academic Press, San Diego.; 2010 Jul 26.

[36]　Riaz M, Kamran M, Fang Y, *et al.* Arbuscular mycorrhizal fungi-induced mitigation of heavy metal phytotoxicity in metal contaminated soils: A critical review. J Hazard Mater 2021; 402: 123919.
[http://dx.doi.org/10.1016/j.jhazmat.2020.123919] [PMID: 33254825]

[37]　Jain R, Raghukumar S, Tharanathan R, Bhosle NB. Extracellular polysaccharide production by thraustochytrid protists. Mar Biotechnol (NY) 2005; 7(3): 184-92.
[http://dx.doi.org/10.1007/s10126-004-4025-x] [PMID: 15909227]

[38]　Short SM. The ecology of viruses that infect eukaryotic algae. Environ Microbiol 2012; 14(9): 2253-

71.
[http://dx.doi.org/10.1111/j.1462-2920.2012.02706.x] [PMID: 22360532]

[39]   French RK, Holmes EC. An ecosystems perspective on virus evolution and emergence. Trends Microbiol 2020; 28(3): 165-75.
[http://dx.doi.org/10.1016/j.tim.2019.10.010] [PMID: 31744665]

[40]   Islam W, Noman A, Naveed H, Huang Z, Chen HYH. Role of environmental factors in shaping the soil microbiome. Environ Sci Pollut Res Int 2020; 27(33): 41225-47.
[http://dx.doi.org/10.1007/s11356-020-10471-2] [PMID: 32829437]

[41]   Thomas A, Tripathi RK, Yadu LK. A laboratory investigation of soil stabilization using enzyme and alkali-activated ground granulated blast-furnace slag. Arab J Sci Eng 2018; 43(10): 5193-202.
[http://dx.doi.org/10.1007/s13369-017-3033-x]

[42]   O'Sullivan D, Murray BJ, Ross JF, Webb ME. The adsorption of fungal ice-nucleating proteins on mineral dusts: a terrestrial reservoir of atmospheric ice-nucleating particles. Atmos Chem Phys 2016; 16(12): 7879-87.
[http://dx.doi.org/10.5194/acp-16-7879-2016]

[43]   Smercina DN, Evans SE, Friesen ML, Tiemann LK. To fix or not to fix: controls on free-living nitrogen fixation in the rhizosphere. Appl Environ Microbiol 2019; 85(6): e02546-18.
[http://dx.doi.org/10.1128/AEM.02546-18] [PMID: 30658971]

[44]   Keenleyside CB, Beaufoy G, Alison J, *et al.* Technical Annex 4: Building ecosystem resilience. Report to Welsh Government (Contract C210/2016/2017).

[45]   Loreti E, Perata P. The many facets of hypoxia in plants. Plants 2020; 9(6): 745.
[http://dx.doi.org/10.3390/plants9060745] [PMID: 32545707]

[46]   Li Y, Niu W, Zhang M, Wang J, Zhang Z. Artificial soil aeration increases soil bacterial diversity and tomato root performance under greenhouse conditions. Land Degrad Dev 2020; 31(12): 1443-61.
[http://dx.doi.org/10.1002/ldr.3560]

[47]   Parr JF, Epstein E, Willson GB. Composting sewage sludge for land application. Agric Environ 1978; 4(2): 123-37.
[http://dx.doi.org/10.1016/0304-1131(78)90016-4]

[48]   da Silva CARDOSO A, Junqueira JB, Reis RA, Ruggieri AC. How do greenhouse gas emissions vary with biofertilizer type and soil temperature and moisture in a tropical grassland? Pedosphere 2020; 30(5): 607-17.
[http://dx.doi.org/10.1016/S1002-0160(20)60025-X]

[49]   Miransari M. Soil microbes and the availability of soil nutrients. Acta Physiol Plant 2013; 35(11): 3075-84.
[http://dx.doi.org/10.1007/s11738-013-1338-2]

[50]   Muthukumar A, Raj TS, Prabhukarthikeyan SR, Kumar RN, Keerthana U. Pseudomonas and Bacillus: A biological tool for crop protection. New and Future Developments in Microbial Biotechnology and Bioengineering 2022 Jan 1 (pp. 145-158). Elsevier.
[http://dx.doi.org/10.1016/B978-0-323-85577-8.00006-8]

[51]   García-Hernández JL, David Valdez-Cepeda R, Murillo-Amador B, *et al.* Compositional nutrient diagnosis and main nutrient interactions in yellow pepper grown on desert calcareous soils. J Plant Nutr Soil Sci 2004; 167(4): 509-15.
[http://dx.doi.org/10.1002/jpln.200320370]

[52]   Bryant JA, Lamanna C, Morlon H, Kerkhoff AJ, Enquist BJ, Green JL. Microbes on mountainsides: Contrasting elevational patterns of bacterial and plant diversity. Proc Natl Acad Sci USA 2008; 105(Suppl 1) (Suppl. 1): 11505-11.
[http://dx.doi.org/10.1073/pnas.0801920105] [PMID: 18695215]

[53]   Fierer N, McCain CM, Meir P, *et al.* Microbes do not follow the elevational diversity patterns of

plants and animals. Ecology 2011; 92(4): 797-804.
[http://dx.doi.org/10.1890/10-1170.1] [PMID: 21661542]

[54]   Fierer N, Jackson RB. The diversity and biogeography of soil bacterial communities. Proc Natl Acad
       Sci USA 2006; 103(3): 626-31.
       [http://dx.doi.org/10.1073/pnas.0507535103] [PMID: 16407148]

[55]   Chu D, Gao CS, De Barro P, Zhang YJ, Wan FH, Khan IA. Further insights into the strange role of
       bacterial endosymbionts in whitefly, *Bemisia tabaci*: Comparison of secondary symbionts from
       biotypes B and Q in China. Bull Entomol Res 2011; 101(4): 477-86.
       [http://dx.doi.org/10.1017/S0007485311000083] [PMID: 21329550]

[56]   Dubey A, Malla MA, Khan F, *et al.* Soil microbiome: a key player for conservation of soil health
       under changing climate. Biodivers Conserv 2019; 28(8-9): 2405-29.
       [http://dx.doi.org/10.1007/s10531-019-01760-5]

[57]   Formánek P, Rejšek K, Vranová V. Effect of elevated CO2, O3, and UV radiation on soils.
       ScientificWorldJournal 2014; 2014(1): 1-8.
       [http://dx.doi.org/10.1155/2014/730149] [PMID: 24688424]

[58]   Zeeshan M, Prasad SM. Differential response of growth, photosynthesis, antioxidant enzymes and
       lipid peroxidation to UV-B radiation in three cyanobacteria. S Afr J Bot 2009; 75(3): 466-74.
       [http://dx.doi.org/10.1016/j.sajb.2009.03.003]

[59]   Dong J, Wan G, Liang Z. Accumulation of salicylic acid-induced phenolic compounds and raised
       activities of secondary metabolic and antioxidative enzymes in Salvia miltiorrhiza cell culture. J
       Biotechnol 2010; 148(2-3): 99-104.
       [http://dx.doi.org/10.1016/j.jbiotec.2010.05.009] [PMID: 20576504]

[60]   Frey B, Kremer J, Rüdt A, Sciacca S, Matthies D, Lüscher P. Compaction of forest soils with heavy
       logging machinery affects soil bacterial community structure. Eur J Soil Biol 2009; 45(4): 312-20.
       [http://dx.doi.org/10.1016/j.ejsobi.2009.05.006]

[61]   Jansson JK, Hofmockel KS. Soil microbiomes and climate change. Nat Rev Microbiol 2020; 18(1):
       35-46.
       [http://dx.doi.org/10.1038/s41579-019-0265-7] [PMID: 31586158]

[62]   Drake JE, Gallet-Budynek A, Hofmockel KS, *et al.* Increases in the flux of carbon belowground
       stimulate nitrogen uptake and sustain the long-term enhancement of forest productivity under elevated
       CO2. Ecol Lett 2011; 14(4): 349-57.
       [http://dx.doi.org/10.1111/j.1461-0248.2011.01593.x] [PMID: 21303437]

[63]   Kandeler E, Mosier AR, Morgan JA, *et al.* Response of soil microbial biomass and enzyme activities
       to the transient elevation of carbon dioxide in a semi-arid grassland. Soil Biol Biochem 2006; 38(8):
       2448-60.
       [http://dx.doi.org/10.1016/j.soilbio.2006.02.021]

[64]   French S, Levy-Booth D, Samarajeewa A, Shannon KE, Smith J, Trevors JT. Elevated temperatures
       and carbon dioxide concentrations: effects on selected microbial activities in temperate agricultural
       soils. World J Microbiol Biotechnol 2009; 25(11): 1887-900.
       [http://dx.doi.org/10.1007/s11274-009-0107-2]

[65]   Carson JK, Gleeson DB, Clipson N, Murphy DV. Afforestation alters community structure of soil
       fungi. Fungal Biol 2010; 114(7): 580-4.
       [http://dx.doi.org/10.1016/j.funbio.2010.04.008] [PMID: 20943169]

[66]   Zogg GP, Zak DR, Ringelberg DB, White DC, MacDonald NW, Pregitzer KS. Compositional and
       functional shifts in microbial communities due to soil warming. Soil Sci Soc Am J 1997; 61(2): 475-
       81.
       [http://dx.doi.org/10.2136/sssaj1997.03615995006100020015x]

[67]   Jansson JK, Taş N. The microbial ecology of permafrost. Nat Rev Microbiol 2014; 12(6): 414-25.

[http://dx.doi.org/10.1038/nrmicro3262] [PMID: 24814065]

[68] McHugh TA, Compson Z, van Gestel N, *et al.* Climate controls prokaryotic community composition in desert soils of the southwestern United States. FEMS Microbiol Ecol 2017; 93(10): fix116.
[http://dx.doi.org/10.1093/femsec/fix116] [PMID: 28961955]

[69] Bogati K, Walczak M. The impact of drought stress on soil microbial community, enzyme activities and plants. Agronomy (Basel) 2022; 12(1): 189.
[http://dx.doi.org/10.3390/agronomy12010189]

[70] Lau JA, Lennon JT. Rapid responses of soil microorganisms improve plant fitness in novel environments. Proc Natl Acad Sci USA 2012; 109(35): 14058-62.
[http://dx.doi.org/10.1073/pnas.1202319109] [PMID: 22891306]

[71] Upton RN, Bach EM, Hofmockel KS. Belowground response of prairie restoration and resiliency to drought. Agric Ecosyst Environ 2018; 266: 122-32.
[http://dx.doi.org/10.1016/j.agee.2018.07.021]

[72] Abdul Rahman NSN, Abdul Hamid NW, Nadarajah K. Effects of abiotic stress on soil microbiome. Int J Mol Sci 2021; 22(16): 9036.
[http://dx.doi.org/10.3390/ijms22169036] [PMID: 34445742]

[73] Furtak K, Grządziel J, Gałązka A, Niedźwiecki J. Prevalence of unclassified bacteria in the soil bacterial community from floodplain meadows (fluvisols) under simulated flood conditions revealed by a metataxonomic approachss. Catena 2020; 188: 104448.
[http://dx.doi.org/10.1016/j.catena.2019.104448]

[74] Rodriguez-Heredia M, Djian-Caporalino C, Ponchet M, *et al.* Protective effects of mycorrhizal association in tomato and pepper against Meloidogyne incognita infection, and mycorrhizal networks for early mycorrhization of low mycotrophic plants. Phytopathol Mediterr 2020; 59(2): 377-84.

[75] Chen H, Chen X, Gu H, *et al.* GmHKT1;4, a novel soybean gene regulating Na^+/K^+ ratio in roots enhances salt tolerance in transgenic plants. Plant Growth Regul 2014; 73(3): 299-308.
[http://dx.doi.org/10.1007/s10725-014-9890-3]

[76] Sharifi M, Ghorbanli M, Ebrahimzadeh H. Improved growth of salinity-stressed soybean after inoculation with salt pre-treated mycorrhizal fungi. J Plant Physiol 2007; 164(9): 1144-51.
[http://dx.doi.org/10.1016/j.jplph.2006.06.016] [PMID: 16919369]

[77] Zhang YF, Wang P, Yang YF, Bi Q, Tian SY, Shi XW. Arbuscular mycorrhizal fungi improve reestablishment of Leymus chinensis in bare saline-alkaline soil: Implication on vegetation restoration of extremely degraded land. J Arid Environ 2011; 75(9): 773-8.
[http://dx.doi.org/10.1016/j.jaridenv.2011.04.008]

[78] Wong MH. Environmental geochemical cycles of persistent toxic substances and emerging chemicals of concern. Proceedings of the International Conference on Environmental Forensics. 17-21.Singapore. 2013; pp.

[79] Khan AG. Role of soil microbes in the rhizospheres of plants growing on trace metal contaminated soils in phytoremediation. J Trace Elem Med Biol 2005; 18(4): 355-64.
[http://dx.doi.org/10.1016/j.jtemb.2005.02.006] [PMID: 16028497]

[80] Thijs S, Sillen W, Weyens N, Vangronsveld J. Phytoremediation: State-of-the-art and a key role for the plant microbiome in future trends and research prospects. Int J Phytoremediation 2017; 19(1): 23-38.
[http://dx.doi.org/10.1080/15226514.2016.1216076] [PMID: 27484694]

[81] Doty SL, Freeman JL, Cohu CM, *et al.*. Enhanced Degradation of TCE on a Superfund Site Using Endophyte-Assisted Poplar Tree Phytoremediation. Environ Sci Technol 2017; 51(17): 10050-8.

[82] Arslan M, Imran A, Khan QM, Afzal M. Plant–bacteria partnerships for the remediation of persistent organic pollutants. Environ Sci Pollut Res Int 2017; 24(5): 4322-36.
[http://dx.doi.org/10.1007/s11356-015-4935-3] [PMID: 26139403]

[83]   Foo JL, Ling H, Lee YS, Chang MW. Microbiome engineering: Current applications and its future. Biotechnol J 2017; 12(3): 1600099.
        [http://dx.doi.org/10.1002/biot.201600099] [PMID: 28133942]

[84]   Zolla G, Badri DV, Bakker MG, Manter DK, Vivanco JM. Soil microbiomes vary in their ability to confer drought tolerance to Arabidopsis. Appl Soil Ecol 2013; 68: 1-9.
        [http://dx.doi.org/10.1016/j.apsoil.2013.03.007]

[85]   Glick BR. Plant growth-promoting bacteria: mechanisms and applications. Scientifica (Cairo) 2012; 2012(1): 963401.
        [PMID: 24278762]

[86]   Lau JA, Lennon JT. Rapid responses of soil microorganisms improve plant fitness in novel environments. Proc Natl Acad Sci USA 2012; 109(35): 14058-62.
        [http://dx.doi.org/10.1073/pnas.1202319109] [PMID: 22891306]

[87]   Lebeis SL. Greater than the sum of their parts: characterizing plant microbiomes at the community-level. Curr Opin Plant Biol 2015; 24: 82-6.
        [http://dx.doi.org/10.1016/j.pbi.2015.02.004] [PMID: 25710740]

[88]   Tripathi BM, Edwards DP, Mendes LW, *et al.* The impact of tropical forest logging and oil palm agriculture on the soil microbiome. Mol Ecol 2016; 25(10): 2244-57.
        [http://dx.doi.org/10.1111/mec.13620] [PMID: 26994316]

[89]   Ge J, Li D, Ding J, Xiao X, Liang Y. Microbial coexistence in the rhizosphere and the promotion of plant stress resistance: A review. Environ Res 2023; 222: 115298.
        [http://dx.doi.org/10.1016/j.envres.2023.115298] [PMID: 36642122]

[90]   Wang Z, Hu X, Solanki MK, Pang F. A synthetic microbial community of plant core microbiome can be a potential biocontrol tool. J Agric Food Chem 2023; 71(13): 5030-41.
        [http://dx.doi.org/10.1021/acs.jafc.2c08017] [PMID: 36946724]

[91]   Chen Y, Yao Z, Sun Y, *et al.* Current studies of the effects of drought stress on root exudates and rhizosphere microbiomes of crop plant species. Int J Mol Sci 2022; 23(4): 2374.
        [http://dx.doi.org/10.3390/ijms23042374] [PMID: 35216487]

[92]   Hartmann M, Six J. Soil structure and microbiome functions in agroecosystems. Nat Rev Earth Environ 2022; 4(1): 4-18.
        [http://dx.doi.org/10.1038/s43017-022-00366-w]

[93]   Herrero M, Stuckey DC. Bioaugmentation and its application in wastewater treatment: A review. Chemosphere 2015; 140: 119-28.
        [http://dx.doi.org/10.1016/j.chemosphere.2014.10.033] [PMID: 25454204]

[94]   Bletz MC, Loudon AH, Becker MH, *et al.* Mitigating amphibian chytridiomycosis with bioaugmentation: characteristics of effective probiotics and strategies for their selection and use. Ecol Lett 2013; 16(6): 807-20.
        [http://dx.doi.org/10.1111/ele.12099] [PMID: 23452227]

[95]   Winkler MKH, Meunier C, Henriet O, *et al.* An integrative review of granular sludge for the biological removal of nutrients and recalcitrant organic matter from wastewater. Chem Eng J 2018; 336: 489-502.
        [http://dx.doi.org/10.1016/j.cej.2017.12.026]

[96]   Li M, Liang P, Li Z, *et al.* Fecal microbiota transplantation and bacterial consortium transplantation have comparable effects on the re-establishment of mucosal barrier function in mice with intestinal dysbiosis. Front Microbiol 2015; 6: 692.
        [http://dx.doi.org/10.3389/fmicb.2015.00692] [PMID: 26217323]

Part 2: Role of Microbiomes in Restoring Polluted Ecosystem and Environment

Microbial-assisted Bioremediation: A Greener Approach for Restoration of Heavy Metal-contaminated Soil

Yogesh Dashrath Naik[1], Rohit Das[2], Santosh Kumar[2], Konderu Niteesh Varma[3], S.T.M. Aravindharajan[3] and Viabhav Kumar Upadhayay[4,*]

[1] Department of Agricultural Biotechnology and Molecular Biology, Dr. Rajendra Prasad Central Agricultural University, Pusa, Samastipur, Bihar, India

[2] Department of Microbiology, School of Life Sciences, Sikkim University, Gangtok - 737102, Sikkim, India

[3] Division of Microbiology, ICAR-Indian Agricultural Research Institute, New Delhi, India

[4] Department of Microbiology, College of Basic Sciences & Humanities, Dr. Rajendra Prasad Central Agricultural University, Pusa, Samastipur, Bihar, India

Abstract: Heavy metals (HMs) pollution is a major environmental concern, posing serious threats to human health and ecological systems. Anthropogenic activities have increased the levels of HMs in the environment, and their pollution is a major issue. Exposure to high levels of these metals can have harmful effects on human health, and they can also damage soil structure, diminish microbial biodiversity, and inhibit plant growth and development. In addition, traditional remediation methods for HMs contaminated soil are often expensive and negatively impact the environment. In recent years, microbial-assisted bioremediation has emerged as a promising and eco-friendly alternative for HM remediation. This approach utilizes microorganisms to transform, immobilize, or detoxify HMs, making them less harmful and more accessible for removal. This chapter highlights the eco-friendly use of microorganisms, the mechanisms that contribute to the bioremediation of HMs, and their potential use in the future.

Keywords: Bioleaching, Biosorption, Bioaccumulation, Biomineralization, Environment, Heavy metals, Microorganisms, Microbial-assisted bioremediation, Mmicrobial diversity, Plant growth, Soil.

* **Corresponding author Viabhav Kumar Upadhayay:** Department of Microbiology, College of Basic Sciences & Humanities, Dr. Rajendra Prasad Central Agricultural University, Pusa, Samastipur, Bihar, India;
E-mail: viabhav.amu@gmail.com

Shiv Prasad, Govindaraj Kamalam Dinesh, Murugaiyan Sinduja, Velusamy Sathya, Ramesh Poornima & Sangilidurai Karthika (Eds.)

INTRODUCTION

Heavy metals (HMs) are generally defined as metallic elements and metalloids that have a high atomic weight and density, typically above 5 g/cm^3. From the biological and environmental science perspective, the term 'heavy metals' is often used to describe a group of metals and metalloids that are toxic to plants and animals even at low concentrations because these metals can accumulate in living tissue and interfere with biological processes, leading to a range of HM health problems and environmental damage. HMs are naturally occurring elements found in soil, water, and rocks. However, human activities like industrialization, urbanization, mining, smelting, burning of fossil fuels, and use of pesticides and fertilizers have led to increased levels of these metals in the environment, particularly in soil, water, and air [1, 2]. The HMs of particular concern due to their toxicity and prevalence in the environment include chromium, mercury, arsenic, cadmium, lead, nickel, copper, and zinc [3]. The physicochemical properties of HMs, such as their ubiquity, non-biodegradability, toxicity, accumulation, and persistence, have made their pollution a significant global concern [4]. As a result, there is growing recognition of the link between HM contamination and public health. The toxicity of HMs to living organisms depends on various factors, including the type of metal, the concentration, and the duration of exposure [5]. High levels of these metals can harm human health, such as neurological damage, cancer, and respiratory problems [6]. HMs can damage soil structure, diminish microbial biodiversity, and inhibit plant growth and development, reducing crop production and quality [2]. However, various microbial species and plants have developed unique coping mechanisms to deal with the toxicity of HM in polluted soils [2, 4]. These adaptations can range from changes in their cell structure and physiology to changes in their gene expression and metabolism. These adaptations can be used for bioremediation, which uses microorganisms and plants to remove or detoxify HMs from polluted soils [7]. Physicochemical techniques such as excavation, soil washing, and chemical stabilization can effectively remove or immobilize HMs from contaminated soil. However, these techniques are often expensive, energy-intensive, and generate toxic waste, making them unsuitable for large-scale remediation projects.

Additionally, these methods may not be environmentally friendly or inappropriate for soils with low metal contamination levels. As a result, biological remediation strategies, such as phytoremediation and bioremediation, are becoming increasingly popular due to their cost-effectiveness, sustainability, and ability to restore soil health. Bioremediation technologies based on microorganisms have become increasingly popular in recent years due to their effectiveness in removing environmental pollutants [8]. Microorganisms such as bacteria, fungi, and algae can transform, immobilize, or detoxify HMs through various mechanisms [2, 9].

The process involves various mechanisms, including bioleaching, biosorption, bioaccumulation, and biomineralization [10]. Bioleaching involves using microorganisms to solubilize and mobilize HMs from the environment, making them more accessible for removal. Biosorption involves binding HMs to the surface of microbial cells, which can then be easily removed from the environment. Bioaccumulation involves the accumulation of HMs within microbial cells, usually in specialized compartments such as vacuoles, which can then be easily removed from the environment. Finally, biomineralization involves the conversion of HMs into less toxic forms, such as metal sulfides or metal carbonates, which are then immobilized in the environment. The current chapter highlights the eco-friendly use of microorganisms and their mechanisms that contribute to the bioremediation of HMs.

HEAVY METALS: SOURCES AND TOXICITY

HMs are a group of elements with a high density (greater than 5 g/cm^3 or a specific gravity at least five times greater than water) and tend to be toxic to living organisms in high concentrations [2]. HMs can be divided into two main categories: essential and non-essential HMs. Some HMs are necessary for living organisms in small amounts for growth and development; these are called essential HMs, and examples include iron, copper, zinc, and manganese [2, 11]. Exposure to high levels of non-essential HMs, like lead, mercury, cadmium, and arsenic, can harm human health. They are not required for normal biological function and can be toxic in large amounts. High concentrations of these metals can cause damage to the nervous system, liver, and kidneys, increasing the risk of cancer [12]. Natural processes and anthropogenic activities are responsible for the entry of HMs into the soil. However, compared to natural processes, anthropogenic activities are more critical for causing HM pollution [13]. Some soil types have naturally high concentrations of HMs due to the parent materials from which they are formed. These soils are often found in areas with high levels of volcanic activity or mineral deposits, which can contain high levels of HMs such as lead, cadmium, arsenic, and mercury [14]. For instance, San Joaquin Valley in California is where naturally occurring geological vents have elevated selenium levels in the soil [13].

HMs are primarily derived from the parent materials from which the soil was formed. Approximately 95% of the Earth's crust comprises igneous rocks, with the remaining 5% comprising sedimentary rocks [13]. Generally, basaltic igneous rocks are abundant in HMs like cadmium, copper, cobalt, and nickel, whereas shale typically has higher concentrations of lead, copper, zinc, manganese, and cadmium [15]. HMs contained in rocks can enter the soil system through various natural processes such as weathering, erosion, leaching, and volcanic activity

[16]. The accumulation of one or more HMs may be due to an anthropogenic disturbance in the HM geochemical cycle [13, 17]. In today's world, urbanization, industrialization, and recent advances in the agricultural sector have contributed significantly to HM soil contamination [18]. In addition, anthropogenic activities such as mining, smelting, burning of fossil fuels, application of pesticides, fertilizer application, improper waste management, *etc.*, also increased the concentration of HMs in the soil environment [1].

HMs are not biodegradable, and their elevated concentration in soil adversely affects the soil's physiochemical and biological properties. The high persistence of HMs in soils allows the entry of HMs into the food chain, which results in severe health hazards for living beings [19]. Moreover, HMs affect soil structure, soil fertility, soil microbial population, biogeochemical cycling of nutrients, soil vegetation, *etc.* The soil enzymatic activities are also affected by the response of HMs due to alterations in the composition and size of microbial communities and their functional activities [2]. HMs harm microorganisms by affecting microbial metabolic processes, respiration, enzymatic activity, and denitrification [2, 20]. The extent of HMs' effect on soil biological activities depends on numerous factors such as texture, clay content, pH, organic matter, chemical form, and speciation of metals and inorganic anions and cations [21]. The drastic effect of HMs on plants is well documented, as various biochemical and physiological processes in plants are affected in response to HMs. The higher concentration of HMs hampers plant growth by affecting seed germination, photosynthesis, stomatal conductance, water balance, the electron transport system, CO_2 assimilation, *etc.*, and results in plant death [2]. An elevated level of HMs impedes cytoplasmic enzymatic activity and damages cell structure due to oxidative stress, consequently affecting plant metabolism and growth [22]. Overall, HMs negatively impact the crop, manifesting in a decline in crop yield. HMs accumulated in the soil can reach other compartments such as rivers, groundwater, and crops, thus threatening humans and animals [2]. When HMs enter living organisms, they can bind to various biomolecules such as enzymes, DNA, and proteins, forming stable and toxic compounds. This can result in the malfunction of cells and hinder biological processes, leading to a range of health problems.

Lead, arsenic, cadmium, and chromium are particularly toxic HMs as they have been linked to mutagenic, carcinogenic, and genotoxic effects [23]. For example, lead exposure has been linked to damage to the nervous system, developmental delays, and behavioral problems in children [24]. Arsenic exposure can cause skin lesions, cancer, and cardiovascular disease, while cadmium exposure has been linked to kidney damage and lung cancer [25]. Chromium exposure can cause lung cancer, skin irritation, and respiratory problems [26].

APPROACHES FOR HEAVY METAL REMOVAL

The prevalence of HM contamination has significantly increased due to the heightened demand and excessive use of various metals and related products. While these metals occur naturally in the environment, pollution arises when human or natural activities cause their concentration to exceed safe levels in soil, air, and water [27]. The rising concentration of HMs in the environment threatens human and animal health due to enhanced anthropogenic and natural activities. With the increased demand for metals and related products, the problem of HM contamination has expanded. Therefore, the need for effective HM removal, clean-up, and recovery methods is more significant than ever before. Traditional strategies for managing HM contamination involved on-site excavation or disposal in landfills. However, these approaches simply transferred the pollution from one location to another without effectively addressing the issue [28]. Many physical, chemical, and biological treatments and bioremediation techniques have been employed to address this issue. Various prominent approaches to heavy metal removal are depicted in Fig. (1).

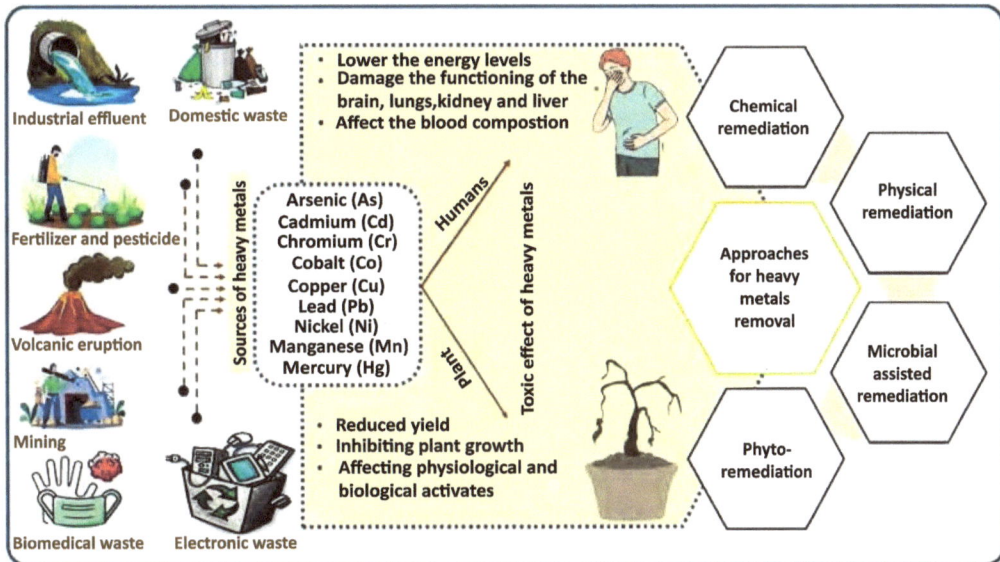

Fig. (1). Comprehensive overview of heavy metal sources impacting plants and humans, along with various remediation approaches

Physical Remediation

Various physical approaches have been developed to remediate HM contamination in soils due to the persistence and toxicity of HMs. Soil replacement was a predominant technique before 1984, replacing contaminated soil with uncontaminated soil. This approach improved the functionality of polluted soil, but the HMs in the restored soil required further treatment or disposal elsewhere [29]. Soil isolation is another physical approach that separates contaminated soil from uncontaminated soil. However, removing HMs through soil isolation requires additional auxiliary engineering procedures [30]. Soil electrokinetic remediation is a recent cutting-edge physical approach that is cost-effective for the bioremediation of HMs. This approach uses an electric field gradient to remove HMs from contaminated soil *via* electromigration, electromigration separation, or electric seepage.

Additionally, techniques such as electrokinetic chemical joint remediation, electrokinetic microbe united remediation, and electrokinetic oxidation/reduction joint remediation have been utilized to remediate contaminated sites. These methods use electrokinetics to mobilize contaminants, promote microbial activity, and facilitate oxidation/reduction reactions, resulting in more effective and comprehensive remediation [31]. Vitrification is a physical technique used for the bioremediation of HMs, which involves treating contaminated soil with high temperatures to form vitreous solids. This process reduces the mobility of HMs in soil, but it is not a commonly used method for metal bioremediation due to its complexity [32]. Physical approaches for HM remediation can be expensive and time-consuming. Soil replacement, soil isolation, and electrokinetic remediation require specialized equipment and engineering procedures, which can increase costs [31]. Vitrification, while effective, is a complex process that requires specialized equipment and high temperatures, further adding to the overall expense [32]. Therefore, implementing physical approaches for HM remediation may be limited by their high costs and the need for specialized expertise and equipment.

Chemical Remediation

Techniques for immobilization, encapsulation, and soil cleaning are the major components of chemical remediation. By adding immobilizing mediators to polluted soil, immobilization procedures reduce the mobility, bioaccessibility, and bioavailability of HMs in the soil. Moreover, complexation, adsorption, and precipitation reactions immobilize HMs in contaminated soil. These procedures assist in transferring HMs from soil solutions to solid particles, reducing their bioavailability and soil movement [33]. Immobilization is mostly accomplished in the soil by adding organic and inorganic amendments. Clay, phosphate, cement, minerals, zeolites, and organic amendments are some of the most popular additions. When contaminated soil is encapsulated, it is combined with other materials like concrete/asphalt and lime, which makes it immobile to stop contamination. Many binding materials are utilized to construct solid blocks, but cement is typically favored due to its affordability, accessibility, and adaptability [34]. Soil washing is a method that uses various extractants and reagents to filter metals from contaminated soil. The goal is to remove the metal from the soil completely. This approach is an effective alternative to traditional methods for soil remediation [35].

Phytoremediation

Phytoremediation, also called green remediation, botany-remediation, vegetative remediation, or agroremediation, involves using plants to remove pollutants from soil and water with high concentrations of hazardous HMs, rendering them harmless by either breaking down or adsorbing the contaminants. Since plants are living organisms utilized for cleaning up polluted areas, this type of remediation is considered a form of biological remediation. Phytoremediation is an effective, non-invasive, and environmentally friendly alternative to HM exclusion. In order to remediate HMs through phytoremediation, various techniques are employed, including phytoextraction, rhizodegradation, phytostabilization, phytovolati-lization, and rhizofiltration [36]. Phytoextraction, or phytoaccu-mulation, is how plant roots absorb different HMs. This method utilizes the ability of plant roots to absorb, translocate, and deposit HMs from the soil to the aboveground harvestable portions of the plant. Once phytoextraction is complete, the plant is harvested and incinerated to generate electricity and recycle metals from the ash.

Phytoextraction is a promising technique for removing metals from polluted environments. Although most plants cannot live in highly contaminated areas, it is crucial to note that this method works best for contaminated sites affected by low to moderate levels of HMs [37]. Plants uptake or absorb HMs from the contaminated soil and transport them through the xylem. The process of turning the HMs into less dangerous vapors or volatile forms and releasing them into the atmosphere through plant transpiration is known as phytovolatilization [36]. Due to their extreme volatility, this technique mainly extracts metals like selenium and mercury. Metals are released into the air as biomolecules during phytovolatilization. For example, converting Hg^{2+} to Hg^0 by plants like *Arabidopsis thaliana* raises mercury volatility. Some plants, including *Chara canescens* and *Brassica juncea*, can absorb HMs from contaminated soil and convert them into gaseous or volatile forms. Phytovolatilization's disadvantage is that it renders pollutants less harmful rather than totally eliminating them. In addition, fruits and other edible plant components may include pollutants [38].

Plants can restrict the mobility or bioavailability of metals during phytostabilization by collecting them within the root or close to the rhizosphere, fixing them in the substrate, or both. Increased metal accumulation in roots may further reduce metal mobility in silt [39]. The fundamental mechanism behind phytostabilization is influenced by several elements, including microorganisms in the rhizosphere, root exudates, metal ion binding to cell walls, metal ion chelation by metal-binding molecules, and metal ion sequestration within vacuoles. The core mechanism behind phytostabilization is the mobility of trace elements in the rhizosphere, which is governed by several soil variables, including pH, organic matter, texture, redox potential, temperature, and microorganisms. Sorption, precipitation, complexation, or metal valence reduction can all result in phytostabilization [40]. The biological processes of plant microbes are used in phytoremediation to remove, lessen, or change toxins from contaminated media such as water, soil, and air. Comparing this developing technology to the present chemical and physical-chemical clean-up techniques, it is more cost-efficient, energy-efficient, and environmentally benign. Rhizofiltration, a mechanism utilized in the phytoremediation process, is frequently used to clean up contaminated groundwater resources [2].

Rhizofiltration employs plant roots to concentrate and precipitate pollutants on or within the root system by adsorbing them from the surrounding root zone (rhizosphere). Exudates from the roots of plants, or secondary metabolites, are subject to biogeochemical reactions that might precipitate pollutants on the roots or into the water body. According to the type of plant, contaminant species, and concentration, pollutants will then bind to the roots, move inside the roots, or go to the phyllosphere (plant organs above ground) [41]. For contaminated surface water, *in-situ* rhizofiltration can be utilized; for contaminated groundwater, *ex-situ* rhizofiltration, which uses an engineered tank system to deliver contaminated water to the plants, can be employed [42]. One of the best ways to effectively remove organic contaminants from soil bodies, such as polycyclic aromatic hydrocarbons (PAHs), is rhizodegradation. It results from intricate interactions between bacteria and plants in the rhizosphere. This heavily depends on the harmonious coexistence of the plant and microbial communities in the rhizosphere and the nearby non-rhizosphere soils. Many bacterial strains of the rhizosphere have been shown to digest a wide range of PAHs; most of these were found in contaminated soil, whereas just a small number were found in non-contaminated soil. It is an exciting field of research for the present and future of bioremediation technologies to understand the intricate molecular communications that occur in rhizosphere zones and to take advantage of these communications to improve results in eliminating contaminants like PAHs [43].

Microbial-assisted Remediation

Microorganisms are all around us and are proven to benefit the environment and people in various ways. They can help with the problem of metal toxicity. Several physical and chemical restoration techniques have been created over the years, but most are expensive, ineffective or produce secondary contaminants. In this case, bioremediation can help because it is inexpensive, very effective, and does not create secondary contaminants [12]. Microbes are known to survive under environmental stress much better than plants and animals because they may quickly mutate and adapt. A small number of HMs in the presence of microorganisms has no discernible impact on the presence of the HM. However, a significant concentration of HM is needed for any discernible alterations in microbial metabolism and expression patterns. It suggests that the availability and amplitude of HM ions determine how a microorganism reacts to HM [44]. Microbial remediation aims to reduce the bioavailability of HMs by immobilizing them *in situ*. The interaction between bacteria and HMs is a complex process that depends on various factors, such as the type and concentration of the metal, the medium's composition, and the microbe species. Microbes can change the toxicity of HMs by converting them from one oxidation state or organic complex to another. This transformation makes the HM less hazardous and water-soluble, causing precipitation and reducing bioavailability. Microbes also remove HMs from the site by utilizing them for growth and development. Microbes employ a variety of tactics, including the oxidation, immobilization, change, binding, and volatilization of HMs, to accomplish the above goals. Some bacteria use enzymatic metabolic pathways to break down HMs into a less dangerous form.

Microorganisms with plant growth-promoting traits can play a vital role in alleviating HM stress in plants. These microorganisms can employ various tolerance mechanisms to transform HMs into bioavailable or soluble forms, which are easier for plants to uptake [9]. Some microorganisms can also biosorb or accumulate HMs, reducing their concentration in the soil. In addition, microorganisms can produce extracellular polymeric substances (EPS) and siderophores, which can chelate HMs and prevent them from causing damage to plant cells. The presence of these compounds can also facilitate the movement of HMs through the soil, making them more accessible to plants. In addition to these direct mechanisms, microorganisms can indirectly promote plant growth in the presence of HMs by producing enzymes that break down organic matter, which can release nutrients such as nitrogen and phosphorus, essential for plant growth. Microorganisms can also solubilize inorganic minerals such as phosphate, zinc, and potassium, making them more available to plants.

Furthermore, some microorganisms can fix atmospheric nitrogen, which is essential for plant growth, reducing the need for synthetic nitrogen fertilizers that can contribute to HM pollution. Overall, using microorganisms with plant growth-promoting traits is a promising approach to alleviate HM stress in plants, and it can potentially contribute to sustainable agriculture practices. The bioremediation of HMs involves various microbes, including bacteria, fungi, yeast, and algae. Nevertheless, bacteria, such as *Bacillus sp.* and *Pseudomonas sp.,* are the most powerful among them [45]. Aspergillus sp., *Streptoverticullum sp., Trichoderma sp.,* and *Saccharomyces sp.* are a few examples of fungi and yeast with significant biosorption potential. A study carried out by Khanna *et al.* (2019) [46] showed that *Pseudomonas aeruginosa* and *Burkholderia gladioli* were able to enhance plant growth in the seedlings of Lycopersicon esculentum (tomato) under cadmium stress. The study by Mukherjee *et al.* (2019) [47] showed that *Halomonas* sp. Exo1, a type of rhizobacteria, produced exopolysaccharides (EPS) and sequestered arsenic, which helped improve the growth of rice plants under salinity and HM stress. The findings of Gontia-Mishra *et al.* (2016) [48] suggest that the use of plant growth-promoting bacteria (*Enterobacter ludwigii* HG 2) and *Klebsiella pneumoniae* (HG 3) could be an effective strategy for enhancing crop productivity, especially in environments contaminated with mercury. Microorganisms have various mechanisms for the bioremediation of HMs, including adsorption, biosorption, changing the redox state of metals, biotransformation, biomineralization, and bioleaching [49].

MECHANISTIC VIEW OF MICROBIAL-ASSISTED BIOREMEDIATION OF HEAVY METALS

Microbial bioremediation is a promising approach for removing HMs from contaminated environments due to its effectiveness, cost efficiency, and sustainability. The process involves various mechanisms, including bioleaching, biosorption, bioaccumulation, and biomineralization (Fig. **2**) [10, 50]. Therefore, understanding the mechanism of bioremediation of HMs is crucial to developing effective strategies to remediate contaminated environments. Furthermore, knowledge of the underlying mechanisms involved in the removal and detoxification of HMs by microorganisms can aid in optimizing remediation strategies, selecting suitable microorganisms, and predicting potential outcomes of remediation efforts (Table **1**). Therefore, understanding the mechanisms involved in the bioremediation of HMs can help develop sustainable and efficient remediation technologies.

Biosorption

Biosorption is a passive process by which HM ions are adsorbed onto the surfaces of microbial cells or extracellular polymeric substances (EPS) [65]. The EPS produced by bacteria, such as polysaccharides, lipids, and proteins, can serve as a binding site for HMs. This mechanism can involve different types of interactions, such as electrostatic interaction, ion exchange, and complexation [66]. The adsorption of HM ions on the microbial surface is primarily governed by the electrostatic attraction between the charged groups on the microbial surface and the oppositely charged metal ions [67]. Several studies have reported the potential of different microorganisms for the biosorption of HMs, such as *Bacillus licheniformis* for chromium (VI) biosorption from contaminated soil [53], *Pseudomonas putida* for nickel biosorption from industrial wastewater [68], and *Aspergillus niger* for lead biosorption from aqueous solutions [69, 70]. These studies highlight the potential of biosorption as a promising approach for the bioremediation of HM-contaminated soil or environments.

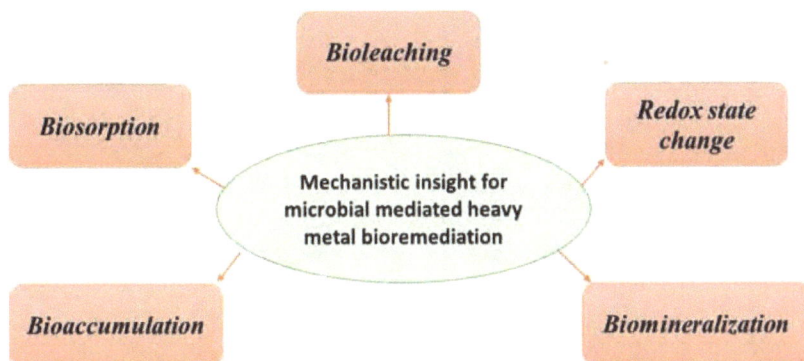

Fig. (2). Mechanistic view of microbial-assisted heavy metal bioremediation.

Table 1. Summarizing the microorganisms involved in the different mechanism of bioremediation and their target heavy metals.

Mechanism	Microorganisms used for bioremediation	Targeted heavy metals	References
Biosorption	*Lysinibacillus fusiformis*	Cadmium, Chromium, Arsenic, Lead, and Nickel	[51]
	Penicillium notatum	Cadmium	[52]
	Aspergillus niger	Nickel	
	Pseudomonas aeruginosa	Chromium and Copper	
	Bacillus licheniformis	Lead	[53]
Bioleaching	*Thiobacillus ferrooxidans*	Copper and Nickel	[54]
	Aspergillus niger	Arsenic	[55]
By redox state change	*Bacillus amyloliquefaciens*	Chromium	[56]
	Sulfate-reducing bacteria	Nickel, Copper, Zinc, Magnesium, and Iron	[57]

(Table 1) cont.....

Mechanism	Microorganisms used for bioremediation	Targeted heavy metals	References
Bioaccumulation	*Corynebacterium glutamicum*	Arsenic	[58]
	Bacillus simplex UFLA CESB127, B. subtilis UFLA SCF590 and *Acetobacter tropicalis UFLA DR6.2*	Iron	[59]
	Sulfate-reducing bacteria	Technetium	[60]
Biomineralization	Phosphate solubilizing bacteria	Lead	[61]
	Ureolytic bacteria	Cadmium and Nickel	[62, 63]
	Dissimilatory metal-reducing bacteria	Iron	[64]

Bioleaching

Bioleaching is a bioremediation approach involving microorganisms to mobilize HMs in contaminated soils, making them more available for removal or transformation [53]. Microorganisms such as bacteria and fungi produce organic acids and enzymes that break down minerals in the soil, releasing bound HMs into solution. The released HMs can then be removed using various physical and chemical processes. A study by Sur *et al.* (2022) [54] investigated the potential of bioleaching for the remediation of HMs from contaminated soil. The study demonstrated that the bacterial strain *Thiobacillus ferrooxidans* effectively mobilized HMs in soil with high removal efficiencies. A mechanistic view of microbial-assisted heavy metal bioremediation is provided in Fig. (**2**).

Another study by Petkova [55] investigated the potential of a fungal strain, *Aspergillus niger*, for the remediation of arsenic-contaminated soil. The study showed that *A. niger* could mobilize arsenic in the soil effectively. Overall, bioleaching has the potential to be a promising strategy for the bioremediation of HM-contaminated soils, and further research is needed to explore its full potential and optimize its use in different environments. A summary of the microorganisms involved in the different mechanisms of bioremediation and their target heavy metals is provided in Table **1**.

By Redox State Change

Microbes play an important role in bioremediating hazardous metals in the environment by transforming their redox status into a less harmful form. Certain bacteria utilize metals like iron (III) or metalloids as electron acceptors to catalyze this process. Researchers have extensively studied the enzymatic transformation of toxic Cr(VI) to nontoxic Cr(III) and insoluble hydroxide to better understand this microbial remediation process. Up-flow anaerobic bed reactors containing sulfate-reducing bacteria (SRB) have also been developed to remove metals such as Ni(II), Cu(II), Zn(II), Mg(II), Al(III), and Fe(III) from contaminated water bodies, with up to 95% efficiency achieved in the first 78 weeks [57]. SRBs can remove hazardous metals through several mechanisms, such as sulfide formation, sulfate reduction, and HM precipitation. SRB effectively immobilizes HM ions and increases sulfide concentration in contaminated water by reducing sulfate and producing insoluble metal sulfides. In addition to SRB, the bacterium *Bacillus amyloliquefaciens* has also shown promising results in reducing and eliminating Cr(VI) in aerobic environments. This bacterium utilizes glucose as an energy source and enzymatically transforms Cr(VI) into less toxic forms [56]. The enzyme arsenite oxidase is capable of catalyzing the oxidation of the highly toxic and soluble form of arsenite (As III) into a less toxic and less soluble form known as arsenate (As V). Arsenite oxidase can function in both anaerobic and aerobic environments, and it is synthesized by various microorganisms such as bacteria, archaea, and fungi [10].

Bioaccumulation

Bioaccumulation is an active process by which microbes actively take up HMs from their surroundings and accumulate them within their cells. Bioaccumulation is effective for the removal of HMs with high concentrations or in soluble forms. Several microbial species, such as *Pseudomonas*, *Bacillus*, and *Aspergillus*, have been reported to have bioaccumulation potentials for different HMs, such as chromium, nickel, and zinc [71]. This mechanism can involve different transporters and ion pumps that enable microbes to concentrate HMs up to several orders of magnitude higher than their surrounding environment [72, 73]. TCDB 1. A is a transporter protein composed of a single α-helix that can facilitate the passive diffusion of HMs across the inner membrane according to their concentration gradient. These channels are typically energy-independent and do not require the proton-motive force or nucleoside triphosphates such as ATP and GTP to transport their substrates [74].

Additionally, porins (TCDB 1. B) constitute a significant class of transporters that use β-barrels to create translational pathways across the outer membrane of gram-negative bacteria [75]. The homotetramer glycerol facilitators (GlpF) from various organisms, including *Escherichia coli*, *Corynebacterium diptheriae*, *Streptomyces coelicolor*, and the homolog Fps1 from *Saccharomyces cerevisiae*, have been utilized for As^{3+} uptake [73, 76, 77]. These findings suggest that transporters could also be a promising target for improving the bioaccumulation of HMs in other microorganisms.

Biomineralization

Biomineralization is the process of transforming HMs into insoluble minerals, which can reduce their bioavailability and toxicity. Transformations of HMs involve a range of processes, including oxidation, reduction, methylation, and demethylation [50, 78]. Various microbial enzymes and metabolic pathways, such as sulfide production, carbonate precipitation, and phosphate precipitation, can mediate this mechanism. The study by Li *et al.* (2013) [79] investigated the biomineralization process of HMs by six metal-resistant bacterial strains isolated from soil. These bacteria were found to produce the enzyme urease, which increased the soil pH and resulted in the mineralization of HM ions present in the soil water. As a result, the HMs were converted to crystalline carbonate minerals, leading to their efficient removal from the environment. Several microbial species, such as SRB, iron-oxidizing bacteria, and cyanobacteria, have been reported to have biomineralization potential for different HMs, such as mercury, arsenic, and uranium [80, 81].

FACTORS AFFECTING MICROBIAL BIOREMEDIATION

Microorganisms can quickly and effectively treat or reduce pollutants to a less hazardous state due to their diminutive size, enabling them to come into contact with contaminants. However, the triumph of microbial remediation hinges on various factors, including the type of microorganisms employed, the characteristics of the pollutants, and the chemical and geological properties of the polluted site. In addition, the efficiency of bioremediation carried out by microorganisms can be significantly affected by several factors, including pH, temperature, soil composition, moisture content, air/oxygen availability, redox potentials, nutrient availability, dissolved oxygen levels and solubility, nutrient diffusion, mass transfer, water solubility, chemical composition, and concentrations of HMs [10, 82]. Considering all of these factors is critical for attaining successful microbial remediation. The pH affects the microbial-mediated bioremediation process. Each microorganism has an optimal pH range in which it can effectively remediate heavy metals. The redox and solubility of heavy metals are also affected by pH. This can result in heavy metals with different valence states, which can be toxic to microbes and impair the bioremediation process. Some microorganisms are more tolerant to acidic and alkaline environments than others. These microbes can still remediate heavy metals in suboptimal conditions. However, adjusting the pH at contaminated sites can be a valuable strategy for improving the bioremediation process. The absorption amplitude of HMs is primarily governed by ambient temperature, exerting a profound influence on the growth and proliferation of microbes. Microorganisms exhibit distinct temperature preferences, with '*T. ferrooxidans*', '*T. acidophilus*', and '*T.

tepidarius' classified as mesophilic bacteria, whereas *'Sulfolobus solfatataricus'* and *'Acidianus brierleyi'* thrive as highly thermophilic bacteria. Temperature emerges as a critical factor in the adsorption process of heavy metals, as higher temperatures expedite the diffusion rate of adsorbates within the boundary layer [10]. Moreover, elevated temperatures elevate the solubility of heavy metals, ultimately augmenting their bioavailability. Furthermore, microbial activities are intrinsically influenced by temperature, culminating in their zenith within an optimal range. Consequently, this engenders a conducive milieu for microbial metabolism and enzyme activity and facilitates the efficacious implementation of bioremediation strategies.

Nutrients found in polluted environments, such as nitrogen, phosphate, sulfur, iron, and potassium, play a crucial role in stimulating and supporting robust microbial growth, cellular metabolism, and the proliferation of microorganisms [83]. These nutrients are fundamental for sustaining life and assist microorganisms in producing essential enzymes to degrade contaminants. Moreover, the cost of remediation is a pivotal factor that determines the feasibility of bioremediation strategies, emphasizing the need for cost-effective approaches. The nature of pollutants also significantly impacts the bioremediation process, encompassing their physical states (solid, semisolid, liquid, or volatile), as well as their toxicity and composition, including organic and inorganic pollutants, heavy metals, polycyclic aromatic hydrocarbons, pesticides, and chlorinated solvents [84]. Furthermore, the characteristics of the polluted area play a crucial role in determining the effectiveness of bioremediation efforts. Adequate moisture content, particularly the availability of water, emerges as a primary factor facilitating biological growth and efficient bioremediation. Additionally, microbial diversity plays a critical role, with a wide range of microorganisms capable of biodegrading various contaminants, such as *Acinetobacter, Aeromonas, Bacilli, Chlorobacteria, Corynebacteria, Flavobacteria, Mycobacteria, Pseudomonas, Streptomyces, Macrobenthos*, and aquatic plants like *E. crassipes* and *L. hoffmeisteri*, contributing to the degradation of turbidity and chemical domestic wastewater [84, 85]. Oxygen also serves a vital function in the initial breakdown of hydrocarbons at contaminated sites, being utilized in both aerobic and anaerobic bioremediation processes [84].

FUTURE OUTLOOK AND CHALLENGES

Bioremediation is a field that shows great promise for cleaning up the environmental pollution. Effective remediation of contaminated soil with HM requires careful assessment, selection of appropriate remediation technologies, and ongoing monitoring to ensure that the contaminants have been sufficiently removed. However, it is crucial to ensure that the bioremediation process does not cause further harm to the environment by selecting suitable microorganisms and monitoring their performance. The future of bioremediation is likely to be shaped by advancements in genetic engineering, which can produce genetically modified microorganisms tailored to specific pollutants [86]. Artificial intelligence can also significantly optimize bioremediation processes, predict performance, and monitor progress [87].

Bioremediation is a promising approach; however, despite its potential, there are several challenges associated with bioremediation both in the present and in the future. One of the primary challenges is the limited understanding of microbial ecology and the complexity of environmental systems. This makes it difficult to predict the effectiveness of bioremediation strategies and select the most appropriate microorganisms for the task. Another challenge is the limited availability of suitable microorganisms and the potential for genetic modification to increase their efficacy. Additionally, environmental factors such as temperature, pH, and nutrient availability can impact the success of bioremediation efforts.

CONCLUSION

In conclusion, microbial-assisted bioremediation is a promising approach for removing HMs from contaminated environments. In addition, the effectiveness of microbial-assisted bioremediation may be influenced by factors such as pH, temperature, nutrient availability, and other contaminants. Therefore, it is crucial to consider these factors and optimize the conditions for bioremediation to achieve maximum efficacy. Despite these challenges, microbial-assisted HM bioremediation holds excellent potential for the remediation of contaminated environments and offers a sustainable solution for HM pollution. Further research is needed to fully understand these mechanisms and how they can be manipulated to improve bioremediation strategies. Genetic engineering of microbial strains may also be necessary to enhance their ability to detoxify HMs from contaminated sites. Overall, microbial-assisted HM bioremediation holds great potential for the remediation of contaminated environments and offers a sustainable solution for HM pollution.

REFERENCES

[1]   Gautam K, Sharma P, Dwivedi S, *et al.* A review on control and abatement of soil pollution by heavy metals: Emphasis on artificial intelligence in recovery of contaminated soil. Environ Res 2023; 225: 115592.
[http://dx.doi.org/10.1016/j.envres.2023.115592] [PMID: 36863654]

[2]   Raklami A, Meddich A, Oufdou K, Baslam M. Plants—microorganisms-based bioremediation for heavy metal clean-up: recent developments, phytoremediation techniques, regulation mechanisms, and molecular responses. Int J Mol Sci 2022; 23(9): 5031.
[http://dx.doi.org/10.3390/ijms23095031] [PMID: 35563429]

[3]   Mitra S, Chakraborty AJ, Tareq AM, *et al.* Impact of heavy metals on the environment and human health: Novel therapeutic insights to counter the toxicity. J King Saud Univ Sci 2022; 34(3): 101865.
        [http://dx.doi.org/10.1016/j.jksus.2022.101865]

[4]   Pande V, Pandey SC, Sati D, Bhatt P, Samant M. Microbial interventions in bioremediation of heavy metal contaminants in agroecosystem. Front Microbiol 2022; 13: 824084.
        [http://dx.doi.org/10.3389/fmicb.2022.824084] [PMID: 35602036]

[5]   Priya AK, Gnanasekaran L, Dutta K, Rajendran S, Balakrishnan D, Soto-Moscoso M. Biosorption of heavy metals by microorganisms: Evaluation of different underlying mechanisms. Chemosphere 2022; 307(Pt 4): 135957.
        [http://dx.doi.org/10.1016/j.chemosphere.2022.135957] [PMID: 35985378]

[6]   Munir N, Jahangeer M, Bouyahya A, *et al.* Heavy metal contamination of natural foods is a serious health issue: a review. Sustainability (Basel) 2021; 14(1): 161.
        [http://dx.doi.org/10.3390/su14010161]

[7]   Arantza S-J, Hiram M-R, Erika K, Chávez-Avilés MN, Valiente-Banuet JI, Fierros-Romero G. Bio- and phytoremediation: plants and microbes to the rescue of heavy metal polluted soils. SN Appl Sci. 2022;4(2).

[8]   Yin K, Wang Q, Lv M, Chen L. Microorganism remediation strategies towards heavy metals. Chem Eng J (Lausanne. 2019; 360:1553-63.
        [http://dx.doi.org/10.1016/j.cej.2018.10.226]

[9]   Upadhayay VK, Maithani D, Dasila H, Taj G, Singh AV. Microbial services for mitigation of biotic and abiotic stresses in plants.Advanced microbial techniques in agriculture, environment, and health management. Elsevier 2023; pp. 67-81.
        [http://dx.doi.org/10.1016/B978-0-323-91643-1.00003-X]

[10]   Khan A, Sharma RS, Panthari D, Kukreti B, Singh AV, Upadhayay VK. Bioremediation of heavy metals by soil-dwelling microbes: an environment survival approach Advanced Microbial Techniques in Agriculture, Environment, and Health Management. Elsevier 2023; pp. 167-90.

[11]   Reddy SS, Kumar P, Dwivedi P. Heavy metal transporters, phytoremediation potential, and biofortification. Plant Metal and Metalloid Transporters, Singapore: Springer. Nat Singap 2022; 387-405.

[12]   Balali-Mood M, Naseri K, Tahergorabi Z, Khazdair MR, Sadeghi M. Toxic mechanisms of five heavy metals: mercury, lead, chromium, cadmium, and arsenic. Front Pharmacol 2021; 12: 643972.
        [http://dx.doi.org/10.3389/fphar.2021.643972] [PMID: 33927623]

[13]   Li C, Zhou K, Qin W, *et al.* A review on heavy metals contamination in soil: effects, sources, and remediation techniques. Soil Sediment Contam 2019; 28(4): 380-94. [doi].
        [http://dx.doi.org/10.1080/15320383.2019.1592108]

[14]   Sarwar N, Imran M, Shaheen MR, *et al.* Phytoremediation strategies for soils contaminated with heavy metals: Modifications and future perspectives. Chemosphere 2017; 171: 710-21.
        [http://dx.doi.org/10.1016/j.chemosphere.2016.12.116] [PMID: 28061428]

[15]   Martínez-Alcalá I, Bernal MP. Environmental impact of metals, metalloids, and their toxicity.Metalloids in plants. John Wiley & Sons, Ltd 2020; pp. 451-88.
        [http://dx.doi.org/10.1002/9781119487210.ch21]

[16]   Lodhi RS, Das S, Zhang A, Das P. Nanotechnology for the remediation of heavy metals and metalloids in contaminated water.Water Pollution and Remediation: Heavy Metals. Springer International Publishing 2021; pp. 177-209.
        [http://dx.doi.org/10.1007/978-3-030-52421-0_7]

[17]   Uddin MM, Zakeel MCM, Zavahir JS, Marikar FMMT, Jahan I. Heavy metal accumulation in rice and aquatic plants used as human food: A general review. Toxics 2021; 9(12): 360.
        [http://dx.doi.org/10.3390/toxics9120360] [PMID: 34941794]

[18]   Ahmed T, Noman M, Rizwan M, Ali S, Shahid MS, Li B. Recent progress on the heavy metals ameliorating potential of engineered nanomaterials in rice paddy: a comprehensive outlook on global food safety with nanotoxicity issues. Crit Rev Food Sci Nutr 2021; 1-15. [doi].
        [PMID: 34554039]

[19]   Bhat SA, Bashir O, Ul Haq SA, *et al.* Phytoremediation of heavy metals in soil and water: An eco-friendly, sustainable and multidisciplinary approach. Chemosphere 2022; 303(Pt 1): 134788.
        [http://dx.doi.org/10.1016/j.chemosphere.2022.134788] [PMID: 35504464]

[20]   Diaconu M, Pavel LV, Hlihor RM, *et al.* Characterization of heavy metal toxicity in some plants and microorganisms—A preliminary approach for environmental bioremediation. N Biotechnol 2020; 56: 130-9.
        [http://dx.doi.org/10.1016/j.nbt.2020.01.003] [PMID: 31945501]

[21]   Aponte H, Meli P, Butler B, *et al.* Meta-analysis of heavy metal effects on soil enzyme activities. Sci Total Environ 2020; 737(139744): 139744.
        [http://dx.doi.org/10.1016/j.scitotenv.2020.139744] [PMID: 32512304]

[22]   Bortoloti GA, Baron D. Phytoremediation of toxic heavy metals by Brassica plants: A biochemical and physiological approach. Environ Adv 2022; 8: 100204.
        [http://dx.doi.org/10.1016/j.envadv.2022.100204]

[23]   Mishra S, Bharagava RN, More N, Yadav A, Zainith S, Mani S, *et al.* Heavy metal contamination: An alarming threat to environment and human health Environmental Biotechnology. For Sustainable Future, Singapore: Springer Singapore 2019; pp. 103-25.

[24]   Al osman M, Yang F, Massey IY. Exposure routes and health effects of heavy metals on children. Biometals 2019; 32(4): 563-73.
        [http://dx.doi.org/10.1007/s10534-019-00193-5] [PMID: 30941546]

[25]   Rahaman MS, Rahman MM, Mise N, *et al.* Environmental arsenic exposure and its contribution to human diseases, toxicity mechanism and management. Environ Pollut 2021; 289(117940): 117940.
        [http://dx.doi.org/10.1016/j.envpol.2021.117940] [PMID: 34426183]

[26]   Shin DY, Lee SM, Jang Y, *et al.* Adverse human health effects of chromium by exposure route: A comprehensive review based on toxicogenomic approach. Int J Mol Sci 2023; 24(4): 3410.
        [http://dx.doi.org/10.3390/ijms24043410] [PMID: 36834821]

[27]   Alengebawy A, Abdelkhalek ST, Qureshi SR, Wang MQ. Heavy metals and pesticides toxicity in agricultural soil and plants: Ecological risks and human health implications. Toxics 2021; 9(3): 42.
        [http://dx.doi.org/10.3390/toxics9030042] [PMID: 33668829]

[28]   Alsafran M, Saleem MH, al Jabri H, Rizwan M, Usman K. Principles and applicability of integrated remediation strategies for heavy metal removal/recovery from contaminated environments. J Plant Growth Regul 2023; 42: 3419-40.

[29]   Raffa CM, Chiampo F, Shanthakumar S. Remediation of metal/metalloid-polluted soils: A short review. Appl Sci (Basel) 2021; 11(9): 4134.
        [http://dx.doi.org/10.3390/app11094134]

[30]   Abdullahi A, Lawal MA, Salisu AM. Heavy metals in contaminated soil: source, accumulation, health risk and remediation process. Bayero J Pure Appl Sci 2021; 14(1): 1-12.
        [http://dx.doi.org/10.4314/bajopas.v14i1.1]

[31]   Vocciante M, Dovì V, Ferro S. Sustainability in ElectroKinetic remediation processes: A critical analysis. Sustainability (Basel) 2021; 13(2): 770.
        [http://dx.doi.org/10.3390/su13020770]

[32]   Trifunović V. Vitrification as a method of soil remediation. Zaštita materijala. 2021;62(3):166-79.
        [http://dx.doi.org/10.5937/zasmat2103166T]

[33]   Lu Y, Cheng J, Wang J, *et al.* Efficient remediation of cadmium contamination in soil by

functionalized biochar: recent advances, challenges, and future prospects. Processes (Basel) 2022; 10(8): 1627.
[http://dx.doi.org/10.3390/pr10081627]

[34]   Popescu SM, Zheljazkov VD, Astatkie T, Burducea M, Termeer WC. Immobilization of Pb in contaminated soils with the combination use of diammonium phosphate with organic and inorganic amendments. Horticulturae 2023; 9(2): 278.
[http://dx.doi.org/10.3390/horticulturae9020278]

[35]   Zhang H, Xu Y, Kanyerere T, Wang Y, Sun M. Washing reagents for remediating heavy-meta--contaminated soil: A review. Front Earth Sci (Lausanne) 2022; 10: 901570.
[http://dx.doi.org/10.3389/feart.2022.901570]

[36]   Yan A, Wang Y, Tan SN, Mohd Yusof ML, Ghosh S, Chen Z. Phytoremediation: A promising approach for revegetation of heavy metal-polluted land. Front Plant Sci 2020; 11: 359.
[http://dx.doi.org/10.3389/fpls.2020.00359] [PMID: 32425957]

[37]   Sladkovska T, Wolski K, Bujak H, Radkowski A, Sobol Ł. A review of research on the use of selected grass species in removal of heavy metals. Agronomy (Basel) 2022; 12(10): 2587.
[http://dx.doi.org/10.3390/agronomy12102587]

[38]   Babu SMOF, Hossain MB, Rahman MS, *et al.* Phytoremediation of toxic metals: A sustainable green solution for clean environment. Appl Sci (Basel) 2021; 11(21): 10348.
[http://dx.doi.org/10.3390/app112110348]

[39]   Parihar JK, Parihar PK, Pakade YB, Katnoria JK. Bioaccumulation potential of indigenous plants for heavy metal phytoremediation in rural areas of Shaheed Bhagat Singh Nagar, Punjab (India). Environ Sci Pollut Res Int 2021; 28(2): 2426-42.
[http://dx.doi.org/10.1007/s11356-020-10454-3] [PMID: 32888151]

[40]   Skuza L, Szućko-Kociuba I, Filip E, Bożek I. Natural molecular mechanisms of plant hyperaccumulation and hypertolerance towards heavy metals. Int J Mol Sci 2022; 23(16): 9335.
[http://dx.doi.org/10.3390/ijms23169335] [PMID: 36012598]

[41]   Kristanti RA, Ngu WJ, Yuniarto A, Hadibarata T. Rhizofiltration for removal of inorganic and organic pollutants in groundwater: a review. Biointerface Res Appl Chem 2021; 11(4): 12326-47.
[http://dx.doi.org/10.33263/BRIAC114.1232612347]

[42]   Available from: https://admin.indiawaterportal.org/articles/bioremediation-its-applicatio-s-contaminated-sites-india-state-art-report-ministry

[43]   Poornachander Rao M, Yerra A, Satyaprasad K. Rhizodegradation of phenanthrene, anthracene and pyrene by augmenting Bacillus cereus and Bacillus subtilis strains. Int J Adv Res (Indore) 2021; 9(2): 89-96.
[http://dx.doi.org/10.21474/IJAR01/12422]

[44]   Funari R, Shen AQ. Detection and characterization of bacterial biofilms and biofilm-based sensors. ACS Sens 2022; 7(2): 347-57.
[http://dx.doi.org/10.1021/acssensors.1c02722] [PMID: 35171575]

[45]   Martínez-Guijarro R, Paches M, Romero I, Aguado D. Sources, Mobility, Reactivity, and Remediation of Heavy Metal(loid) Pollution: A Review. Adv Environ Eng Res 2021; 2(4): 033.
[http://dx.doi.org/10.21926/aeer.2104033]

[46]   Khanna K, Jamwal VL, Gandhi SG, Ohri P, Bhardwaj R. Metal resistant PGPR lowered Cd uptake and expression of metal transporter genes with improved growth and photosynthetic pigments in *Lycopersicon esculentum* under metal toxicity. Sci Rep 2019; 9(1): 5855.
[http://dx.doi.org/10.1038/s41598-019-41899-3] [PMID: 30971817]

[47]   Mukherjee P, Mitra A, Roy M. Halomonas rhizobacteria of *Avicennia marina* of Indian Sundarbans promote rice growth under saline, and heavy metal stresses through exopolysaccharide production. Front Microbiol 2019; 10: 1207.

[http://dx.doi.org/10.3389/fmicb.2019.01207] [PMID: 31191507]

[48]   Gontia-Mishra I, Sapre S, Sharma A, Tiwari S. Alleviation of mercury toxicity in wheat by the interaction of mercury-tolerant plant growth-promoting rhizobacteria. J Plant Growth Regul 2016; 35(4): 1000-12.
[http://dx.doi.org/10.1007/s00344-016-9598-x]

[49]   Dell'Anno F, Brunet C, van Zyl LJ, *et al.* Degradation of hydrocarbons and heavy metal reduction by marine bacteria in highly contaminated sediments. Microorganisms 2020; 8(9): 1402.
[http://dx.doi.org/10.3390/microorganisms8091402] [PMID: 32933071]

[50]   Medfu Tarekegn M, Zewdu Salilih F, Ishetu AI. Microbes used as a tool for bioremediation of heavy metal from the environment. Cogent Food Agric 2020; 6(1): 1783174.
[http://dx.doi.org/10.1080/23311932.2020.1783174]

[51]   Jibrin AM, Oyewole OA, Yakubu JG, Hussaini A, Egwim EC. Heavy metals biosorption by urease producing Lysinibacillus fusiformis 5B. Eur J Biol Res 2020; 10(4): 326-35.

[52]   Oyewole OA, Zobeashia SSLT, Oladoja EO, Raji RO, Odiniya EE, Musa AM. Biosorption of heavy metal polluted soil using bacteria and fungi isolated from soil. SN Appl Sci 2019; 1: 1-8.

[53]   Sarankumar RK, Selvi A, Murugan K, Rajasekar A. Electrokinetic (EK) and bio-electrokinetic (BEK) remediation of hexavalent chromium in contaminated soil using alkalophilic bio-anolyte. Indian Geotechnical Journal 2020; 50(3): 330-8.
[http://dx.doi.org/10.1007/s40098-019-00366-6]

[54]   Sur IM, Micle V, Hegyi A, Lăzărescu AV. Extraction of metals from polluted soils by bioleaching in relation to environmental risk assessment. Materials (Basel) 2022; 15(11): 3973.
[http://dx.doi.org/10.3390/ma15113973] [PMID: 35683266]

[55]   Petková K. Potencial of aspergillus niger in bioremediation of contaminated soils. Int Multidiscip Sci Geoconf SGEM 2013; 1: 757.
[http://dx.doi.org/10.5593/SGEM2013/BE5.V1/S20.100]

[56]   Das S, Mishra J, Das SK, *et al.* Investigation on mechanism of Cr(VI) reduction and removal by *Bacillus amyloliquefaciens*, a novel chromate tolerant bacterium isolated from chromite mine soil. Chemosphere 2014; 96: 112-21.
[http://dx.doi.org/10.1016/j.chemosphere.2013.08.080] [PMID: 24091247]

[57]   Newsome L, Falagán C. The microbiology of metal mine waste: Bioremediation applications and implications for planetary health. GeoHealth 2021;5:e2020GH000380.

[58]   Kelly CR, Cristina FS, Whasley FD, Rosane FS. Bioaccumulation of Fe^{3+} by bacteria isolated from soil and fermented foods for use in bioremediation processes. Afr J Microbiol Res 2014; 8(26): 2513-21.
[http://dx.doi.org/10.5897/AJMR2014.6838]

[59]   Mateos LM, Ordóñez E, Letek M, Gil JA. *Corynebacterium glutamicum* as a model bacterium for the bioremediation of arsenic. Int Microbiol 2006; 9(3): 207-15.
[PMID: 17061211]

[60]   Henrot J. Bioaccumulation and chemical modification of Tc by soil bacteria. Health Phys 1989; 57(2): 239-45.
[http://dx.doi.org/10.1097/00004032-198908000-00001] [PMID: 2547734]

[61]   Zhang K, Xue Y, Xu H, Yao Y. Lead removal by phosphate solubilizing bacteria isolated from soil through biomineralization. Chemosphere 2019; 224: 272-9.
[http://dx.doi.org/10.1016/j.chemosphere.2019.02.140] [PMID: 30825853]

[62]   Arias D, Cisternas L, Rivas M. Biomineralization mediated by ureolytic bacteria applied to water treatment: a review. Crystals (Basel) 2017; 7(11): 345.
[http://dx.doi.org/10.3390/cryst7110345]

[63]   Xu G, Li D, Jiao B, *et al.* Biomineralization of a calcifying ureolytic bacterium *Microbacterium* sp. GM-1. Electron J Biotechnol 2017; 25: 21-7.
[http://dx.doi.org/10.1016/j.ejbt.2016.10.008]

[64]   Zachara JM, Kukkadapu RK, Fredrickson JK, Gorby YA, Smith SC. Biomineralization of poorly crystalline Fe(III) oxides by dissimilatory metal reducing bacteria (DMRB). Geomicrobiol J 2002; 19(2): 179-207.
[http://dx.doi.org/10.1080/01490450252864271]

[65]   Torres E. Biosorption: a review of the latest advances. Processes (Basel) 2020; 8(12): 1584.
[http://dx.doi.org/10.3390/pr8121584]

[66]   Henao GS, Herrera GT. Heavy metals in soils and the remediation potential of bacteria associated with the plant microbiome. Front Environ Sci 2021; 15.

[67]   Padilla-Ortega E, Leyva-Ramos R, Mendoza-Barron J. Role of electrostatic interactions in the adsorption of cadmium(II) from aqueous solution onto vermiculite. Appl Clay Sci 2014; 88-89: 10-7.
[http://dx.doi.org/10.1016/j.clay.2013.12.012]

[68]   Vélez JMB, Martínez JG, Ospina JT, Agudelo SO. Bioremediation potential of *Pseudomonas* genus isolates from residual water, capable of tolerating lead through mechanisms of exopolysaccharide production and biosorption. Biotechnol Rep (Amst) 2021; 32: e00685.
[http://dx.doi.org/10.1016/j.btre.2021.e00685] [PMID: 34765463]

[69]   Jianlong W, Xinmin Z, Decai D, Ding Z. Bioadsorption of lead(II) from aqueous solution by fungal biomass of Aspergillus niger. J Biotechnol 2001; 87(3): 273-7.
[http://dx.doi.org/10.1016/S0168-1656(00)00379-5] [PMID: 11334669]

[70]   Tian D, Jiang Z, Jiang L, *et al.* A new insight into lead (II) tolerance of environmental fungi based on a study of *Aspergillus niger* and *Penicillium oxalicum*. Environ Microbiol 2019; 21(1): 471-9.
[http://dx.doi.org/10.1111/1462-2920.14478] [PMID: 30421848]

[71]   Srichandan H, Mohapatra RK, Parhi PK, Mishra S. Bioleaching: A bioremediation process to treat hazardous wastes Soil Microenvironment for Bioremediation and Polymer Production. 2019; pp. 115-29.
[http://dx.doi.org/10.1002/9781119592129.ch7]

[72]   Hlihor RM, Apostol LC, Gavrilescu M. Environmental bioremediation by biosorption and bioaccumulation: principles and applications. Enhancing clean-up of environmental pollutants. Biol Approaches 2017; 1: 289-315.

[73]   Diep P, Mahadevan R, Yakunin AF. Heavy metal removal by bioaccumulation using genetically engineered microorganisms. Front Bioeng Biotechnol 2018; 6: 157.
[http://dx.doi.org/10.3389/fbioe.2018.00157] [PMID: 30420950]

[74]   Saier MH Jr. Transport protein evolution deduced from analysis of sequence, topology and structure. Curr Opin Struct Biol 2016; 38: 9-17.
[http://dx.doi.org/10.1016/j.sbi.2016.05.001] [PMID: 27270239]

[75]   Reddy BL, Saier MH Jr. Properties and phylogeny of 76 families of bacterial and eukaryotic organellar outer membrane pore-forming proteins. PLoS One 2016; 11(4): e0152733.
[http://dx.doi.org/10.1371/journal.pone.0152733] [PMID: 27064789]

[76]   Shah D, Shen MWY, Chen W, Da Silva NA. Enhanced arsenic accumulation in Saccharomyces cerevisiae overexpressing transporters Fps1p or Hxt7p. J Biotechnol 2010; 150(1): 101-7.
[http://dx.doi.org/10.1016/j.jbiotec.2010.07.012] [PMID: 20638426]

[77]   Singh S, Kang SH, Lee W, Mulchandani A, Chen W. Systematic engineering of phytochelatin synthesis and arsenic transport for enhanced arsenic accumulation in *E. coli*. Biotechnol Bioeng 2010; 105(4): 780-5.
[http://dx.doi.org/10.1002/bit.22585] [PMID: 19845016]

[78] Dhami NK, Reddy MS, Mukherjee A. Biomineralization of calcium carbonates and their engineered applications: a review. Front Microbiol 2013; 4: 314.
[http://dx.doi.org/10.3389/fmicb.2013.00314] [PMID: 24194735]

[79] Li M, Cheng X, Guo H. Heavy metal removal by biomineralization of urease producing bacteria isolated from soil. Int Biodeterior Biodegradation 2013; 76: 81-5.
[http://dx.doi.org/10.1016/j.ibiod.2012.06.016]

[80] Miot J, Benzerara K, Morin G, *et al.* Iron biomineralization by anaerobic neutrophilic iron-oxidizing bacteria. Geochim Cosmochim Acta 2009; 73(3): 696-711.
[http://dx.doi.org/10.1016/j.gca.2008.10.033]

[81] Paganin P, Alisi C, Dore E, *et al.* Microbial diversity of bacteria involved in biomineralization processes in mine-impacted freshwaters. Front Microbiol 2021; 12: 778199.
[http://dx.doi.org/10.3389/fmicb.2021.778199] [PMID: 34880845]

[82] Kumar L, Bharadvaja N. Microbial remediation of heavy metals Microbial Bioremediation & Biodegradation. Singapore: Springer Singapore 2020; pp. 49-72.
[http://dx.doi.org/10.1007/978-981-15-1812-6_2]

[83] Atagana HI, Haynes RJ, Wallis FM. Optimization of soil physical and chemical conditions for the bioremediation of creosote-contaminated soil. Biodegradation 2003; 14(4): 297-307.
[http://dx.doi.org/10.1023/A:1024730722751] [PMID: 12948059]

[84] Sayqal A, Ahmed OB. Advances in heavy metal bioremediation: An overview. Appl Bionics Biomech 2021; 2021: 1-8.
[http://dx.doi.org/10.1155/2021/1609149] [PMID: 34804199]

[85] Mangunwardoyo W, Sudjarwo T, Patria MP. Bioremediation of effluent wastewater treatment plant Bojongsoang Bandung Indonesia using consortium aquatic plants and animals. International Journal of Research and Reviews in Applied Sciences 2013; 14(1): 150-60.

[86] Nath D, Kumari V, Laik R, Mukhopadhyay R. Genetically engineered microorganisms: A promising approach for bioremediation.Omics for environmental engineering and microbiology systems. CRC Press 2022; pp. 453-67.
[http://dx.doi.org/10.1201/9781003247883-23]

[87] Altowayti WAH, Shahir S, Othman N, *et al.* The role of conventional methods and artificial intelligence in the wastewater treatment: A comprehensive review. Processes (Basel) 2022; 10(9): 1832.
[http://dx.doi.org/10.3390/pr10091832]

Part 3: Role of Microbiomes in Agriculture Crop Growth and Development

Role of Microbes and Microbiomes in Biofertilizer Production and as Plant Growth Promoters

Nikul B. Chavada[1,*] and **Ramesh Poornima**[2]

[1] *Om College of Science, Bheshan Highway, Junagadh- 362001, Gujarat, India*

[2] *Department of Environmental Sciences, Tamil Nadu Agricultural University, Coimbatore, India*

Abstract: In 2050, 8.3 billion people will live on Earth, and 70 to 100% more food will be needed. Food and its products are available through agricultural practices. Soil biological systems play an essential role in food production. However, it is a complex process that leads to the stability of crop production and the maintenance of soil health. Healthy food with eco-friendly agriculture practices is required to sustain the soil ecosystem globally. Additionally, the continued depletion of the Earth's natural resources and the increasing use of harmful chemical fertilizers are significant concerns for agriculture's future. Biofertilizers are gaining popularity as a viable alternative to unsafe chemical fertilizers in the pursuit of sustainable agriculture. Biofertilizers have an important role in enhancing crop output and preserving long-term soil fertility, both of which are critical for fulfilling global food demand. Microbes can interact with agricultural plants to improve their resistance, growth, and development. Nitrogen, phosphorus, potassium, zinc, and silica are the fundamental elements needed for crop growth, yet they are normally present in insoluble or complex forms. Certain microbes dissolve them and make them accessible to plants.

Keywords: Biofertilizer, Biological N_2 fixing bacteria, PGPR (Plant Growth Promoting Rhizobacteria), PSB.

INTRODUCTION

Farmers are applying various crop nutrition tactics to assist and fulfill the increasing demand for food caused by the world's population growth. According to FAO predictions, agricultural product consumption will climb to 60% by 2030 [1]. One of the primary issues of the twenty-first century is increasing output while protecting the environment [2]. Fertilizers have been widely utilized to boost crop output from agricultural land. Increased use of chemical fertilizers in agriculture may help a country become self-sufficient in food production, but

* **Corresponding author Nikul B. Chavada:** Om College of Science, Bheshan Highway, Junagadh- 362001, Gujarat, India; E-mail: nikulfriends8@gmail.com

Shiv Prasad, Govindaraj Kamalam Dinesh, Murugaiyan Sinduja, Velusamy Sathya, Ramesh Poornima & Sangilidurai Karthika (Eds.)

chemicals are harmful to both the environment and living beings. Furthermore, chemical fertilizers are costly, have an impact on the soil, impair its water-holding capacity and fertility, induce nutrient imbalances in the soil, and result in unsustainable levels of water contamination [1]. Biofertilizers, on the other hand, are environmentally benign, cost-effective, non-toxic, and simple to apply; they assist in sustaining agricultural land soil structure and biodiversity. As a result, they are an excellent replacement for chemical fertilizers [3].

In the period 2022-2027, the biofertilizer market is expected to grow 13.81%, according to a new report by IMARC Group (Bio-Fertilizer Market, forecast 2029, CAGR report by the United States biofertilizer market), where the India biofertilizer market is expected to grow 15.89% (CAGR report) [4]. Growth controllers, genetically modified crops (GM crops), and tolerant varieties have been applied to improve crop resistance to salt stress [5 - 7]. These methods have been restricted due to their cost, labor work, and environmental threats. The soil needs several essential minerals to carry out plant growth. Very significant minerals are nitrogen, phosphorus, and potassium. Other 13 minerals as microelements are required for plant development and yield production [8 - 10]. They describe two categories based on their quantitative requirement: 1) macronutrients and 2) micronutrients. Nitrogen and phosphorus are necessary for plant growth and development; potassium is the third essential plant nutrient [11]. After nitrogen and phosphorus, five million tonnes of potash are required every year in the world, but in India, the commercial source is not available for potassic fertilizer production, so it is imported from other countries [12, 13].

Biofertilizers, also known as microbial inoculants, are organic materials formed from plant roots and root zones that include particular microorganisms. They have been demonstrated to increase plant growth and yield by 10-40% [14]. When applied to the seed, plant surface, or soil, these bioinoculants colonize the rhizosphere and the interior of the plant, stimulating plant development. By supplying nutrients to the soil, they not only boost soil fertility and crop yield but also protect the plant against pests and diseases. They have been found to improve root system development, lengthen its life, destroy hazardous elements, boost seedling survival, and shorten the time to bloom [15]. Another advantage is that after 3-4 years of continuous usage of biofertilizers, there is no need for their application since parental inoculum is adequate for development and multiplication [16]. The plant needs 17 critical ingredients for effective growth and development. Nitrogen (N), phosphorus (P), and potassium (K) are three of the most important. Several microorganisms, including phosphate-solubilizing bacteria, moulds, and fungi, and nitrogen-fixing soil bacteria and cyanobacteria, are routinely utilized as biofertilizers [17]. Similarly, phytohormone-generating bacteria are utilized in the manufacture of biofertilizers. They give the plant growth-promoting chemicals such as amino acids, indole acetic acid (IAA), and vitamins, thereby improving soil fertility and plant productivity.

Biofertilizers contain efficient bio-inoculants that improve the nutrient quality of plants and regulate their physiological properties [14, 18]. Biofertilizers are produced by soil-helpful living microorganisms [19]. Synthetic fertilizers can be applied to plants to improve plant growth. In soil systems, chemical fertilizers can cause leaching, volatilization, acidification, and denitrification if not appropriately managed [20 - 23]. Synthetic fertilizers and pesticides have caused several environmental issues (greenhouse effect, ozone layer depletion, and water acidification). The application of bio-fertilizers and biopesticides can overcome these issues. In addition to being natural and beneficial, both are eco-friendly for the user. Community waste and sewage sludge are alternative sources of organic fertilizer. It is an inexpensive and attractive source of nutrients [21, 24]. However, due to the presence of heavy metals and some poisonous chemicals, it cannot be used as an ideal organic fertilizer; heavy metals produce adverse effects on crop growth, so they cannot be an appropriate source of fertilizer [20].

Biofertilizers are grouped into different types on the basis of their functions and mode of action (Fig. **1**). The commonly used biofertilizers are nitrogen fixer (N-fixer), potassium solubilizer (K-solubilizer), phosphorus solubilizer (P-solubilizer), and plant growth promoting rhizobacteria (PGPR). In one gram of fertile soil, up to 10^{10} bacteria can be present, with a live weight of 2000 kg ha^{-1} [25]. Soil bacteria could be cocci (sphere, 0.5 µm), bacilli (rod, 0.5–0.3 µm), or spiral shaped (1–100 µm). The presence of bacteria in the soil depends upon the physical and chemical properties of the soil, organic matter, and phosphorus contents, as well as cultural activities. However, nutrient fixation and plant growth enhancement by bacteria are key components for achieving sustainable agriculture goals in the future. Microbes also facilitate various nutrient cycles in the ecosystem.

The microbial strains used as biofertilizer and their effect on crop yield are given in Table **1**. Burris [26] confirmed that atmospheric nitrogen is fixed biologically in legumes. A French agriculturist suggested that legumes are higher in cereals plants; they can play a vital role in biological nitrogen fixation and may deliver nitrogen to plants. Bacteria, fungi, and blue-green algae are used for biofertilizer applications [19]. They are applied to the rhizosphere of plants to improve their soil ecosystem. Efficient strains are isolated from soil (other environments) and are suitable for soil environments and climate [27 - 29]. These strains are multiplied in the laboratory to produce a suitable mass of isolates and then provide farmers with suitable carriers like peat and lignite powder to have a good shelf life for their farm.

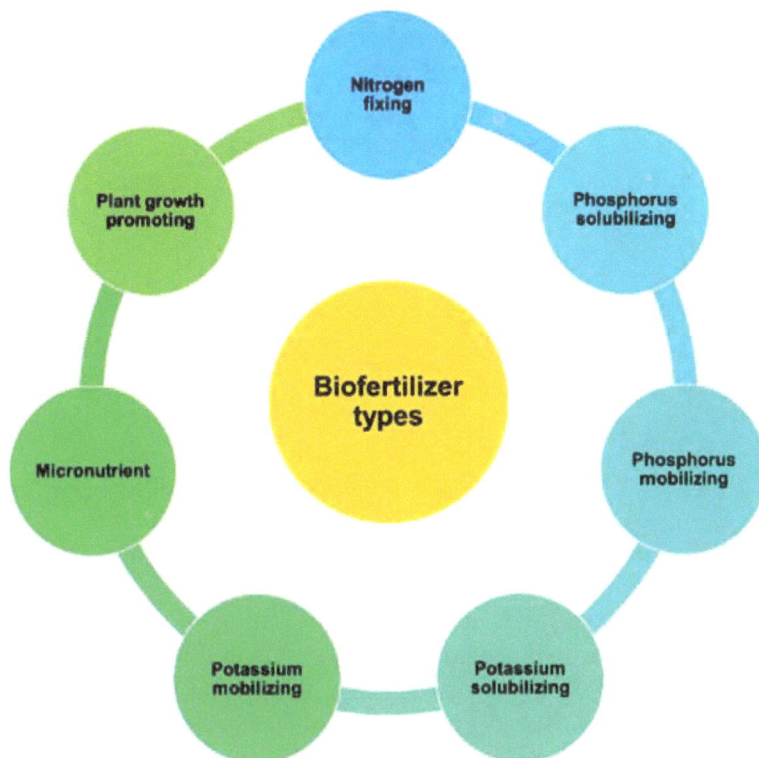

Fig. (1). Types of biofertilizers.

Table 1. Microbial strains used as biofertilizer and their effect on crop yield.

S. No.	Strain	Crops	Benefit in yield production	Comments
1.	Azotobacter	Free-living organism Applying on non-legume crops	Rise 15% yield –adds 25 kg N/ha	Controls illnesses in plant
2.	Azospirillum	Free-living organisms applied on non-legume crops	It may have a 10-20% yield increase	Improves plant growth
3.	Rhizobium strains	It can apply legumes like crops,	Increase 10-35% crop yield production.	Biological nitrogen fixation by legumes
4.	Phosphate Solubilizers (PSB)	Apply to all plants	Increase 5-30% yield	Apply with rock phosphate
5.	Microhizae (VAM)	Several trees, plants, root fungi	It can enhance the uptake of minerals and water for the plant	Generally applied during seeding

Microorganisms can increase soil's biochemical characteristics [22] and can be used as a substitute for chemical fertilizer; bioinoculants can improve soil ecosystems. Microbes add nutrients through natural biological mechanisms like nitrogen fixation, phosphorus solubilizing, and producing growth-promoting hormones [30, 31]. Microbes can convert complex organic and inorganic nutrients into simple nutrients readily available for plants. The proper application of biofertilizers can improve 20 – 30% of crop yields. There are two reasons for using biofertilizers in agriculture practice. First, they can increase crop productivity and reduce the application of chemical fertilizers, which cause pollution in the soil ecosystem, therefore, biofertilizers are considered economically and environmentally healthy for the soil ecosystem. This chapter describes a variety of nitrogen-fixing bacteria, phosphate solubilizing bacteria (PSB), zinc solubilizer bacteria, and fungi as biofertilizers [32 - 34].

NITROGEN-FIXING MICROORGANISMS AS BIOFERTILIZERS

Nitrogen (N_2) is the most limiting nutrient for plant development. Although the atmosphere contains around 80% of free nitrogen, most plants are unable to utilize it. Nitrogen-fixing bacteria play a crucial role in agriculture by converting atmospheric nitrogen into a form that can be readily used by plants [35]. Nitrogen fixation can supply 300-400 kg N/ha/year while increasing crop production by 10-50%. N-fixation accounts for up to 25% of total nitrogen in plants. While leguminous crops are well-known for their symbiotic relationship with nitrogen-fixing bacteria, it is less common for cereals to form such associations. However, some cereals, such as rice, can benefit from associative nitrogen fixation, where non-leguminous plants associate with certain bacteria to fix nitrogen.

Nitrogen-fixing bacteria are active at the plant root region in the soil [36]. The multimeric enzyme complex nitrogenase plays a vital role in reducing nitrogen to ammonia [37]. These nitrogen-fixing microbes are generally categorized as blue-green algae, free-living bacteria (*Azospirillum* and *Azotobacter*), and symbionts (*Frankia, Rhizobium,* and *Azolla*). The N_2-fixing bacteria associated with legumes include *Rhizobium, Azorhizobium, Mesorhizobium, Allorhizobium, Bradyrhizobium,* and *Sinorhizobium,* whereas microbes involved with non-leguminous plants include *Alcaligenes, Achromobacter, Acetobacter, Arthrobacter, Azomonas, Beijerinckia, Clostridium, Bacillus, Enterobacter, Erwinia, Desulfovibrio, Corynebacterium, Derxia, Campylobacter, Mycobacterium, Herbaspirillum, Klebsiella, Lignobacter, Rhodo-pseudomonas,* Rhodospirillum, Xanthobacter, Methylosinus, and *Mycobacterium.* The major N-fixing bacteria and their potential are given in Fig. (**2**).

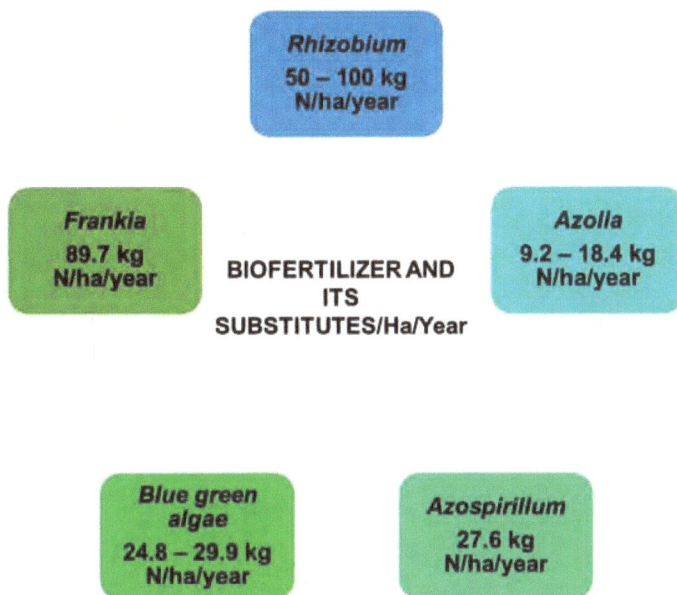

Fig. (2). Major nitrogen-fixing bacteria.

Rhizobium is a symbiotic bacterium that belongs to the *Rhizobiaceae* family group [21, 38]. It is a symbiotic microorganism and is associated with legumes of plant roots, used as bioinoculant. It can depend on the availability of strain for the legume [39]. It can colonize plant roots like tumor growths called root nodules, which work for ammonia production and fix nitrogen with legumes in the soil. *Azospirillum* is a heterotrophic bacterial group belonging to the family *Spirilaceae,* which fixates approximately 15-35 kg/ha of nitrogen and manufactures growth-promoting organic compounds. *Azospirillum* can be used as a biofertilizer for maize-like millet plants [40]. This genus of bacteria has been extensively studied for its ability to enhance plant growth and nitrogen availability.

Azospirillum forms a beneficial association with the roots of various plants, including legumes and some cereals, promoting nitrogen fixation and nutrient uptake. *Azotobacter* is mainly found in neutral or alkaline soil. It is an aerobic, free-living, and heterotrophic bacteria belonging to family *Azotobacteriaceae. A. chroococcum* is a commonly found species in soils. *Azotobacter* can produce natural antimicrobial substances that serve as anti-fungal substances. In the case of leguminous crops like soybeans, peas, and lentils, a different type of nitrogen-fixing bacteria called Rhizobium forms nodules on the roots, where the bacteria convert nitrogen gas into ammonium, which is then used by the plants for growth. This symbiotic relationship is highly specific to legumes. It is worth noting that while nitrogen-fixing bacteria can be beneficial for plant growth and reduce the need for synthetic nitrogen fertilizers, the effectiveness of such associations with cereals varies. Scientists continue to research and explore ways to enhance nitrogen fixation in non-leguminous crops like cereals to reduce the environmental impact of agriculture and improve crop productivity.

Azotobacter is a nitrogen-fixing diazotrophic free-living bacterium that plays an important part in the nitrogen cycle due to its numerous metabolic activities [41]. Azotobacter may manufacture vitamins such as thiamine and riboflavin. It is a member of the *Azotobacteriaceae* family and is utilized as a biofertilizer for non-leguminous plants such as rice, sweet potato, cotton, sugarcane, and sweet sorghum. Dutta and Singh [42] found that *Azotobacter* inoculation increased seed production in rapeseed and mustard. It fixes about 30 kg/N/year and is mostly used in sugarcane crops, increasing cane production by 25-50 tons/hectare and sugar content by 10-15%. Azotobacter may be found in both acidic and alkaline soils.

A. chroococcum is the most common species discovered in soil, although *A. insignis, A. vinelandii, A. macrocytogenes,* and *A. beijerinckii* are also present. Blue-green alga (Cyanobacteria), on the other hand, is a phototropic alga that can produce auxin, indole acetic acid, and gibberellic acid-like plant growth-promoting hormones and fix approximately 15-25 kg N/ha in rice fields, so it is referred to as "paddy organisms". Cyanobacteria or blue-green algae (BGA) groups can fix atmospheric nitrogen. *Nostoc* and many other cyanobacterial groups can be bioinoculants [43]. *Frankia* is another nitrogen-fixing actinomycete, and it is competent for infecting and nodulating more than eight woody family plants. It is an actinorhizal plant used in land retrieval for timber and fuel wood manufacture [44] *Frankia* can fix nitrogen and is estimated to be similar to rhizobia activity [31], but limited information is available regarding the potential use of Frankia as a bioinoculant.

PHOSPHATE SOLUBILIZING AND MOBILIZING MICROBES AS BIOFERTILIZERS

Plants contain around 0.2% phosphorus by dry weight, and it is an important nutrient for plant growth and development. Plant growth can be affected by phosphorus, an essential growth-limiting nutrient [32], by operating key metabolic activities in plants [33]. About 95-99% of P is present in the soil in an unavailable form, so plants cannot efficiently utilize it directly from the soil [34]. P plays a critical biochemical role in energy storage, energy allocation, and other perilous biological processes in the living plant. Organic acid and phosphatase enzymes can convert unavailable/insoluble phosphate compounds to solubilized form and make them available for plants. Phosphate-solubilizing bacteria (PSB) use many processes to convert insoluble phosphates, such as H_2PO_4 and HPO_4, into soluble forms, including chelation, organic acid generation, and ion exchange reactions.

Phosphate-solubilizing bacteria account for 1-50% of microbial populations, while fungi account for just 0.1-0.5% of phosphate-solubilizing activity. The PSB can produce metabolites with carboxyl (ketogluconic) and hydroxyl (gluconic) groups that chelate the cation attached to the phosphate and convert it to a soluble form that plants can use. The released acids also lower soil pH and dissolve bound phosphate to make it accessible to plants [45]. Microorganisms employ the proton-extrusion process to solubilize phosphate in addition to the chemical technique [46]. The PSB provides phosphate as well as other trace elements like Fe and Zn, which eventually improve plant development. They also produce the enzyme that destroys infections, protecting the plant from diseases.

Pseudomonas, Bacillus, Aereobacter, Flavobacterium and *Erwinia Rhizobium, Burkholderia, Achromobacter, Agrobacterium, and Micrococcus* group of bacterial genera were reported with this ability. Soil microbiota release phosphate from organic and inorganic compounds and add it to the soil through solubilization [47]. The organization of soil phosphorus increases crop growth and reduces P damage from soils. PSB soil inoculums have improved plant growth and yield [48]. A phosphate solubilizing microorganism (PSM) belongs to the *Bacillus* genus. Different phosphorus-containing mediums can be used to grow several bacteria, which serve a vital function in providing phosphorus [49]. Plant roots can absorb insoluble inorganic phosphate from the bulk soil by mobilizing it from their nearby soil mineral matrix [18]. Bacterial and fungal species can solubilize inorganic phosphates, known as phosphate solubilizers; they exist in plants' rhizosphere in insufficient numbers and compete with other microbial species present in the rhizosphere [19].

Phosphate solubilizing bacteria (PSB) apply individually or combined with other microorganisms, producing higher effects on crop growth and biomass production [20]. Developing successful microbial inoculants is a significant technical challenge [50]. The application of rhizobia will have a dual beneficial effect on P mobilization and nitrogen fixation [22]. Phosphate uptake by plant roots, PSB can release rock P value in soil by converting unavailable P to available phosphate. Microorganisms produce organic acids and other compounds that favor the phosphate solubilization activity of microorganisms (PSM). PSB-secreting auxins-like hormone enhances plant growth promotion activities. IIA and cytokinin-like substances can improve plants' metabolism activity. The most crucial soil bacterial communities are *Pseudomonas*, Bacillus, Enterobacter, and *Klebsiella* strains; however, *Pseudomonas* sp. comparatively has higher phosphorous uptake efficiency. Biofertilizers contain microbe ecotype diversity and their tolerance to environmental stress. In agriculture, biofertilizers are essential for sustainable practices. Biofertilizers are the most promising natural compounds that can improve microbial activities in soil ecosystems.

Phosphate-mobilizing microorganisms have the ability to mobilize immobile forms of phosphorous. They move insoluble phosphate from the soil layers to the root cortex. For instance, Arbuscular mycorrhiza, a phosphate-mobilizing fungus, enters the roots and increases the surface area of the roots while also stimulating metabolic processes and absorbing nutrients. Phosphate mobilizers have been reported to be phosphorus-solubilizing bacteria (PSB). Under ideal conditions, they have the capacity to solubilize/mobilize 30-50 kg P_2O_5/ha, increasing crop production by 10-20% [51].

POTASSIUM SOLUBILIZING AND MOBILIZING MICROBES AS BIOFERTILIZERS

After nitrogen and phosphorus, potassium (K) is the second most plentiful and significant plant nutrient. Although K is plentiful in soil, only 1-2% is accessible to plants, with the remainder existing as mineral K that cannot be absorbed by plants. As a result, soil solution K must be replenished on a constant basis [9]. It is essential for the growth and development of plants. If sufficient nutrients are not provided, the plants will grow slowly, have poorly formed roots, generate little seeds, and provide low yields. A variety of bacterial and fungal strains have been found to employ several processes to convert insoluble K into soluble K, including chelation, acid generation, complexolysis, acidolysis, and exchange reactions.

Aspergillus niger, Cladosporium, Arthrobacter spp., Sphingomonas aminobacter, and *Bacillus spp.* are examples of potassium-solubilizing biofertilizers. *B. mucilaginosus* and *B. edaphicus* have been shown to improve both solubilization and mobilization. When *B. mucilaginosus* was inoculated into soil, it increased the oil content and biomass of groundnuts by 35.4% and 25%, respectively, while also increasing K and P availability. Research published recently found that a potassium-solubilizing strain of *Bacillus pseudomycoides* increased K absorption in tea plants grown in mica waste-treated soil by increasing potassium availability [52]. Some fungi, such as *Aspergillus* and *Penicillium*, have the ability to solubilize and mobilize K from organic and inorganic sources. Thus, the involvement of K solubilizers is critical for guaranteeing a consistent supply of K to agricultural plants. These also have a good effect on the availability of other necessary elements in the soil and hence play a significant role in soil sustainability [53].

SULFUR-OXIDIZING MICROBES

Sulfur-oxidizing microbes are microorganisms that can use sulfur complexes, such as hydrogen sulfide, thiosulfate, and elemental sulfur, as an energy source to perform chemosynthesis [54]. These microbes are habitats in various surroundings, including hot springs, hydrothermal vents, and sulfur springs. Some examples of sulfur-oxidizing microbes include bacteria like *Thiobacillus, Acidithiobacillus, and Beggiatoa*, and some archaea like *Acidianus* and *Sulfolobus* [55]. Sulfur-oxidizing microbes can also remove sulfur compounds from contaminated soil and water. Additionally, they are being studied for their potential use in industrial processes, such as biomining and biodesulfurization. Sulfur-oxidizing microbes can use sulfur compounds as their energy source for metabolism. These microbes are found in various environments, including hot springs, deep-sea vents, and sulfuric acid pools.

One of the most well-known groups of sulfur-oxidizing microbes is the *Thiotrichales*, commonly found in marine environments. These bacteria can oxidize various sulfur compounds, including elemental sulfur and thiosulfate, to produce energy. They play a significant part in the sulfur cycle, changing sulfur compounds into forms that other organisms can use. Another group of sulfur-oxidizing microbes is the *Acidithiobacillales*, commonly found in acidic environments such as mine drainage. These bacteria can tolerate high levels of acidity and are essential in biomining, where they are used to extract metals from ores. Overall, sulfur-oxidizing microbes are essential for the cycling of sulfur in the environment and have applications in various industries, such as mining and wastewater treatment [56, 57]. Sulfur-oxidizing microbes indirectly contribute to plant growth and nutrient availability through their activity in the soil. Here is an overview of their mechanism as biofertilizers:

- *Sulfur Oxidation:* Sulfur-oxidizing microbes have the ability to convert elemental sulfur (S^0) or sulfide compounds (such as hydrogen sulfide) into sulfate (SO_4^{2-}). This process is known as sulfur oxidation and is carried out by enzymes produced by these microbes.
- *Sulfur Mineralization:* By oxidizing sulfur compounds, sulfur-oxidizing microbes convert sulfur into sulfate forms that can be easily absorbed by plants. Sulfate is an important form of sulfur that serves as a vital nutrient for plant growth and development.
- *Enhanced Sulfur Availability:* The conversion of sulfur into sulfate by sulfur-oxidizing microbes increases the availability of sulfur in the soil. Sulfur is an essential macronutrient for plants and plays a crucial role in various metabolic processes, including protein synthesis, enzyme activity, and chlorophyll

production.

- *Plant Uptake:* Once sulfur is in the sulfate form, plants can readily take it up through their roots. Sulfate is transported into plant cells and incorporated into organic molecules, contributing to the growth and health of the plant.
- *Soil Fertility Improvement:* Sulfur-oxidizing microbes indirectly enhance soil fertility by increasing the bioavailability of sulfur. Adequate sulfur levels in the soil promote optimal plant growth, improve nutrient balance, and can positively influence crop yield and quality.

It is important to note that the effectiveness of sulfur-oxidizing microbes as biofertilizers can vary depending on several factors, including soil conditions, microbial populations, and crop requirements. Additionally, other factors such as pH, temperature, and the presence of other microorganisms can influence the activity and effectiveness of sulfur-oxidizing microbes in the soil. To harness the benefits of sulfur-oxidizing microbes, it may be necessary to assess the specific needs of the crop and soil conditions. Soil amendments or inoculants containing these microbes can be applied to enhance sulfur availability and promote plant growth, particularly in sulfur-deficient soils.

ZINC SOLUBILIZING MICROBES

Zinc solubilizing microbes that are capable of converting insoluble forms of zinc into a bioavailable form that can be efficiently uptake by plants. These microbes can significantly contribute to soil fitness and plant growth by improving zinc availability, an essential plant micronutrient. These microbes produce organic acids that can chelate or bind with the zinc ions, making them more soluble and available to plants. Zinc solubilizing microbes can be applied to soils as biofertilizers, alone or in combination with other PGPR or mycorrhizal fungi. This can lead to improved plant growth and yields, as well as increased soil fertility and sustainability. Zinc-solubilizing microbes are essential for sustainable agriculture and can help reduce the use of chemical fertilizers. Some examples of zinc-solubilizing bacteria include Pseudomona*s, Bacillus*, and *Rhizobium* species. These bacteria are commonly found in plants' rhizosphere, enhancing their zinc uptake. Overall, zinc solubilizing microbes play a significant role in improving the availability of zinc in soil for uptake, and they have potential applications in agriculture for enhancing crop production and reducing fertilizer use. Zinc (Zn) solubilizing microbes can act as biofertilizers by enhancing plant's zinc availability. Here is an overview of their mechanism as biofertilizers:

- *Zinc Solubilization:* Zinc solubilizing microbes produce organic acids, enzymes, or other compounds that facilitate the solubilization of insoluble zinc compounds in the soil. These compounds can include zinc oxides, zinc hydroxides, or zinc-containing minerals that are not easily accessible to plants.

- *Organic Acid Production:* Organic acids, such as gluconic acid, citric acid, and oxalic acid, are commonly produced by zinc solubilizing microbes. These acids have chelating properties, meaning they can bind to zinc ions and form soluble complexes, making zinc more available for plant uptake.
- *Zinc Mobilization:* The organic acids released by zinc solubilizing microbes help mobilize zinc from the soil particles, releasing it into the soil solution. This process increases the concentration of soluble zinc, making it more accessible to plant roots.
- *Plant Uptake:* Once zinc is solubilized and available in the soil solution, plants can take it up through their roots. Zinc is an essential micronutrient for plants and plays a critical role in various physiological processes, including enzyme activity, hormone synthesis, and chlorophyll production.
- *Plant Growth Promotion*: Adequate zinc availability through the activity of zinc-solubilizing microbes can promote plant growth, development, and overall health. Zinc is involved in various metabolic pathways and is necessary for optimal crop yield, quality, and resistance to certain diseases and stresses.

Applying zinc solubilizing microbes as biofertilizers can enhance zinc availability in zinc-deficient soils, benefiting zinc-demanding crops. However, it's important to consider factors such as soil pH, organic matter content, and the specific microbial strains used, as these factors can influence the effectiveness of zinc solubilization and subsequent plant uptake. Biofertilizers containing zinc solubilizing microbes can be applied to soils either through seed treatment, soil inoculation, or foliar application. These applications aim to increase the population and activity of zinc-solubilizing microbes, leading to improved zinc availability and uptake by plants. Proper diagnosis of zinc deficiency in soils and crops is essential to determine the need for zinc supplementation and the appropriate application method.

ARBUSCULAR FUNGI

Mamatha and Bagyaraj [58] reported that about 90% of plant roots are associated with different types of fungi. This association is known as Mycorrhiza or fungus root. Plant roots associate with fungi with their hyphae in soil. Fungi have scarce nutrients that are transported to the plant [58]. The fungus can enter the root of plant cells by a specified structure called arbuscules; it is highly branched and develops a widespread hyphal network external to the plant root and the established corporal link between the soil and root ecosystem. Fungal hypha can absorb minerals from the soil and transport them to the plant's root. Osorio and Habte [59] studied that P-solubilizing microorganisms and AM fungi can play an essential role in the growth of Leucaena *leucocephala* (Lam). They screened isolates based on their ability to solubilize rock phosphate in a culture medium [59]. Interaction between *Mortierella sp* and mycorrhizal fungi can raise P uptake and help in plant growth. AM symbiosis increased microorganism levels in the maize rhizosphere soil with improved soil enzyme activity. It can improve MBC (microbial biomass C), DOC (dissolved organic carbon, and ROC (root organic culture) contents by carbon mineralization. AM fungi were established to progress the soil ecosystem [60].

Arbuscular mycorrhizal fungi (AMF) play a crucial role in nutrient uptake and enhance plant growth, making them valuable biofertilizers. Here's an overview of the mechanisms by which arbuscular fungi act as biofertilizers:

- *Nutrient uptake and transport:* Arbuscular fungi have a vast network of fine hyphae that extend into the soil, effectively increasing the surface area available for nutrient absorption. These hyphae can penetrate the soil more extensively than plant roots, allowing them to access nutrients from a larger soil volume. The fungi absorb nutrients, such as phosphorus, nitrogen, potassium, and micronutrients, and transport them to the plant roots [54].
- *Phosphorus solubilization:* One of the key benefits of arbuscular fungi is their ability to solubilize phosphorus (P) in the soil. They produce enzymes called phosphatases that break down organic and inorganic forms of phosphorus, making them more available for plant uptake. This is especially beneficial in soils with low phosphorus availability, as the fungi enhance the plant's ability to acquire this essential nutrient [61].
- *Nitrogen fixation:* Although arbuscular fungi cannot fix atmospheric nitrogen like some bacteria, they indirectly contribute to nitrogen availability in the soil. By improving nutrient uptake and plant growth, these fungi can enhance the efficiency of nitrogen fertilizers applied to the soil. This reduces the dependency on synthetic nitrogen fertilizers and mitigates their potential environmental impacts.
- *Enhancing water uptake:* The extensive hyphal network of arbuscular fungi improves soil structure and increases water-holding capacity. The hyphae create channels that allow water to infiltrate and move through the soil more effectively. As a result, plants associated with these fungi have improved drought tolerance and better access to soil moisture, even in water-limited conditions.
- *Disease resistance:* Arbuscular fungi can help plants combat various soil-borne diseases. They stimulate the plant's immune system and produce antimicrobial compounds that inhibit the growth of pathogenic microorganisms. Additionally, the colonization of plant roots by these fungi creates a physical barrier that prevents pathogens from directly accessing the root system [47, 62].

Overall, arbuscular fungi act as biofertilizers by enhancing nutrient uptake, solubilizing phosphorus, improving water availability, and promoting disease resistance in plants. Their mutualistic relationship with plants benefits both parties and has significant implications for sustainable agriculture and ecological restoration.

PLANT GROWTH-PROMOTING RHIZOBACTERIA

PGPR refers to a group of free-living rhizosphere bacteria that colonize plant roots and have a beneficial influence on plant development. They function as biofertilizers by stimulating plant growth and development, enabling biotic and abiotic stress tolerance, and aiding in soil mineralization by decomposing organic materials. The plant benefits from PGPR inoculation in a variety of ways. They improve plant resistance to drought, salt, and biotic stress. They improve seed germination, soil fertility, and growth by releasing phytohormones such as IAA, auxins, gibberellin, ethylene, and others. They have the ability to alter plant secondary metabolites as well as bioremediation of heavy metals and pollutants. PGPR also includes microbes like *Arthrobacter, Agrobacterium, Azotobacter, Alcaligenes, Actinoplanes, Acinetobacter, Frankia, Bacillus, Rhizobium, Pseudomonas*, Streptomyces, Micrococcus, Enterobacter, Xanthomonas, Serratia, Cellulomonas, and *Flavobacterium* [63].

MECHANISM AND MODES OF ACTION OF BIOFERTILIZERS

Biofertilizers are biological agents that contain beneficial microorganisms, such as bacteria, fungi, or algae, which contribute to plant growth and enhance soil fertility. The mechanisms and modes of action of biofertilizers can vary depending on the type of microorganism involved. Here are some common mechanisms and modes of action of biofertilizers:

- *Nitrogen fixation:* Certain bacteria, such as *Rhizobium spp., Azotobacter spp.,* and *Azospirillum spp.,* have the ability to fix atmospheric nitrogen and convert it into a form that plants can utilize. These nitrogen-fixing bacteria form symbiotic associations with plant roots (*e.g.,* legumes) or colonize the rhizosphere. They convert atmospheric nitrogen into ammonia or ammonium, which can be readily absorbed by plants, thus reducing the reliance on synthetic nitrogen fertilizers.
- *Phosphorus solubilization:* Some microorganisms, such as phosphate-solubilizing bacteria (PSB) and mycorrhizal fungi, play a crucial role in solubilizing bound forms of phosphorus (P) in the soil, making it more available to plants [57, 64]. PSB produce organic acids and enzymes that break down insoluble phosphorus compounds, releasing soluble forms that can be taken up by plants. Arbuscular mycorrhizal fungi (AMF) form mutualistic associations

with plant roots and enhance phosphorus uptake by extending their hyphae into the soil and improving nutrient absorption [61].

- *Production of growth-promoting substances:* Many biofertilizers, including certain bacteria and fungi, produce growth-promoting substances such as phytohormones (*e.g.*, auxins, cytokinins, and gibberellins) and enzymes (*e.g.*, cellulases and proteases). These substances stimulate plant growth, root development, and nutrient uptake, leading to improved plant vigor and productivity.

- *Disease suppression:* Some biofertilizers possess biocontrol properties and can suppress soil-borne plant diseases. Certain bacteria, such as Bacillus spp. and *Pseudomonas* spp., produce antibiotics or antifungal compounds that inhibit the growth of plant pathogens. They can also compete with pathogens for nutrients and space, reducing their population and disease incidence.

- *Improving soil structure and nutrient cycling:* Biofertilizers can enhance soil structure and nutrient cycling processes. For example, certain bacteria and fungi produce extracellular enzymes that decompose organic matter, releasing nutrients and improving soil fertility. They also promote the aggregation of soil particles, improving soil structure and water infiltration.

- *Enhanced nutrient availability and uptake:* Biofertilizers can improve the availability and uptake of essential nutrients, including nitrogen, phosphorus, potassium, and micronutrients. They can solubilize nutrients, convert them into plant-available forms, and facilitate their uptake by plants through root colonization and the release of organic acids and chelating agents.

Overall, biofertilizers function through various mechanisms and modes of action to enhance nutrient availability, stimulate plant growth, improve soil fertility, suppress diseases, and promote sustainable agricultural practices. Their use can contribute to reduced reliance on synthetic fertilizers, increased crop yields, and improved environmental sustainability.

BIOFERTILIZERS AND THEIR APPLICATION METHODS IN AGRICULTURAL CROP

- Biofertilizers are beneficial microorganisms that can be applied to agricultural crops to enhance soil fertility and plant growth. Here are some commonly used biofertilizers and their application methods:

- Rhizobium biofertilizers: Rhizobium bacteria form a symbiotic relationship with leguminous plants (*e.g.*, soybeans, peas, and lentils) and fix atmospheric nitrogen in specialized structures called nodules on the plant roots. Rhizobium biofertilizers are commonly applied as seed inoculants or seedling root dips. The seeds are coated with the inoculant or immersed in a suspension of the bacteria before planting.

- Azotobacter and Azospirillum biofertilizers: These bacteria are free-living nitrogen-fixing bacteria that colonize the rhizosphere of various crops. They enhance nitrogen availability and promote plant growth. Azotobacter and Azospirillum biofertilizers are typically applied as seed treatments or through soil drenching. The seeds are coated with the biofertilizer or mixed with a suspension before sowing. Soil drenching involves applying the biofertilizer directly to the soil around the plant roots.
- Phosphate-solubilizing bacteria (PSB) biofertilizers: PSB solubilize bound forms of phosphorus in the soil, making it more available to plants. PSB biofertilizers can be applied as seed treatments, seedling root dips, or soil amendments. Seeds are coated with the biofertilizer or immersed in a PSB suspension before planting. Alternatively, the biofertilizer can be mixed with compost or applied directly to the soil during land preparation.
- Arbuscular mycorrhizal fungi (AMF) biofertilizers: AMF form mutualistic associations with plant roots and enhance nutrient uptake, particularly phosphorus. AMF biofertilizers are typically applied as soil amendments during land preparation or as seed coatings. The biofertilizer is mixed with soil or placed in close proximity to the plant roots during planting [50].
- Azolla biofertilizer: Azolla is a floating aquatic fern that harbors a symbiotic nitrogen-fixing cyanobacterium called Anabaena azollae. It can be used as a biofertilizer for flooded rice fields. Azolla is added to the water in rice fields, and as it decomposes, it releases nitrogen and other nutrients into the soil.
- Compost and vermicompost: Composting is a process that involves the decomposition of organic matter by microorganisms, resulting in nutrient-rich compost. Vermicompost is a compost produced through the activity of earthworms. Both compost and vermicompost can be used as biofertilizers. They are generally applied as soil amendments before planting or as top-dressings during crop growth.

The role of biofertilizers in enhancing plant growth and development is given in Fig. (**3**). It is important to note that the specific application methods may vary depending on the crop, local conditions, and biofertilizer product recommendations. Following the manufacturer's instructions and considering the specific requirements of the crop is crucial for successful biofertilizer application.

BIOFERTILIZER REQUIREMENTS AND LIMITATIONS

The application of chemical synthesis fertilizer has a negative impact on soil health. In addition, it may pollute the water body and destroy beneficial soil microorganisms and friendly insects, which are helpful for plant growth and soil fertility. Therefore, the demand for fertilizer will increase every year. Fossil fuels and depleting feedstock will increase the cost of chemical fertilizers, making them unaffordable for small and marginal farmers for agriculture. However, biofertilizer application is economical, more efficient, and productive compared to chemical fertilizer. The government of India has implemented several schemes (promotion and production of biofertilizers since 7th five-year plan). Biofertilizer demand in India is approximately 18,500 tonnes per year. At the same time, the estimated production of Biofertilizers in India (10,000 tonnes per year, 2017-18). Therefore, the Indian government is focused on generating additional demand for biofertilizers and micronutrients through proper extension and promotion.

Fig. (3). Role of biofertilizers in enhancing plant growth and development.

Nonetheless, biofertilizer application has its own set of constraints. Despite the fact that biofertilizers have shown their usefulness in agriculture with promising outcomes over the last 50 years, the expected success has yet to be realized. Several obstacles prevent this technique from being used on a big basis. Some of the possible reasons include poor soil characteristics, competition for a niche, intrusion of soil pollutants and contaminants, the lack of appropriate strain, the lack of sufficient funds and equipment from government and private bodies, lack of skilled and experienced staff in the production unit, extreme climatic conditions, suitable carrier material, insufficient storage, and transport facilities.

CONCLUSION AND PROSPECTS

Organic agriculture practices should be promoted for the sustainable development of agricultural products. Microorganisms or bioinoculants can play a vital role in limiting soil degradation and environmental contamination and may reduce the problems of low productivity. Biofertilizer is an essential aspect of organic farming, owing to its eco-friendly nature, pollution-free, renewable source, and low cost compared to synthetic fertilizer. In order to improve the soil ecosystem and encourage plant yield, biofertilizers are added to the soil. Unlike chemical fertilizers, which are made up of synthetic compounds, biofertilizers are derived from natural sources and work in harmony with the soil and plants. Biofertilizers are considered to be environmentally friendly because they do not cause pollution or harm to the ecosystem, and they can improve soil health and reduce the use of chemical fertilizers. They are widely used in organic farming and are becoming increasingly popular in conventional agriculture as well. However, an understanding of the field environment, soil parameters, and strain host specificity is required for efficient biofertilizer synthesis and application. Recent breakthroughs in molecular biology, biotechnology, genetic engineering, microbial taxonomy, and nanotechnology have aided in the development of biofertilizers with greater efficiency, higher competitive ability, and varied capabilities. Biofertilizers can sustain agricultural output while having a minimal environmental impact and can be a viable alternative to chemical fertilizers. In this subject, further research is needed to study and find soil-specific strains, obtain new insights into biofertilizer.

ACKNOWLEDGEMENTS

The authors would like to thank the support provided by *Om College of Science, Bheshan Highway, Junagadh*, and Tamil Nadu Agricultural University.

REFERENCES

[1]   Nosheen S, Ajmal I, Song Y. Microbes as biofertilizers, a potential approach for sustainable crop production. Sustainability (Basel) 2021; 13(4): 1868.
[http://dx.doi.org/10.3390/su13041868]

[2]   Berg G. Plant–microbe interactions promoting plant growth and health: perspectives for controlled use of microorganisms in agriculture. Appl Microbiol Biotechnol 2009; 84(1): 11-8.
[http://dx.doi.org/10.1007/s00253-009-2092-7] [PMID: 19568745]

[3]   Thomas L, Singh I. Microbial biofertilizers: types and applications. Soc Biol 2019; 55: 1-19.
[http://dx.doi.org/10.1007/978-3-030-18933-4_1]

[4]   Krishijagran.com [website on the Internet]. Indian Fertilizer Sector at a Glance. [cited: 16th Dec 2023]. Available from: https://krishijagran.com/farm-data/indian-fertilizer-sector-at-a-glance/2022

[5]   Verma JP, Yadav J, Tiwari KN, Jaiswal DK. Evaluation of plant growth promoting activities of microbial strains and their effect on growth and yield of chickpea (Cicer arietinum L.) in India. Soil Biol Biochem 2014; 70: 33-7.

[http://dx.doi.org/10.1016/j.soilbio.2013.12.001]

[6]   Hajihashemi S. Agronomic practices Steviol Glycosides. Elsevier 2021; pp. 31-56.
[http://dx.doi.org/10.1016/B978-0-12-820060-5.00002-9]

[7]   Sun B, Gu L, Bao L, *et al.* Application of biofertilizer containing Bacillus subtilis reduced the nitrogen loss in agricultural soil. Soil Biol Biochem 2020; 148: 107911.
[http://dx.doi.org/10.1016/j.soilbio.2020.107911]

[8]   Ansari SH, Ahmed A, Razzaq A, Hildebrandt D, Liu X, Park YK. Incorporation of solar-thermal energy into a gasification process to co-produce bio-fertilizer and power. Environ Pollut 2020; 266(Pt 3): 115103.
[http://dx.doi.org/10.1016/j.envpol.2020.115103] [PMID: 32650303]

[9]   Meena VS, Maurya BR, Verma JP. Does a rhizospheric microorganism enhance K+ availability in agricultural soils? Microbiol Res 2014; 169(5-6): 337-47.
[http://dx.doi.org/10.1016/j.micres.2013.09.003] [PMID: 24315210]

[10]   Goteti PK, Emmanuel LDA, Desai S, Shaik MHA. Prospective zinc solubilizing bacteria for enhanced nutrient uptake and growth promotion in maize (*Zea mays* L.). Int J Microbiol 2013; 2013: 1-7.
[http://dx.doi.org/10.1155/2013/869697] [PMID: 24489550]

[11]   Deshwal VK, Kumar P. Production of Plant growth promoting substance by *Pseudomonas*. J Acad Ind Res 2013; 12: 1-10.

[12]   Malhotra M, Srivastava S. Stress-responsive indole-3-acetic acid biosynthesis by *Azospirillum brasilense* SM and its ability to modulate plant growth. Eur J Soil Biol 2009; 45(1): 73-80.
[http://dx.doi.org/10.1016/j.ejsobi.2008.05.006]

[13]   Kim KY, Jordan D, McDonald GA. *Enterobacter agglomerans*, phosphate solubilizing bacteria, and microbial activity in soil: Effect of carbon sources. Soil Biol Biochem 1998; 30(8-9): 995-1003.
[http://dx.doi.org/10.1016/S0038-0717(98)00007-8]

[14]   Kawalekar JS. Role of biofertilizers and biopesticides for sustainable agriculture. J Bio Innov 2013; 2: 73-8.

[15]   Youssef MMA, Eissa MFM. Biofertilizers and their role in management of plant parasitic nematodes. A review. J of Biotech and Pharma Res 2014; 5: 1-6.

[16]   Bumandalai O, Tserennadmid R. Effect of Chlorella vulgaris as a biofertilizer on germination of tomato and cucumber seeds. Int J Aquat Biol 2019; 7: 95-9.

[17]   Umesha S, Singh PK, Singh RP. Microbial biotechnology and sustainable agriculture Biotechnology for sustainable agriculture. Elsevier 2018; pp. 185-205.

[18]   Milić VM, Jarak MN, Mrkovački NB, *et al.* Microbiological fertilizer use and study of biological activity for soil protection purposes. Zbornik Radova Instituta Za Ratarstvo i Povrtarstvo 2004; pp. 153-69.

[19]   Venkatashwarlu B. Role of bio-fertilizers in organic farming: Organic farming in rain fed agriculture: Central institute for dry land agriculture. Hyderabad Pakistan Pp 2008; pp. 85-95.

[20]   Goldstein AH. Bacterial solubilization of mineral phosphates: Historical perspective and future prospects. Am J Altern Agric 1986; 1(2): 51-7.
[http://dx.doi.org/10.1017/S0889189300000886]

[21]   Vessey JK. Plant growth promoting rhizobacteria as biofertilizers. Plant Soil 2003; 255(2): 571-86.
[http://dx.doi.org/10.1023/A:1026037216893]

[22]   Yosefi K, Galavi M, Ramrodi M, Mousavi SR. Effect of bio-phosphate and chemical phosphorus fertilizer accompanied with micronutrient foliar application on growth, yield and yield components of maize (Single Cross 704). Aust J Crop Sci 2011; 5: 175-80.

[23]   Das K, Katiyar V, Goel R. 'P' solubilization potential of plant growth promoting *Pseudomonas*

mutants at low temperature. Microbiol Res 2003; 158(4): 359-62.
[http://dx.doi.org/10.1078/0944-5013-00217] [PMID: 14717458]

[24]   Aneja KR. Experiments in microbiology, plant pathology, tissue culture and microbial biotechnology. New Age International 2018; pp. 1-632.

[25]   Raynaud X, Nunan N. Spatial ecology of bacteria at the microscale in soil. PLoS One 2014; 9(1): e87217.
[http://dx.doi.org/10.1371/journal.pone.0087217] [PMID: 24489873]

[26]   Burris RH. Biological nitrogen fixation: A scientific perspective. Plant Soil 1988; 108(1): 7-14.
[http://dx.doi.org/10.1007/BF02370094]

[27]   Ashby SF. Some observations on the assimilation of atmospheric nitrogen by a free living soil organism.—Azotobacter chroococcum of Beijerinck. J Agric Sci 1907; 2(1): 35-51.
[http://dx.doi.org/10.1017/S0021859600000988]

[28]   Song T, Mårtensson L, Eriksson T, Zheng W, Rasmussen U. Biodiversity and seasonal variation of the cyanobacterial assemblage in a rice paddy field in Fujian, China. FEMS Microbiol Ecol 2005; 54(1): 131-40.
[http://dx.doi.org/10.1016/j.femsec.2005.03.008] [PMID: 16329979]

[29]   Saadatnia H, Riahi H. Cyanobacteria from paddy fields in Iran as a biofertilizer in rice plants. Plant Soil Environ 2009; 55(5): 207-12.
[http://dx.doi.org/10.17221/384-PSE]

[30]   Schwencke J, Carú M. Advances in actinorhizal symbiosis: host plant-Frankia interactions, biology, and applications in arid land reclamation. A review. Arid Land Res Manage 2001; 15(4): 285-327.
[http://dx.doi.org/10.1080/153249801753127615]

[31]   Torrey JG. Nitrogen fixation by actinomycete-nodulated angiosperms. Bioscience 1978; 28(9): 586-92.
[http://dx.doi.org/10.2307/1307515]

[32]   Wang X, Wang Y, Tian J, Lim BL, Yan X, Liao H. Overexpressing AtPAP15 enhances phosphorus efficiency in soybean. Plant Physiol 2009; 151(1): 233-40.
[http://dx.doi.org/10.1104/pp.109.138891] [PMID: 19587103]

[33]   Shenoy VV, Kalagudi GM. Enhancing plant phosphorus use efficiency for sustainable cropping. Biotechnol Adv 2005; 23(7-8): 501-13.
[http://dx.doi.org/10.1016/j.biotechadv.2005.01.004] [PMID: 16140488]

[34]   Kannapiran E, Ramkumar VS. Isolation of phosphate solubilizing bacteria from the sediments of Thondi coast, Palk Strait, Southeast coast of India. Ann Biol Res 2011; 2: 157-63.

[35]   Weil RR and Brady NC.. Soil phosphorus and potassium. The Nature and Properties of Soils 2002; 643-95.

[36]   Egamberdieva D, Kucharova Z. Cropping effects on microbial population and nitrogenase activity in saline arid soil. Turk J Biol 2008; 32: 85-90.

[37]   Berman-Frank I, Lundgren P, Falkowski P. Nitrogen fixation and photosynthetic oxygen evolution in cyanobacteria. Res Microbiol 2003; 154(3): 157-64.
[http://dx.doi.org/10.1016/S0923-2508(03)00029-9] [PMID: 12706503]

[38]   Afzal A, Ashraf M, Asad SA, Farooq M. Effect of phosphate solubilizing microorganisms on phosphorus uptake, yield and yield traits of wheat (*Triticum aestivum* L.) in rainfed area. Int J Agric Biol 2005; 7: 207-9.

[39]   Azcon R, Barea JM, Hayman DS. Utilization of rock phosphate in alkaline soils by plants inoculated with mycorrhizal fungi and phosphate-solubilizing bacteria. Soil Biol Biochem 1976; 8(2): 135-8.
[http://dx.doi.org/10.1016/0038-0717(76)90078-X]

[40]   Sørensen J, Sessitsch A. Plant-associated bacteria lifestyle and molecular interactions. Modern Soil

Microbiology 2007; 2: 211-36.

[41]   Sahoo RK, Ansari MW, Dangar TK, Mohanty S, Tuteja N. Phenotypic and molecular characterisation of efficient nitrogen-fixing Azotobacter strains from rice fields for crop improvement. Protoplasma 2014; 251(3): 511-23.
[http://dx.doi.org/10.1007/s00709-013-0547-2] [PMID: 24005473]

[42]   Dutta S, Singh MS. Effect of Azotobacter on yield and oil content of rapeseed mustard varieties under Manipur condition. Indian Journal of Hill Farming 2002; 15: 44-6.

[43]   Prasad RC, Prasad BN. Cyanobacteria as a source biofertilizer for sustainable agriculture in Nepal. J Plant Sci Bot Orientalis 2001; 1: 127-33.

[44]   Herrmann L, Lesueur D. Challenges of formulation and quality of biofertilizers for successful inoculation. Appl Microbiol Biotechnol 2013; 97(20): 8859-73.
[http://dx.doi.org/10.1007/s00253-013-5228-8] [PMID: 24037408]

[45]   Itelima JU, Bang WJ, Onyimba IA, Sila MD, Egbere OJ. Bio-fertilizers as key player in enhancing soil fertility and crop productivity. RE:view 2018; 6(3): 73-83.

[46]   Patel D, Goswami D. Phosphorus solubilization and mobilization: mechanisms, current developments, and future challenge. Advances in Plant Microbiome and Sustainable Agriculture: Functional Annotation and Future Challenges 2020:1–20.
[http://dx.doi.org/10.1007/978-981-15-3204-7_1]

[47]   Rodríguez H, Fraga R. Phosphate solubilizing bacteria and their role in plant growth promotion. Biotechnol Adv 1999; 17(4-5): 319-39.
[http://dx.doi.org/10.1016/S0734-9750(99)00014-2] [PMID: 14538133]

[48]   Young CC, Chang CH, Chen LF, Chao CC. Characterization of the nitrogen fixation and ferric phosphate solubilizing bacteria isolated from a Taiwan soil. Journal-chinese Agricultural Chemical Society 1998; 36: 201-10.

[49]   Rizvi A, Ahmed B, Khan MS, Umar S, Lee J. Psychrophilic bacterial phosphate-biofertilizers: a novel extremophile for sustainable crop production under cold environment. Microorganisms 2021; 9(12): 2451.
[http://dx.doi.org/10.3390/microorganisms9122451] [PMID: 34946053]

[50]   Richardson AE. Prospects for using soil microorganisms to improve the acquisition of phosphorus by plants. Funct Plant Biol 2001; 28(9): 897-906.
[http://dx.doi.org/10.1071/PP01093]

[51]   Asoegwu CR, Awuchi CG, Nelson KCT, *et al.* A review on the role of biofertilizers in reducing soil pollution and increasing soil nutrients. Himalayan Journal of Agriculture 2020; 1: 34-8.

[52]   Pramanik P, Goswami AJ, Ghosh S, Kalita C. An indigenous strain of potassium-solubilizing bacteria *Bacillus pseudomycoides* enhanced potassium uptake in tea plants by increasing potassium availability in the mica waste-treated soil of North-east India. J Appl Microbiol 2019; 126(1): 215-22.
[http://dx.doi.org/10.1111/jam.14130] [PMID: 30326179]

[53]   Bahadur I, Maurya BR, Kumar A, Meena VS, Raghuwanshi R. Towards the soil sustainability and potassium-solubilizing microorganisms. Potassium Solubilizing Microorganisms for Sustainable Agriculture 2016; pp. 255-66.
[http://dx.doi.org/10.1007/978-81-322-2776-2_18]

[54]   Corbridge DEC. Phosphorus: chemistry, biochemistry and technology. CRC press 2013; pp. 53-74.

[55]   Freeman WH, Christensen JJ, Hansen LD, *et al.* General references. Annu Rev Biophys Bioeng 3: 95-126. n.d.

[56]   Kpomblekou-A K, Tabatabai MA. Effect of organic acids on release of phosphorus from phosphate rocks1. Soil Sci 1994; 158(6): 442-53.
[http://dx.doi.org/10.1097/00010694-199415860-00006]

[57]   Reyes I, Bernier L, Simard RÃR, Antoun H. Effect of nitrogen source on the solubilization of different inorganic phosphates by an isolate of Penicillium rugulosum and two UV-induced mutants. FEMS Microbiol Ecol 1999; 28(3): 281-90.
[http://dx.doi.org/10.1111/j.1574-6941.1999.tb00583.x]

[58]   G M, D B, S J. Inoculation of field-established mulberry and papaya with arbuscular mycorrhizal fungi and a mycorrhiza helper bacterium. Mycorrhiza 2002; 12(6): 313-6.
[http://dx.doi.org/10.1007/s00572-002-0200-y] [PMID: 12466919]

[59]   Osorio NW, Habte M. Synergistic influence of an arbuscular mycorrhizal fungus and a P solubilizing fungus on growth and P uptake of Leucaena leucocephala in an Oxisol. Arid Land Res Manage 2001; 15(3): 263-74.
[http://dx.doi.org/10.1080/15324980152119810]

[60]   Dong L, Li Y, Xu J, *et al.* Biofertilizers regulate the soil microbial community and enhance *Panax ginseng* yields. Chin Med 2019; 14(1): 20.
[http://dx.doi.org/10.1186/s13020-019-0241-1] [PMID: 31143242]

[61]   Jha A, Sharma D, Saxena J. Effect of single and dual phosphate-solubilizing bacterial strain inoculations on overall growth of mung bean plants. Arch Agron Soil Sci 2012; 58(9): 967-81.
[http://dx.doi.org/10.1080/03650340.2011.561835]

[62]   Glick BR. The enhancement of plant growth by free-living bacteria. Can J Microbiol 1995; 41(2): 109-17.
[http://dx.doi.org/10.1139/m95-015]

[63]   Yadav AN, Verma P, Singh B, Chauhan VS, Suman A, Saxena AK. Plant growth promoting bacteria: biodiversity and multifunctional attributes for sustainable agriculture. Adv Biotechnol Microbiol 2017; 5: 1-16.

[64]   Pérez E, Sulbarán M, Ball MM, Yarzábal LA. Isolation and characterization of mineral phosphate-solubilizing bacteria naturally colonizing a limonitic crust in the south-eastern Venezuelan region. Soil Biol Biochem 2007; 39(11): 2905-14.
[http://dx.doi.org/10.1016/j.soilbio.2007.06.017]

CHAPTER 5

Role of Microbes and Microbiomes in Biotic and Abiotic Stress Management in Agriculture

M. Vijayalakshmi[1,*], Christobel R. Gloria Jemmi[1], G. Ramanathan[2] and S. Karthika[3]

[1] *V.V. Vanniaperumal College for Women, Virudhunagar, Tamil Nadu, India*

[2] *Sri Paramakalyani College, Alwarkurichi, Tirunelveli, Tamil Nadu, India*

[3] *Tamil Nadu Agricultural University, Coimbatore, Tamil Nadu, India*

Abstract: Agriculture is our sensible recreation and the foremost food source for all animals and human beings. It gives laurels to us, but knowingly or unknowingly, agricultural systems face stress, resource quality degradation, and depletion by human activities. Abiotic stresses, such as nutrient deficiency, water logging, extreme cold, frost, heat, and drought, affect agricultural productivity. Biotic factors like insects, weeds, herbivores, pathogens, bacteria, viruses, fungi, parasites, algae, and other microbes limit good-quality products. Climate change leads to more complications when interpreting how plants and microbes interact to protect themselves from stress. Plants need water, carbon, and nutrients to grow. The extreme conditions mentioned restrict the growth of plants. Although plants can sense and exhibit natural mechanisms during stress conditions, increased non-sustainable agricultural practices and other human activities lead to highly stressful conditions for plant growth and yield. While in stressful situations, fungi play an essential role in energy transfer and uptake of nutrients by releasing the adverse effects of stress on plant growth. Many strategies in bacteria and fungi need to be addressed here, including stress conditions such as cysts and spore formation, cell membrane deformation, production of damage repair enzymes, and chemical synthesis to relieve stress. The mechanism of salt tolerance, symbiotic microbes, xenobiotics, and hazardous tolerance genes induces plant growth in unfavorable conditions. In recent days, technological improvements such as gene modification by genetic engineering have shown the potential to enhance the positive effects on agricultural production and products.

Keywords: Microbiomes, stress management, agriculture, climate change, mitigation.

---

* **Corresponding author M. Vijayalakshmi:** V.V. Vanniaperumal College for Women, Virudhunagar, Tamil Nadu, India; E-mail: vijimicrobes@gmail.com

Shiv Prasad, Govindaraj Kamalam Dinesh, Murugaiyan Sinduja, Velusamy Sathya, Ramesh Poornima & Sangilidurai Karthika (Eds.)

INTRODUCTION

In every ecosystem, plants are vital to its survival, as are furry animals and microscopic bacteria. Therefore, soil microflora cannot be sustained without a balanced ecosystem. Soil microbes are vital in restoring soil fertility and breaking down agricultural waste, particularly bacteria. They act as an ecological transformation inducer and are apt to live in geochemically extreme environments. Extremophiles can be divided into oligotrophic, acidophilic, alkaliphilic, thermophilic, psychrophilic, and sulfidophilic [1]. Based on oxygen requirements, these are named aerobes, anaerobes, and facultative. In the modern world, plants adapt to grow in all conditions because of recent technological inventions. However, microorganisms are native to the soil; they can change their potential effect with the least changes.

Bacteria are active in the soil near the roots, called rhizosphere. They are categorized as decomposers, mutualistic bacteria, pathogens, and lithotrophs, and they form micro-aggregates, which are very important for water filtration and holding. These live under starvation conditions or stress. Once it turns into a flourishing state, the crop yield and soil fertility improve. Aeromonas, *Streptomycetes,* and Actinomycetes release "geosmin", which provides an earthy smell. *Azotobacter, Azospirillum,* and *Clostridium* also fix nitrogen without the host plant [2].

Rhizobium removes nitrogen from the atmosphere and converts it into ammonia. *Arthrobacteria* sp. is involved in nitrification by turning ammonia into nitrite and nitrate. *Nitrosomonas sp.* convert ammonia to nitrite, and *Nitrobacter spp.* can convert nitrites into nitrates. However, denitrifying bacteria enable the conversion of nitrate to atmospheric nitrogen. Lithotrophs get their energy from other than carbon compounds. Interestingly, the sulphur-oxidizing and sulphur-reducing bacteria are available in the soil bed. Without bacteria, new plant populations struggle to survive. The food chain begins with solar energy, and decomposers act as the vital components at the end. Algae are universal in soils, permanent ice, snowfields, hot springs, and cold deserts. They also produce oxygen for human consumption and other aerobic microbes in our ecosystems [3]. They are the leading producers of organic compounds and play a significant role in the food chain in aquatic and agricultural areas. Mainly, in agriculture, the algae are used as a biofertilizer and soil stabilizer by providing nitrogen and phosphorus. These are grown worldwide for human consumption as dietary supplements. They can yield carbon in the desert and wild lands with minimal freshwater demands and reduce soil erosion.

The biological nitrogen fixation using blue-green algae is termed Algalization [4]. This will help bring the water, micro, and macronutrients of calcium, potassium, and magnesium into the upper layer. This process will maintain the soil structure and property of nutrient uptake up to 30 - 40 kg N/ha/Season. In the world, 90% of marine plants are algae, and 50% of the agricultural microbes almost belong to the algal community. All the molecules of oxygen we inhale come from algae and plants. Cyanobacteria are usually used in rice paddies everywhere in Asia. Marginal soils, such as saline-alkaline and limestone soils are applied [5].

In the emerging microbiome, the marine algae (red and brown algae) are cast off as organic fertilizers on farmlands, capable of reducing atmospheric nitrogen to ammonia *via* heterocysts, and a superior category of seaweeds produce polymers of agars, alginates, tannins, antibiotics, colorants, pigments, defense ingredients for the plant growth. *Tolypothrix tenuis* is grown and added to the rice field for increased yield [6]. In Thailand, *Chlorella* and *Dunaliella* are commercially available for the treatment of wastewater and have been successfully maintained for 55 years for irrigation purposes. The goal of agriculture is to improve existing cultivars and to develop new cultivars.

CLIMATE CHANGE IMPACTS

Climate and agriculture are interconnected; a slight change in climate affects agriculture adversely. The warning of our atmosphere is now impossible to avoid. Climate change indirectly contributes to ocean acidification, shrinking glaciers, and destroying coral reefs and other aquatic plants and animals. Climate change will significantly affect rural communities around the world. According to the Ministry of Agriculture, approximately 40 million hectares of agricultural land in India was severely damaged by hydro-meteorological disasters in 2022. The glaciers will disappear in 80 years, and their cores show a rich chronological history of unique, frozen microbial life reported by the IPCC - International Panel on Climate Change, Fifth Assessment Report 2014 [7]. They also identified the impacts of climate change on biodiversity, ecosystems, and human communities at global and regional levels and examined the vulnerability and ability of the natural world and human societies to adapt to climate change. Microbes have been found in thawed permafrost for more than 400,000 years.

The temperature in the Arctic has risen by 3 to 5°C, according to the IPCC Special Report on Oceans and the Cryosphere [8]. The temperature affects the yield negatively; the opposite applies to the precipitation conditions. Using the probability ratio (PR) concept, we can measure how much of the change in the probability of extreme events is due to a change in the average climate. The biodiversity loss will lead to the maximum percentage of species at high risk of extinction across forests and land. Drought means the dry land population is exposed to water stress, heat stress, desertification, *etc.*

Extreme heat, floods, sea level rise, and forest fires are alarming effects of climate change. Food security is a problem that crucially affects the costs for adaptation and residual damage to major crops. There are 67.79 million hectares of agricultural land in India that is still not irrigated, as documented in the Ministry of Agriculture and Social Affairs' report 2016. Many small islands share the common characteristics that increase their vulnerability due to their small size, limited fresh water and natural resources, dense populations with poor infrastructure, limited financial and economic resources, and limited human resources. According to EnviStats-India-2022, an analysis of the monsoon trends of India for the past few years, shown in Table **1,** revealed that only 703 districts in India recorded minimum rainfall [9].

Table 1. Percentage deviation of rainfall from the long-period average (Adopted from [9]).

Year	Normal	Large	Deficit	% of Rainfall for the Country as a Whole
2017	25	5	6	95
2018	24	1	11x	91
2019	21	12	3	110
2020	16	15	5	109
2021	20	10	6	99

Today, 3.3 billion people live in highly vulnerable countries around the world. The number of deaths from droughts, storms, and floods was 15 times higher in high-risk countries between 2010 and 2020 than in low-risk countries. Climate change could starve 90 million Indians by 2030 due to reduced agricultural production and disruptions to food supplies, as reported by the International Food Policy Research Institute's (IFPRI) Global Food Policy, 2022. Enhancement of crop production is essential for safeguarding food and national security [9 - 11].

The main subsidized sectors are commercial buildings, industry, electricity power plants, housing, and land transportation. Urbanization increases carbon emissions, which are hotspots unless cities take action to reduce emissions through changes in urban planning, technology, and behavior. Global CO_2 emissions related to urban energy range from 8.8 to 14.3 Gt CO_2 per year, which accounts for 53% to 87% of the CO_2 emissions from the last global energy consumption [12]. Altogether, the cities' CO_2 emissions are almost 70% from manmade fossil fuel, varying depending on land use, energy consumption, and various socio-economic and geographical factors. This data will help scientists and policymakers to decide how to make a plan for utilizing the technologies or prevent behavioral changes in city mitigation. The goals of the Paris Climate Agreement focus on reducing global CO_2 emissions from industrialized or modernized cities. This characteristic varies from city to city, even within the same country, and these differences are due to socio-economic factors in the energy sector.

PERMAFROST PANDEMIC

Many scientists have explained that parts of Antarctica, the Arctic, and Tibet act as natural laboratories to help select organisms that can withstand pressures. The effect of hydrate evaporation is evident in the Antarctic ice sheet, causing upwelling of methane through seawater [13]. When cryonite pits melt, the material can ultimately affect the oceans, feeding nutrients into marine ecosystems. This can lead to massive algal blooms, which remove carbon dioxide from the atmosphere. Green algae and diatoms supported the 'Glacial runoff Hypothesis' that states long-term survival occurred and led to the emergence of some species of Antarctic flora and fauna [14]. These holes are the hotspots of microbial life in glaciers and stated the discovery of more than 300 microorganisms.

Out of these, 70% corresponds to new species in Permafrost. Antarctica is a source of extreme microorganisms. Global climate changes have caused the evacuation of extreme forms of life that were not previously described or found on this continent. Thawing Permafrost releases microorganisms, which are still unknown. As Permafrost melts, ancient microbes such as bacteria and viruses survive in extreme conditions, causing illness to humans and animals. Microbiologist Brent Christner said, 'Not only have we found that things are alive, but there are active ecosystems. Scientists have reported that over 400,000 years old microbes in thawed Permafrost. Thriving communities of red algae are responsible for causing danger to the iceberg and water resources [15].

Dinoflagellates and diatoms can modify pond water color and produce toxins in public water supply areas. All the related fields of study, such as climatology, geology, and microbiology, focus on the threat of microbes revived by the thawing of the Permafrost (Fig. **1**). Permafrost is a current development in the last 10 to 20 years. The real danger is not a thawing of a million years old, but exploiting or digging the layers causes the real disaster [16]. It is crucial to identify the risks of permafrost thawing and to manage them for future sustainability.

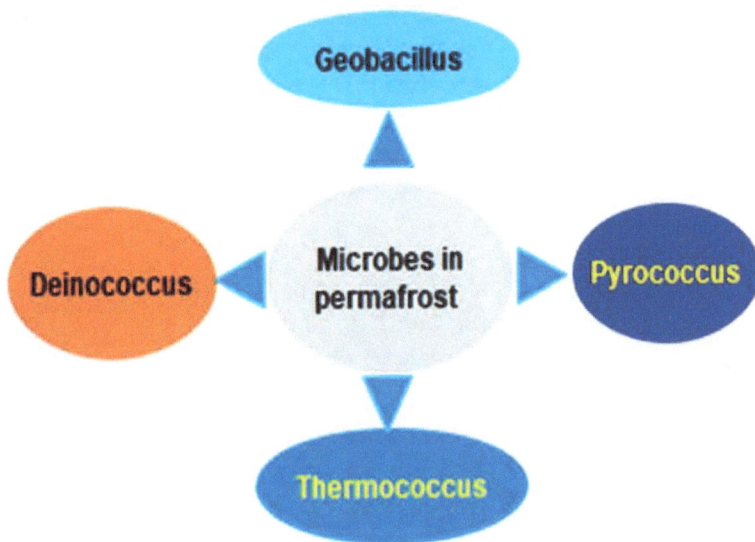

Fig. (1). Microbes in permafrost.

Permafrost is a permanently frozen area in high mountain regions as well as in Earth's higher latitudes of the North and South poles. So many new species are identified from the Permafrost active layer, such as *Thermococcus, Pyrococcus, Geobacillus,* and *Deinococcus,* by 16S rRNA sequencing. Their features are unique compared to other species, *e.g.*, cold and radiation-resistant [17].

The bacterial cells and their spores can survive in the frozen and thawed permafrost. Permafrost microbes need to use cold-protection proteins, as well as hibernation, so that they can survive in extreme conditions. Two main pathways of energy metabolism dominate in cold environments: i) the reduction of CO_2 to CH_4 using H_2 and ii) the fermentation of acetate to CH_4 and CO_2. The regions with suitable temperature and pressure conditions are favorable for the formation of methane hydrate under sedimentary rock areas below the Arctic, Antarctic permafrost, and continental margins.

Thermococccus sp. is a gram-negative, flagellated, hyperthermophilic archaea, requiring $80^0C - 85°C$ and sulfur as a significant electron acceptor [18]. Spores are specialized structures that protect cells from lethal conditions such as radiation, desiccation, toxins, extreme heat, and extreme cold [19]. The mechanism of persistence of thermophiles in psychobiotic conditions may lead to new concepts of microbial physiological plasticity, which have been cultured from 3 to 30^0 C for at least 9 months in cold environments.

Pyrococcus sp. is a hyperthermophilic, coccus-shaped archaeon with polar peritrichous flagella. They are anaerobic, have optimum growth at 90°C, and require a carbon source [20]. The energy utilized by nature in the form of carbon can easily be digested into carbon dioxide and methane. Methanogenesis is one of the predominant processes during anaerobic decomposition of organic matter. Under sea hot vents, *P. abyssi* has been found at a pressure of 200 atm without sunlight and extremely high temperatures.

Geobacillus sp. is an obligate thermophile with peritrichous flagella. They are rod-shaped, gram-positive, aerobic, and an endospore former, with optimum growth at 37 – 78°C and oxygen as an electron acceptor. These are motile and contain a maximum of 51 – 58% hydrolyzed starch, casein, citrate, gelatine, and glucose. *G. stearothermophilus* and *G.thermoleovorans* are dominant species [21].

Deinococcus sp. is gram-positive, mesophilic, and resistant to high doses of gamma UV radiation. Radio-sensitivity of microbes is a complex mechanism that requires additional factors like medium composition, temperature, humidity, *etc.* Glacier ice harbors dormant microbes that do not multiply and have normal metabolic functions. They become reactive when the environment turns to favorable conditions. The microbes will induce climate change by emitting greenhouse gases such as nitrous oxide and methane. In gene cloning strategies, unpredictable or dormant genes are essential to modify the normal plants into transgenic plants with resistance to climate change, stress, drought, and all abiotic factors [22]. The surface layer of permafrost showed positive antigen-antibody reactions against alpha, beta, and gamma proteobacteria, actinobacteria, bacteroides, firmicutes, and some archaeal species. For example, Supraglacial ecosystems are covered with cyanobacteria and photosynthetic algae. These pigment molecules help us adapt to cold and sun radiation by trapping energy, increasing resistance to stress, and providing UV protection.

SURVIVAL STRATEGIES UNDER STRESS FACTORS

The membrane permeability, nutrient transport, and osmotic regulation of plant cells are important factors in soil moisture maintenance. Abiotic stress initiates the synthesis of many proteins, including transcription factors, enzymes, ion channels, and transporters. Many scientists reported that solute enrichment can vary by ions, soluble sugars, organic acids and amino acids [23]. Accumulation of solutes in plants, salinity, drought, and heavy metals (Cu, Cd, and Cr) resistance is associated with stress tolerance, which leads to sustainable growth and high rates of photosynthesis. The production of ROS, osmotic pressure, and ionic effects lead to maintaining ionic homeostasis, solutes enrichment, antioxidant defense, and hormonal regulation in plants. This pressure will amplify global climate change and reduce crop yield by around 30%. More carbon is needed to improve plant metabolism, flower development physiology, and grain-filling efficiency.

Salinity is the excessive concentrations of soluble salts like Na^+, Ca^+, Mg^+ cations, C anions, carbonates, bicarbonates, and sulfates in soil, which inhibit seed germination, create high water potential, leaf necrosis, and chlorosis. Salt stress in plants is due to water and nutrient unavailability, nitric oxide production, antioxidant enzymes, hormonal changes, accumulation of polyols, and other changes. Secondary salinity is caused by logging, planting, saltwater irrigation, and poor drainage. Changes in acidic conditions adversely affect soil nutrients, reducing their availability and causing the loss of normal physiological growth patterns. Defence, repair, and adaptation are the components of the resistance response to stress. Salt tolerance is created by plant pathways that involve molecular, physiological, and metabolic changes. Phytohormones include IAA, Jas, GAs, ET, SL, and BR; they interact with metabolic stimuli to increase stress tolerance in plants. Minerals such as nitrogen (N), phosphorus (P), sulfur (S), and calcium (Ca) have been reported to regulate metabolic acceleration under stress conditions [24].

ADAPTIVE MECHANISMS OF SALT TOLERANCE IN PLANTS

Salt can be found in groundwater at high concentrations. Halophytes are valuable for economic or ecological reasons. Plant adaptation to salinity starts from stress perception by absorbing the salt from the soil and storing it in bladder-shaped cells on the surface of its leaves. Halophytes have several adaptations to avoid or resist stress: salt hairs, salt glands, waxy cuticles, selective ion uptake, salt exclusion from different plant organs, root excretion of salts, and molecular sieves. The adaptive mechanisms of salt tolerance in plants are categorized based on intercellular damage results from osmotic adaptation, cell wall modification, ROS detoxification, vesicular trafficking, homeostasis, vacuolar compartmentalization, and transport proteins [25, 26]. The activity of nutrient elements (Na, Ca & P) in soil decreases as salinity increases, *e.g.*, phosphate availability is reduced in saline soils by its ionic strength; the physiological inactivation of P and its concentration is controlled by the sorption process and by precipitation of Ca-P minerals. Sodium-induced Ca^{2+} deficiencies have been observed in the grass family (e.g., corn, sorghum, rice, wheat, and barley). The Na^+ inhibits the radial movement of calcium from the root epidermis to the root xylem vessels. Yield reduction in plants and crops is not specifically sensitive to Na^+, which reflects the combined effects of nutritional problems with impaired soil and physical conditions. The parameters of soil water content, salt composition, soil biota, soil fertility, and irrigation methods also influence the plants in terms of salt stress. Salt distribution patterns and salt accumulation are interlocked with the effects of root water extraction patterns and the net direction of water flow in the soil. So, managing these variables may reduce the cumulative stresses on a crop and minimize the impact of soil stress.

Damage to Plant Organs of Fruits, Flowers, Leaves, Trunk and Root

The crop yields influence the response of salt concentrations within the root zone through the interactions between salinity and soil, water, and climatic conditions. The salinity affects nutrient ion activities and produces extreme ion ratios of Na^+/Ca^{2+}, Na^+/K^+, and Cl^-/NO_3^- in the soil. The salt-tolerant traits have the following advantages: greater root growth, high efficiency in water uptake, low Na+ permeability, high root pressure, and osmotic adjustments. Nutrient imbalances in plants may result from the effect of salinity on nutrient availability, uptake, or distribution of a nutrient within a plant, increasing the internal plant requirement for nutrient uptake by physiological inactivation. The primary defensive plant response to oxidative stress is the biosynthesis of antioxidants. Salt-stressed plants have antioxidants of flavonoids, ascorbates, tocopherols, carotenoids, and lycopenes [27, 28].

Salt tolerance is based on yield, quality of crops, and size of the fruits, tubers, and edible organs, and it reduces the market value of carrots, cucumber, potatoes, yams and lettuce. Damage to plant organs such as fruits, flowers, leaves, trunk and root can be affected or modified in various ways: (i) High salt concentration can severely affect plant growth and yield due to osmotic, ionic stress on root system; the osmotic stress caused by salt reduces cell expansion in growing tissue and stomatal closure, which helps minimize water loss and plant damage; (ii) Salinity alters flowering time and photosynthetic remobilization; (iii) Salts are distributed and retransformed in the leaf blade instead of mesophyll or the sheath/petiole instead of the epidermis; (iv) This is used to control long-distance transportation and storage of salt in the trunk; (v) Eliminating salt maxima in root-soil solutions, salt removal from xylem, symbiosis-related or PGPR, and changes in root structures; (vi) Total energy gain decreases at higher salinity by slowing growth and photosynthetic rate; (vii) Too much salt has a negative impact on root and shoot length, low salt has a positive effect on longer shoots and roots. Simply, there were more lateral roots with lower salt levels. This type of system will be useful for analyzing the signaling mechanisms of phytohormones with other parameters such as photosynthesis, redox homeostasis, and cell cycle regulation. In general, abiotic stress responses in plants are shown in Fig. (2).

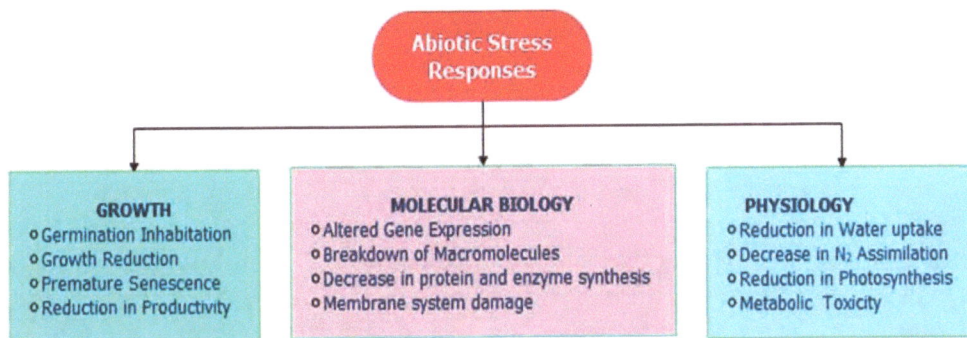

Fig. (2). Abiotic stress responses in plants.

Abiotic stresses are caused by changes in ecological nature, such as precipitation, heat, heavy metals, salinity, light, water, pests, nutrients, ultraviolet radiation, and soil biota. These also affect plants in many ways: (i) when plants are exposed to these stresses, their physiological, morphological, and biochemical processes are affected; (ii) stresses trigger at a molecular level of plants by altered gene expression, breakdown of macromolecules, membrane damage, cellular metabolism, changes in growth rates, crop yields, *etc.*; (iii) Physiological modifications of reduction in water uptake, reduce N_2 assimilation, reduce photosynthesis, and increase the metabolic toxicity in plants; (iv) reduced germination, less growth, premature senescence, and low yield. Abiotic diseases do not show the signs of diseases. Continuous exposure to abiotic stress causes disorders in plants, including wilting of the leaf, abscission of the leaf, decreased leaf region, and decreased water loss through transpiration. Some treatments are used for treating abiotic disorders, such as testing soil and fertilizing effects, identifying and treating pests or diseases, and aerating compacted soil with an air tool to 'uplift' it without damaging roots.

Drought refers to a prolonged lack of water that affects plant growth and survival. Drought stress is a complex action that limits plant development at various stages. There are five drought levels: Abnormally dry, moderate, severe, extreme, and exceptional. The main effects are imbalances in water, osmotic pressure, metabolic activity, and less ATP production. Low humidity can disturb the rate of nutrient uptake and impair enzyme activity. Droughts are divided into mild, moderate, and severe stress conditions. The overall RWC (relative water content) in mild drought stress ranges from 60 – 70%, moderate from 40 – 60%, severe from 0–40%, and over 90% as a control. There are three factors: water withdrawal, water use efficiency, and yield index, which stimulate trait-based breeding and genetic analysis of drought adaptation mechanisms. Drought tolerance depends on how efficiently plants perceive changing environmental conditions and respond to diminishing water availability. A drought-tolerant plant, *Arabidopsis thaliana* has the potential to become an infectious crop by manipulating cell and tissue-specific responses.

Stress mitigation is possible by spraying chemicals like DAP, KCL, Kaolin, GA, IAA, thiourea, and cycocel on the foliar stage [29, 30]. Plants face many biotic stresses caused by microorganisms. These agents cause various types of infections and diseases in crop plants. Nutrients, antitranspirants, plant growth regulators, and soil or rhizosphere microbes can mitigate the adverse effects of water. *Achromonas, Azospirillum, Aeromonas, Bacillus, Enterobacter, Azotobacter, Pseudomonas,* and *Klebsiella* soil-dwelling microbes have been shown to promote plant growth under unfavorable conditions [17]. Genetically modified transgenic plants have proven to be an alternative to normal plants. The exogenous application of ABA reduces the release of ethylene and leaf dropout. After using jasmonic acid, gibberellin can relieve stress, especially salt. Genes involved in stress management are shown in Table **2.**

Table 2. Genes involved in stress management (Modified from [30]).

Stages of Plant Growth	Genes Involved	Stresses Effect
Early flowering	GIGANTEA, ABFs, SOC1	Suppress the seed germination and seedling development
Leaf senescence & chlorophyll contents	SGR1, SGR2, CspA, CspB	Reduce the height and area of the leaf
Stomatal aperture	CLE25, NF-YB1, NF-YB2	Reduce stomatal conductance
Cuticular waxes production	MYB96, OsWR2, TsSHN1	Affect reproductive growth
Carbon assimilation	BRL3, SnRK1, TPS/TPP, TsVP	Poor vegetative growth
Root architecture	BRL3, HvBR11, OsNACs, OsERF71, HVA1, DRO1	Poorly organized
Ethylene sensitivity	Hahb4, ARGOS8, AtOSR1	Slow ripening

GIGANTEA (GI-photoperiodic gene), SOC – suppressor of overexpression of 1-CO1, CLE – CLAVATA3/ Embryo surrounding region related 25 gene, ABF – ABRE binding factors, NF-Y – Nuclear factor, TPS/TPP – Trehalose phosphate synthase/ Trehalose phosphate phosphatase, SnRK1 – SNF related AMPK protein kinases and OsERF71- Oscinnamoyl coenzyme A reductase1. NAC genes such as ANAC019, ANAC055, and ANAC072 are induced by drought, high salinity, cold, and freezing [31].

Four floral promotion pathways have been identified in long-day plants: photoperiod, autonomous, vernalization, and GA pathways. The genes within the photoperiod pathway or long-day pathway play a significant role in controlling flowering time. Gigantea (GI-photoperiodic gene), a clock-associated protein, maintains circadian period length and amplitude and regulates flowering time and hypocotyl growth in response to day length.

Agricultural Techniques for Stress Reduction

Once the soil conditions and moisture conditions become favorable, the following general recommendations are used to reduce the stress in agriculture: (i) selection of suitable crops and varieties, (ii) improved method of seedling, (iii) adaptation of soil and moisture conservation techniques, (iv) enhance soil organic matter, (v) application of foliar nutrition, (vi) use of drip and micro sprinkler irrigation, (vii) wastewater should not be utilized without pre-treatment and (viii) use of absorbent polymer for water absorption and slow release (Table **3**).

Table 3. Important agricultural techniques in foliar nutrition and stress reduction (Adopted from [32]).

Stages of Application	Mechanism of Spraying
Flowering and grain formation stage	Spray DAP + KCl, cycocel
Moisture stress	Foliar spray- inc sulfate, boric acid, ferrous sulfate, kaolin, and urea
Seed hardening time	1% KH_2PO_4 for 6 - 8 hours
Application of N and K fertilizers	*Azospirillum, Rhizobiaceae, Azotobacter, Vesicular arbuscular mycorrhizae,* or *Phosphobacteria* with 25 kg of soil
For leaves	Neem cake 4 grams /liter, triazophos 1.5 ml/L
For mites	Abamectin 0.5ml/L

TOLERANCE OF THE ABIOTIC STRESSES

Plants are constantly modifying their metabolism based on external stimuli. Some are natural (light intensity and relative humidity) and sporadic events (drought, salinity, xenobiotics, cold triggers, and extreme temperatures). A diagrammatic representation of survival strategies under stress is shown in Fig. (**3**). The capability of plants to tolerate salinity stress is usually evaluated by biomass production. The root system or rhizosphere enables the plant to trace non-saline areas for water and minerals. More recent progress has been made under the tolerance mechanism of plants to saline stress and climate change stress to ensure the maintenance of plant growth and feasible crop yields. Survival under adverse conditions requires the development of stress-tolerant genotypes and agricultural regimens that avoid the stress effects of salinity, metalloids, and drought. The several classifications to identify and modify resistance effects are:

• Tools for detecting the effect of salt tolerance in tissue and vacuoles
• Morphological features differentiation in shoots and roots
• Substitute for salt intolerance species with agricultural/horticultural potential (GMMs)

- Biochemical indicators and chemical markers of plant uptake (phytohormones)
- Symbiotic microbes (RIDER)
- Gene expression-related mechanisms (radiation tolerance)

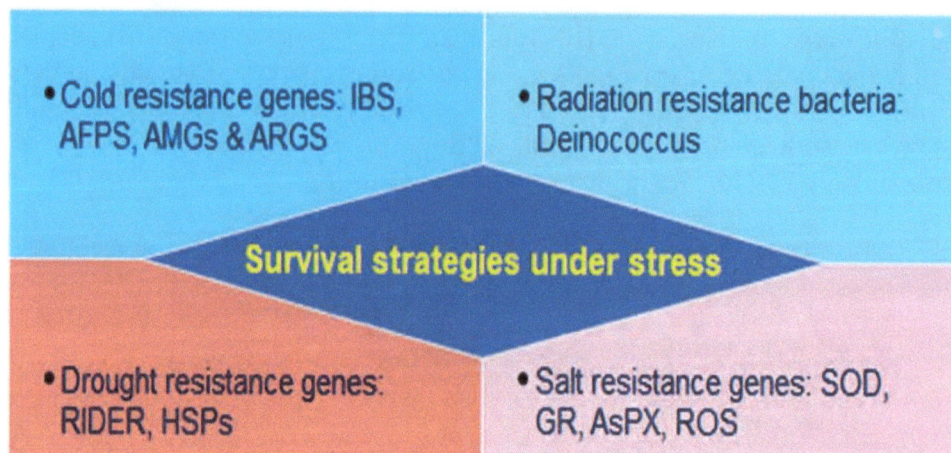

Fig. (3). Diagrammatic representation of survival strategies under stress.

There are many survival strategies under stress; Infection-encoded auxiliary metabolic genes (AMGs) are found in ice cryo peg brine and ocean ice. About 1466 genes were examined and reported in the Global Ocean Viromes 2.0 data set. These qualities are utilized to create flagellar assembly and overcome the injuries caused by low temperatures. The Antifreeze proteins (APs) bind to microscopic crystals, thus preventing them from forming larger crystals and remaining liquid even at -18^0C [33]. These APS or cryoprotectant agents are 200 to 300 times more effective in the IRI (recrystallization inhibition). In Antarctica, *Acinetobacter radioresistens* contain several ARGs in an isolate recovered from the region and genetically modified by recent technology.

Drought tolerance genes are *Rhizobacterium*-induced drought tolerance and resilience (RIDER), and RSPs are also introduced into the plant by genetic engineering. The SOD, GR, and ROS genes have the potential to modify the salinity.

Global water scarcity, population, and climate change are the biggest challenges in agriculture in recent years. This places great responsibility on the development of drought-tolerant cultivars by molecular breeding techniques of Marker-assisted recurrent selection (MARS), Genome-wide selection (GWS), and next-generation drought-tolerant crop production [30]. Integrating the concept of genetic phenotype into plant improvement requires attention to phenotypic analysis and Eco physiological modeling through the novel concept of 'Genetic genomics'.

EMERGING MICROBIOME IN SOIL BIOTA

Although microbial clades are loosely recognizable taxonomically, little is known about their genetics. The physiological and ecological role has been termed 'microbial dark matter' due to new research and innovation. Genetic resistance forms the backbone of integrated management. The genetic hypothesis promotes sustained resistance by linking multiple resistance genes with the process of identifying new resistance genes. Stress-tolerant genotypic processes focus on biotechnological, genomic and agronomic approaches. The plant transformation in stress tolerance has been applied in crops of *Arachis sp., Cajanus sp., Cicer sp., Glycine max, and Lupinus sp. Pisum sativum, Vigna sp.,etc.* Many proteins involved in the modification of transgenic plants (aquaporins, endogenous vacuolar protein, p5cs, Arabidopsis antiporter, salt) are hypersensitive to salt stress.

Chinese F1 rice hybrids and sunflowers have identical male sterility genes with short stems. Similarly, resistance to his *Phytophthora infestans* was introduced into potatoes to withstand weather conditions [34]. *The Ama1* gene is isolated from Amaranthus to develop protein-rich potatoes by gene modification. OXDC (Oxalate decarboxylase) & C-5 sterol desaturase (FvC5SD) genes have been isolated from the edible fungus *Collybia velutipes*. Other strategies for silencing the plant gene, *e.g.*, polygalacturonase, induce the shelf life of fruits and vegetables [35]. The GM papaya carries the papaya ringspot virus envelope protein (PRSV). *Rhizobacterium*-induced drought tolerance and resilience (RIDER) is optimal for microbial-mediated plant responses [36]. Soil-borne Aeromonas, Azospirillum, and Bacillus bacteria promote plant growth under adverse conditions. Exogenous reporter genes GFP, beta-glucuronidase (GUS), luciferase (LUC), yellow fluorescent protein (YEP), and red fluorescent protein (RFP), as well as the endogenous reporter gene phytoene desaturase (PDS) were used for RNAi assays and screening transgenics from non-transgenic plants. The PDS gene leads to an albinism or dwarf phenotype by depleting chlorophyll, carotenoids, and gibberellins. The usage of CRISPR 9 (cas9) -mediated genome editing, transcriptional regulation by dcas9 and Cas13a, fine-tuning of miRNAs, DNA/RNA editing, and genetic use restriction technologies (GURTs) also provides clarification to avoid transgene spread. Moreover, applied and agricultural research is vital for developing new plants with negligible adverse effects on non-target organisms. The management process of abiotic stresses is shown in Fig. (**4**).

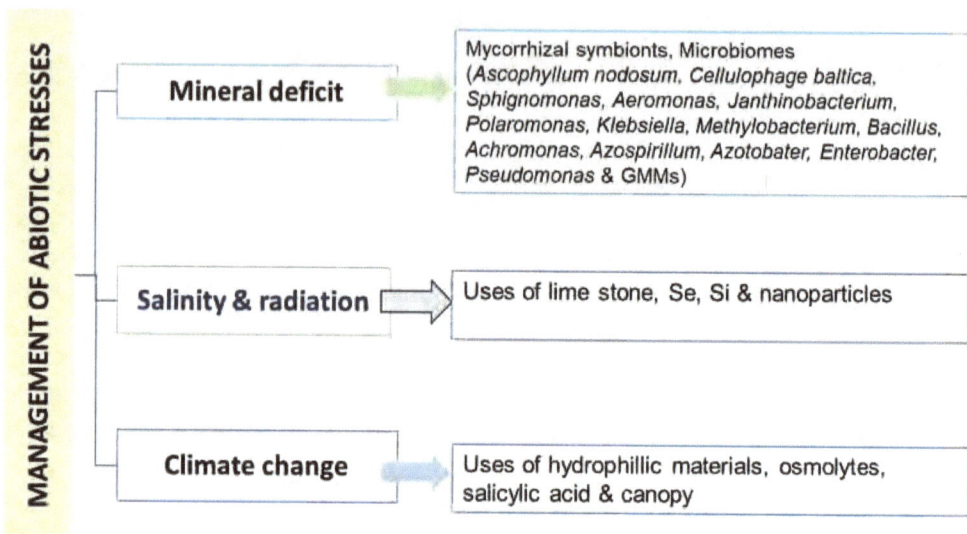

Fig. (4). Management process of abiotic stresses.

Agronomic strategies have innovations in crop tolerance, are static for a long time, are continuously revised, and also require genetic improvements. Plants adapt to low water availability by reducing the leaf area, increasing the root biomass, and reducing the ripening speed. The basic rules for management are: (i) Soil water reservoir must be refilled; (ii) Excess water is removed from the field; (iii) Impact of rain and runoff effect. Water stress depends on the ability of plants to undergo physiological, biochemical, and morphological changes to enhance water use efficiency. The physiology of crops is regulated by soil water availability and ecological conditions. Mineral deficiency is eliminated by microbial symbionts, microbiomes, GMMs, *etc*. Salinity and radiation effects are reduced by using lime stones, selenium, and silicon nanoparticles. Climatic change is easily suppressed by hydrophilic materials, osmolytes, salicylic acid, and canopy. Positive effects have been reported for the application of PGPR (Plant growth-promoting bacteria) to increase salinity tolerance. The mitigation effect is measured by the sodium adsorption rate (SAR); it depends on the concentration of sodium, calcium, and magnesium.

Polaromonas is a psychrotolerent, cold-active bacterium of 9 types that produces four toxins from just 2 genes, ICE1 and DREB2 (Dehydration Responsive Element binding proteins). These cold virus stress reaction proteins are produced at the transcriptional level. ROS-diminishing genes are also present in these bacteria, which capture reactive oxygen and induce psychrophilic and ultraviolet radiation resistance [37]. This is typical in many abiotic stresses like salinity, temperature limits, and dry seasons. This genome editing (GEd) benefits by introducing modifications in legume plants and producing minor genome changes. Crop varieties with improved abiotic stress tolerance are shown in Fig. (**5**).

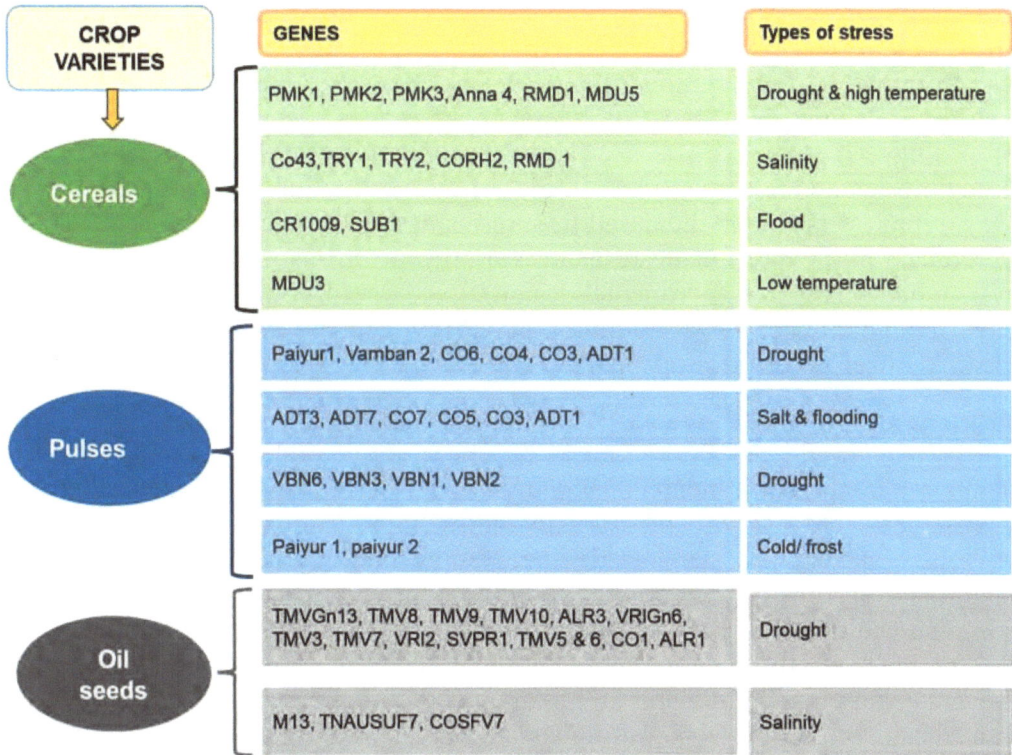

CROP VARIETIES	GENES	Types of stress
Cereals	PMK1, PMK2, PMK3, Anna 4, RMD1, MDU5	Drought & high temperature
	Co43,TRY1, TRY2, CORH2, RMD 1	Salinity
	CR1009, SUB1	Flood
	MDU3	Low temperature
Pulses	Paiyur1, Vamban 2, CO6, CO4, CO3, ADT1	Drought
	ADT3, ADT7, CO7, CO5, CO3, ADT1	Salt & flooding
	VBN6, VBN3, VBN1, VBN2	Drought
	Paiyur 1, paiyur 2	Cold/ frost
Oil seeds	TMVGn13, TMV8, TMV9, TMV10, ALR3, VRIGn6, TMV3, TMV7, VRI2, SVPR1, TMV5 & 6, CO1, ALR1	Drought
	M13, TNAUSUF7, COSFV7	Salinity

Fig. (5). Crop varieties with improved abiotic stress tolerance (Adapted from [38]).

Crop production depends on drought, intensity of light, and duration of exposure of seed to the soil. Various stresses, such as cold, salinity, and flooding, also interfere with the growth rate. Drought negatively affects fertility and grain mass and also reduces yield. Heat mitigation strategies are followed in different crops. Seed inoculation with rhizobacteria, bacterial seed treatment, and application of Mg, ascorbic acid, sulfur, calcium chloride, and salicylic acid also mitigate heat stress. Phytohormones of jasmonic acid, salicylic acid, naphthylphthalamic acid, aminoethoxyvinyl glycine, and gibberellic acid are beneficial for waterlogging mitigation effects on crops. There are countless applications like SOD (Superoxide dismutase), glutathione reductase (GR), and ascorbate peroxidase (AsPX) for utilizing this kelp (*Ascophyllum nodosum*). CO4,3,5,7, TRY1,2, ADT3,5,1, MDU3, RMD1, and TMV3,7,8,9,10 genes are genetically modified and introduced in transgenic plants. These plants can withstand the dormant state for a prolonged period. Micro-level climate risk assessment, major crop-based production, and real-time monitoring are used for long-term and short-term management. Various reports propose that the enemy of stress impacts of algal concentrates might be connected with cytokine action. Cytokine transforming pressure is induced by preventing the reactive oxygen species (ROS) formation or by repressing xanthine oxidation [30].

The commercial formulation of seasol from seaweed has also improved the freezing tolerance and osmotic tolerance in grapes. There are many reports proposing that algal extracts or seaweeds might be helpful tools for improving drought tolerance in citrus trees. In the field investigation of winter, barley has shown improvement and increased frost resistance. This type of gene is artificially introduced in plants to make resistant varieties or dormant plants by transgenic system modification. Experimental design, tissue or clone tissue cell type-specific have a high impact on agricultural plants by genome editing technology to identify novel genes in crop plants.

EFFICACY OF NANOMATERIALS ON STRESSES

Plants can hold environmental change by modifying their genes. Plants safeguard themselves from osmotic pressure by delivering different osmolytes such as polyols, trehalose, proline, and glycine. Nanotechnology involves agriculturally vital nano fertilizers to flourish agronomic productivity and efficiency and diminish ecological pressure. Nanofertilizers are eco-friendly and moderate the loss of nutrients from agricultural products. They offer continuous support to prevent contamination of water bodies and the environment using nanotechnology in a sustainable future. Nanopesticides are recycled to enhance harvest efficacy, shielding plants from harmful factors.

Polymer nanopesticides have gained attention for preparing biodegradable and biocompatible nanogels, nanospheres, and nanofibers. Nanobug spray is nanostructured alumina (NSA), a nano-designed material for killing bugs. Nano formulation, nanocarrier, or transporter of chitosan liposome acts as the vehicle to control the delivery or release of active compound present in its core and is used to protect the satisfactory measure of the compound during the transition from hatching to the development phase of bugs. The UN-CBD embraced the proposal of the Worldwide agreement to preserve plants and animals in April 2010. The Activity Plan for Biodiversity (BAP) is a universally perceived program focused on plant species and the climate and was implemented in 2009 among 191 nations. AMBIO has financed several global programs like SCAR, EBA (Evolution and Biodiversity in the Antarctic), and MERGE-IPY. MERGE aims to study microbiological and ecological responses in polar areas [39]. Many researchers have shown that training implies high-yielding varieties that are resistant to biotic and abiotic stresses, nourishing quality improvement with enormous seed and size of the fruit and matured growth. The effects of nanomaterials on abiotic stresses [40] are shown in Fig. (**6**).

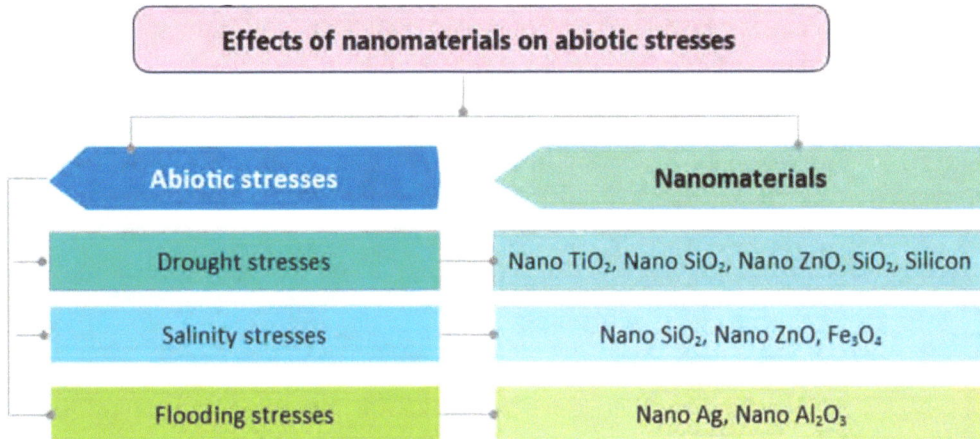

Fig. (6). Effects of nanomaterials on abiotic stresses.

Phytotechnology is the synthesis of different types of nanoparticles in an eco-friendly, simple, stable, and cost-effective method for the development of stress-resilient plants. Natural nanoparticles are native components of biological systems of nanoclay, liposomes, exosomes, magnetosomes, and viruses. The biosynthesis of these particles in plants is analyzed by factors including metal ion concentration, concentration of plant extracts, temperature, and pH. The abiotic stresses are treated by using Nano Tio_2, Nano Sio_2, Nano ZnO, Silicon, Nano Fe_3O_4, Nano Ag, and Nano Al_2O_3. Nanoparticles of platinum, iron, silver, gold, copper, selenium, and zinc can be synthesized within the plant parts, including stem, leaves, flowers, bark, and roots. These particles are stimulated by phytochemicals of alkaloids, flavonoids, phenolics, saponins, and steroids. The stem extract of plants like *Coleus aromaticus, Salvadora persica,* and *Momordica charantia* has silver nanoparticles; leaves of *Diospyros kaki* produce platinum nanoparticles, and *Vitex negundo* leaves have copper silver particles. *Zingiber officinale* produces gold from root extracts. *Syzygium aromaticum* buds can also synthesize silver, gold, copper, palladium nanoparticles. In recent days, bacteria and algae such as *Sargassum tenerrimum, Turbinaria conoides, Laminaria japonica, Bacillus, Pseudomonas, E. coli,* and *Serratia* have been used for the synthesis of silver nanoparticles., Potato virus X and Hepatitis viruses act as nanocarriers and nanoconjugates. In plants, these nanoparticles have the effect of reducing secondary metabolites and cytotoxic metabolites and regulating the enzymatic and non-enzymatic antioxidants, leading to lowered ROS and maintaining higher nitrogen assimilation and ionic homeostasis. Administration of these nanoparticles at 50 to 80 ppm can be efficient in various plants. Tio_2 nanoparticles enhance carbohydrate metabolism and elevate the negative stress effects. Nanomaterials are mostly involved in seed germination, stress reduction, and growth factor enhancement for plants, as well as cell organization, inducing photosynthesis rate, and increasing the biomass level and root and shoot length.

Plant pathogenic fungi of *Bipolaris sorokiniana* and *Magnaporthe grisea* are removed or eliminated by Ag NPs (Silver nanoparticles), *Alternaria alternata* and *Botrytis cinerea* can be suppressed by CuNPs and AgNPs. *Rhizopus stolonifer, Fusarium oxysporum and Mucor plumbeus* have an impact against ZnO NPs and MgO NPs. Carbon-based NPs, Ti-based NPs, Ce-based NPS, Mg-based NOs, Si-based NPs, Mn-based nano pesticides, and nano fertilizers are accessible for plant growth [41]. A half-breed material encapsulated pesticides of zineb and mancozeb, which is more intense against *A. alternata.* Nanosensors are a fantastic asset. Nanotubes, nanowires, nanoparticles, or nanocrystals are used to upgrade or optimize signal transduction by sensing elements.

Zhong and co-workers attempted to develop new strategies to eliminate the pollutants on glacier ice core surfaces [42]. The Byrd Polar and Climate Research Center has prepared to conserve more than 7000 meters of ice core sections due to the awareness of this hazard. Recently, multi-omics systems covering genomics, transcriptomics, proteomics, and phenomics have assimilated studies on plant-microbe interaction. As per this report, antimicrobial compounds are molecules widely found in life forms of psychrotolerans.

. Farmers and agronomists can concentrate on plants and their developmental obstruction. Plants can adjust to their environment by introducing resistance and stress-enduring. On the off chance that it is not helpful, plants need some advancement inside their genome. Nanofertilizers and biofertilizers give resistance to crops against stress and simultaneously help them recover. Plants have optional properties to available or inactive precursors by host enzymes in response to pathogen attack or tissue damage.

CONCLUSION

'Honour your ecological community and environmental people group' because all changes like environmental and climate variability and natural fluctuations may occur due to human activities. The ozone-diminishing substances lead to climate extremes and the total amount of carbon dioxide we emit. There are three climate change adaptation approaches: social awareness program, ecosystem-based adaptation, and new technologies and infrastructure, which can help in climate resilience and water-related coastal risks. WWF's Climate Savers Program set an objective to reduce fossil fuel by-products and provoked urban communities to use sun-based energy to improve air quality and supply safe water. As climate change worsens, dangerous weather fluctuations frequently happen on the planet. Modern industrial organizations, commodities, or businesses are responsible for reducing their contribution to climate change. With expanding improvement, climate change will achieve different permafrost-related impacts on vegetation, water quality, geo-risks, and occupations. This way, we can reduce emissions and avoid the worst effect on plant stress management. Exploring the microbes is not an easy errand, but our machinery life needs extraordinary power to heal everything in the future. This final warning from nature will reoccur repeatedly because of ignorance. We should be ready for the climate change era.

ACKNOWLEDGEMENTS

The authors thank the V. V. Vanniaperumal College for Women, Virudhunagar, Tamil Nadu, India, and Tamil Nadu Agricultural University, Coimbatore, Tamil Nadu, India, for their continual encouragement and unflinching support.

REFERENCES

[1]   Engel AS. Microbes.Encyclopedia of caves. Academic Press 2019; pp. 691-8.
        [http://dx.doi.org/10.1016/B978-0-12-814124-3.00083-2]

[2]   Snow T. Ecosystem metabolisms and functions: an eco-literacy framework and model for designers. FormAkademisk - forskningstidsskrift for design og designdidaktikk 2020; 13(4): 4.
        [http://dx.doi.org/10.7577/formakademisk.3370]

[3]   Abdel-Raouf N , Al-Homaidan AA, Ibraheem IB. Agricultural importance of algae. Afr J Biotechnol 2012; 11(54): 11648-58.
        [http://dx.doi.org/10.5897/AJB11.3983]

[4]   Joshi H, Shourie A, Singh A. Cyanobacteria as a source of biofertilizers for sustainable agriculture.Advances in cyanobacterial biology. Academic Press 2020; pp. 385-96.
        [http://dx.doi.org/10.1016/B978-0-12-819311-2.00025-5]

[5]   Fernández-Valiente E, Quesada A. A shallow water ecosystem: rice-fields. The relevance of cyanobacteria in the ecosystem. Limnetica. 2004; 23(1-2):095-107.

[6]   Hegde DM, Dwivedi BS, Sudhakara Babu SN. Biofertilizers for cereal production in India: A review. Indian J Agric Sci 1999; 69(2): 73-83.

[7]   Change IC. Mitigation of climate change. Contribution of working group III to the fifth assessment report of the intergovernmental panel on Climate Change. 2014 Apr; 1454:147.

[8]   Levin K, Boehm S, Carter R. 6 Big findings from the IPCC 2022 report on climate impacts, adaptation and vulnerability. World Resources Institute. 6 Feb 22.

[9]   NSO (2022), EnviStats-India 2022: Vol. I: Environment Statistics, National Statistical Office, Ministry of Statistics & Programme Implementation , Government of India, New Delhi EnviStats – India 2022 (Volume I: Environment Statistics) (ruralindiaonline.org) Accessed on 20 June 2023.

[10]   Smith LC, Ramakrishnan U, Ndiaye A, Haddad L, Martorell R. The importance of Women's status for child nutrition in developing countries: international food Policy Research Institute (IFPRI) research report abstracts 131. Food Nutr Bull 2003; 24(3): 287-8.
        [http://dx.doi.org/10.1177/156482650302400309]

[11]   Georgieva K, Sosa S, Rother B. Global food crisis demands support for people, open trade, bigger

local harvests. International Monetary Fund, blog post. 2022 Sep 30.

[12]   Ray S, Kumar N. Strategies to Lower Carbon Emissions in Industry. Low Carbon Pathways for Growth in India 2018; pp. 65-80.
       [http://dx.doi.org/10.1007/978-981-13-0905-2_7]

[13]   De Wit R, Bouvier T. ' *Everything is everywhere*, but, *the environment selects* '; what did Baas Becking and Beijerinck really say? Environ Microbiol 2006; 8(4): 755-8.
       [http://dx.doi.org/10.1111/j.1462-2920.2006.01017.x] [PMID: 16584487]

[14]   Blamey J. Scientists Discover Antarctic Microbes for Future Biotech Applications Crop biotech update. ISAAA Inc. 2011.

[15]   Miteva V. Bacteria in snow and glacier ice InPsychrophiles: from biodiversity to biotechnology. Berlin, Heidelberg: Springer Berlin Heidelberg 2008; pp. 31-50.

[16]   Marcello R. Mysterious Microbes Turning Polar Ice Pink. Speeding Up Melt, Environment News 2018.

[17]   Kochetkova TV, Toshchakov SV, Zayulina KS, *et al.* Hot in cold: Microbial life in the hottest springs in permafrost. Microorganisms 2020; 8(9): 1308.
       [http://dx.doi.org/10.3390/microorganisms8091308] [PMID: 32867302]

[18]   McTernan PM. Genetic engineering and purification of the hydrogenases from the hyperthermophilic archaeon *Pyrococcus furiosus* 2015.

[19]   Milojevic T, Cramm MA, Hubert CRJ, Westall F. "Freezing" Thermophiles: From One Temperature Extreme to Another. Microorganisms 2022; 10(12): 2417.
       [http://dx.doi.org/10.3390/microorganisms10122417] [PMID: 36557670]

[20]   Huber R, Kristjansson JK, Stetter KO. Pyrobaculum gen. nov., a new genus of neutrophilic, rod-shaped archaebacteria from continental solfataras growing optimally at 100□C. Arch Microbiol 1987; 149(2): 95-101.
       [http://dx.doi.org/10.1007/BF00425072]

[21]   Martins PH, Rabinovitch L, Orem J, *et al.* Biochemical, physiological, and molecular characterisation of a large collection of aerobic endospore-forming bacteria isolated from Brazilian soils. ARPHA Preprints 2022.

[22]   Moreira ERB, Ottoni JR, De Oliveira VM, Passarini MRZ. Potential for resistance to freezing by non-virulent bacteria isolated from Antarctica. An Acad Bras Cienc 2022; 94 (Suppl. 1): e20210459.
       [http://dx.doi.org/10.1590/0001-3765202220210459] [PMID: 35293946]

[23]   Al-Mashakbeh HM. The influence of lithostratigraphy on the type and quality of stored water in mujib reservoir-Jordan. J Environ Prot (Irvine Calif) 2017; 8(4): 568-90.
       [http://dx.doi.org/10.4236/jep.2017.84038]

[24]   El Sebai T, Abdallah M. Role of Microorganisms in Alleviating the Abiotic Stress Conditions Affecting Plant Growth. Advances in Plant Defense Mechanisms 2022 Aug 3. IntechOpen.
       [http://dx.doi.org/10.5772/intechopen.105943]

[25]   Zhao Q, Suo J, Chen S, *et al.* Na_2CO_3-responsive mechanisms in halophyte *Puccinellia tenuiflora* roots revealed by physiological and proteomic analyses. Sci Rep 2016; 6(1): 32717.
       [http://dx.doi.org/10.1038/srep32717] [PMID: 27596441]

[26]   Choudhary S, Wani KI, Naeem M, Khan MMA, Aftab T. Cellular responses, osmotic adjustments, and role of osmolytes in providing salt stress resilience in higher plants: Polyamines and nitric oxide crosstalk. J Plant Growth Regul 2023; 42(2): 539-53.
       [http://dx.doi.org/10.1007/s00344-022-10584-7]

[27]   Rao MJ, Wu S, Duan M, Wang L. Antioxidant metabolites in primitive, wild, and cultivated citrus and their role in stress tolerance. Molecules 2021; 26(19): 5801.
       [http://dx.doi.org/10.3390/molecules26195801] [PMID: 34641344]

[28]   Pinedo-Guerrero ZH, Cadenas-Pliego G, Ortega-Ortiz H, *et al.* Form of silica improves yield, fruit quality and antioxidant defense system of tomato plants under salt stress. Agriculture 2020; 10(9): 367.
[http://dx.doi.org/10.3390/agriculture10090367]

[29]   Wright, A. Antarctic, Microbes, Could Help Unlock Climate Mysteries.

[30]   Martignago D, Rico-Medina A, Blasco-Escámez D, Fontanet-Manzaneque JB, Caño-Delgado AI. Drought resistance by engineering plant tissue-specific responses. Front Plant Sci 2020; 10: 1676.
[http://dx.doi.org/10.3389/fpls.2019.01676] [PMID: 32038670]

[31]   Negrão S, Schmöckel SM, Tester M. Evaluating physiological responses of plants to salinity stress. Ann Bot (Lond) 2017; 119(1): 1-11.
[http://dx.doi.org/10.1093/aob/mcw191] [PMID: 27707746]

[32]   Stress management: Drought, TNAU Agritech portal https://agritech.tnau.ac.in/agriculture/agri_drought management.htmlAccessed on 20 June 2023.

[33]   DeVries AL, Cheng CHC. Antifreeze proteins and organismal freezing avoidance in polar fishes. Fish Physiol 2005; 22: 155-201.
[http://dx.doi.org/10.1016/S1546-5098(04)22004-0]

[34]   Robbins MD, Masud MAT, Panthee DR, Gardner RG, Francis DM, Stevens MR. Marker-assisted selection for coupling phase resistance to tomato spotted wilt virus and Phytophthora infestans (late blight) in tomato. HortScience 2010; 45(10): 1424-8.
[http://dx.doi.org/10.21273/HORTSCI.45.10.1424]

[35]   Jiang CZ, Lu F, Imsabai W, Meir S, Reid MS. Silencing polygalacturonase expression inhibits tomato petiole abscission. J Exp Bot 2008; 59(4): 973-9.
[http://dx.doi.org/10.1093/jxb/ern023] [PMID: 18316316]

[36]   He GH, Xu JY, Wang YX, *et al.* Drought-responsive WRKY transcription factor genes TaWRKY1 and TaWRKY33 from wheat confer drought and/or heat resistance in Arabidopsis. BMC Plant Biol 2016; 16(1): 116.
[http://dx.doi.org/10.1186/s12870-016-0806-4] [PMID: 27215938]

[37]   Hodges DM. Chilling effects on active oxygen species and their scavenging systems in plants. InCrop responses and adaptations to temperature stress 2023 Apr 28 (pp. 53-76). CRC Press.
[http://dx.doi.org/10.1201/9781003421221-2]

[38]   Vijayalakshmi D. Abiotic stresses and its management in agriculture. TNAU Agritech, Coimbatore. 2018:361-87 https://agritech.tnau.ac.in/pdf/11.pdf

[39]   Wilmotte A, Vyverman W, Willems A, *et al.* Antarctic Microbial Biodiversity: the Importance of Geographical and Ecological factors "AMBIO" (SD/BA/01). D/2012/1191/3. 2012.

[40]   Mustafa G, Komatsu S. Toxicity of heavy metals and metal-containing nanoparticles on plants. Biochim Biophys Acta Proteins Proteomics 2016; 1864(8): 932-44.
[http://dx.doi.org/10.1016/j.bbapap.2016.02.020] [PMID: 26940747]

[41]   Bhattacharjee R, Kumar L, Mukerjee N, *et al.* The emergence of metal oxide nanoparticles (NPs) as a phytomedicine: A two-facet role in plant growth, nano-toxicity and anti-phyto-microbial activity. Biomed Pharmacother 2022; 155: 113658.
[http://dx.doi.org/10.1016/j.biopha.2022.113658] [PMID: 36162370]

[42]   Zhong ZP, Tian F, Roux S, *et al.* Glacier ice archives nearly 15,000-year-old microbes and phages. Microbiome 2021; 9(1): 160.
[http://dx.doi.org/10.1186/s40168-021-01106-w] [PMID: 34281625]

Part 4: Role of Microbes in Sustainable Waste Management for Protecting our Ecosystem

Role of Microbes and Microbiomes in Wastewater Treatment for Aquatic Ecosystem Restoration

Suganthi Rajendran[1,*], Sinduja Murugaiyan[2], Poornima Ramesh[3] and **Govindaraj Kamalam Dinesh[4,5,6]**

[1] *JSA College of Agriculture and Technology, Avatti, Cuddalore, Tamil Nadu, India*

[2] *National Agro Foundation, Chennai, Tamil Nadu, India*

[3] *Department of Environmental Sciences, Tamil Nadu Agricultural University, Coimbatore, India*

[4] *Division of Environment Science, ICAR-Indian Agricultural Research Institute (IARI), New Delhi, India*

[5] *Division of Environmental Sciences, Department of Soil Science and Agricultural Chemistry, SRM College of Agricultural Sciences, SRM Institute of Science and Technology, Baburayanpettai-603201, Chengalpattu, Tamil Nadu, India*

[6] *INTI International University, Persiaran Perdana BBN, Putra Nilai, 71800 Negeri Sembilan, Malaysia*

Abstract: Industrial development improves our life quality. Nevertheless, the industries, such as those producing paper and pharmaceutical products, generate large amounts of industrial wastewater. This wastewater contains various pollutants, which are organic and inorganic. Various physical, chemical, and biological methods have been employed to eliminate the pollutants. Both physical and chemical methods involve more capital and produce secondary contaminants. During wastewater treatment, the wastewater microbiome facilitates the degradation of organic matter, reduction of nutrients, and removal of pathogens and parasites. For the purification of water and the preservation of the ecosystem, microbes in wastewater treatment are crucial. However, little is known about how microbial diversity is controlled and for what reasons. The varied microbial community supports flocculation, heterotrophic respiration, nitrification under aerobic conditions, and denitrification under anaerobic conditions. Although recycled water is reinstated for recreational and agricultural use, biomonitoring is vital for assessing treatment effectiveness. Microorganism-based biological treatment is developing as an effective and environmentally friendly method. This chapter thoroughly introduces biological wastewater treatment, growth and kinetics, and different microbial community types that include bacteria and fungus, actinomycetes, algae, plants, and the range of microbial wastewater treatment, among other topics.

* **Corresponding author Suganthi Rajendran:** JSA College of Agriculture and Technology, Avatti, Cuddalore, Tamil Nadu, India; E-mail: suganthi.tamilselvi@gmail.com

Shiv Prasad, Govindaraj Kamalam Dinesh, Murugaiyan Sinduja, Velusamy Sathya, Ramesh
Poornima & Sangilidurai Karthika (Eds.)

Keywords: Aerobic process, Biological treatment, Anaerobic process, Bioreactors, Mycoremediation, Microbes, Phytoremediation.

INTRODUCTION

Microbiomes are crucial in biological wastewater treatment methods such as activated sludge, anaerobic digestion, and bioelectrochemical systems. The microbial population's activity and resilience in the microbiome significantly impact the performance and stability of these activities. However, it is challenging to manage these wastewater treatment processes properly, and it is hard to anticipate their efficiency because of the complex nature of the microbiota (microbial type, role, and interactions) [1]. Biological wastewater treatment is regarded as a biological and chemical process that governs the removal and degradation of organic, inorganic, trace, and recalcitrant contaminants in the wastewater [2]. It is also considered an efficient, reliable, and cost-efficient technology for removing pollutants. The diminution of organic matter in the aquatic ecosystem is primarily mediated by microbial respiration and photochemical degradation [3, 4], which strengthens microbes as a feasible candidate for treating wastewater. In recent decades, the increasing occurrence of trace contaminants altogether, with their persistent, bioaccumulative, and toxic nature, threatens the homeostasis of the ecosystem, which has become a critical concern [5]. The diversified pollutants contain personal care products, pharmaceutically active compounds, steroids, antibiotics, chemicals, endocrine disruptors, pesticides, artificial sweeteners, and metabolites [6, 7]. These emerging contaminants do not yet have many regulations regarding their limits.

The generation of wastewater increases multi-fold because of industries and the overexploitation of freshwater resources. Wastewater treatment is always linked with waste recovery and its optimum utilization, which broadens the amplitude of wastewater treatment, enhancing the quality of the byproducts as an efficient alternative for non-potable purposes. Composition, concentration, volume produced, compatibility to treatment, and environmental impact must be considered before selecting the appropriate treatment for the particular wastewater that ropes microbes in since it can detoxify the discrete pollutants. Microorganisms detoxify pollutants through various mechanisms, and the application of microorganisms in treating the wastewater determines the efficiency of pollutant elimination. This section explores various biological treatment processes, possibilities, and the limitations associated with those treatment processes.

WASTEWATER TREATMENT

Even conventional biological wastewater treatment has specific implications, such as high electricity consumption, high-rate filters, and surplus sludge biomass, which lead to their enhancements. The application of microorganisms at various treatment phases significantly influences wastewater treatment efficiency.

Principles of Biological Wastewater Treatment

Samer [8] stated the following principles of biological wastewater treatment.

- The biological system existing in the wastewater treatment system is susceptible to hydraulic loads. Variation greater than 250% is said to be problematic.
- Temperature largely influences the growth of microorganisms. A reduction of 10°C decreases the rate of biological reactions.
- At doses between 60 and 500 mg L^{-1}, treatment effectiveness is higher. If it is above 500, it can be coupled with an anaerobic process.
- The strength of the wastewater should not be more than 150% or 1000 mg L^{-1} BOD; otherwise, equalization has to be done to ensure the safer strength of wastewater entering the treatment system.
- The carbon: nitrogen: phosphorous ratio in the wastewater should be from 100:20:1 to 100:5:1 for higher treatment efficiency.
- Wastewater should undergo pretreatment

Biological Growth and Kinetics

The biological growth in wastewater treatment plants can be ascribed from the Monod equation [8]: the activity of bacteria and the speed of metabolic processes are influenced by a number of variables. The most crucial criterion is the environment's temperature, pH, dissolved oxygen level within the wastewater, nutritional content, and other contaminants. These beneficial elements are used in fermenters to promote the microbial community's development and reactors' efficiency in wastewater treatment.

$$\mu = \frac{(\lambda S)}{(K_s + S)}$$

Where,

μ stands for the specific growth rate coefficient,

λ for the maximum growth rate coefficient at 0.5 μ_{max},

S for the concentration of the limiting nutrient (BOD and COD), and

Ks stands for the Monod coefficient.

BIOLOGICAL TREATMENT

Treatment of wastewater involves vital biological processes. The biological environment, the kind of biological transformation, and reactor design are typically used to characterize these biological activities. The biological environment determines the process's nature, whether organic or inorganic, and the transition process governs the different pathways for contaminant transformation depending on its source, such as organic or inorganic. These pathways include destabilization, flocculation, and adsorption onto organic matter. Reactor configurations comprise various physical setups or systems primarily intended for biological intervention and entail various principles. Removing organic pollutants that can be oxidized and coagulating non-settlable solids with microorganisms form the basis of biological treatment. Wastewater treatment generally falls into two categories: aerobic and anaerobic wastewater treatment. The primary disparity between the aerobic and anaerobic processes is clearly illustrated in Fig. (**1**). The type of microorganism involved in the process determines the output of the selected treatment process.

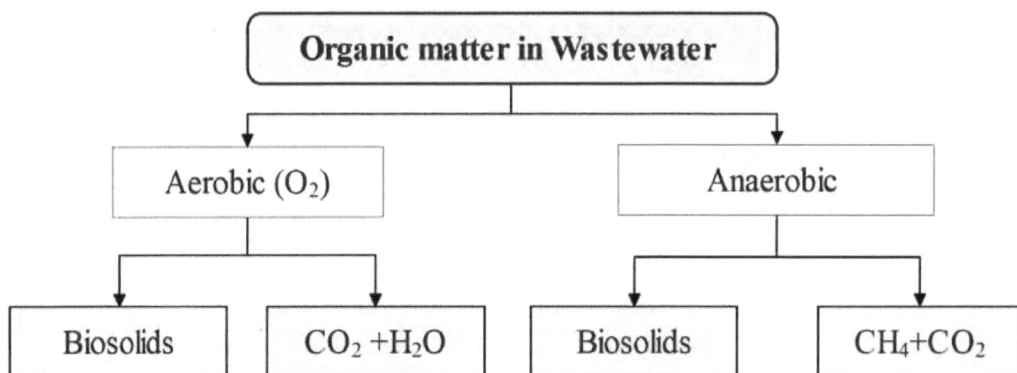

Fig. (1). Fate of organic matter under aerobic and anaerobic degradation process.

Aerobic treatment involves pollutant degradation by aerobes in the presence of molecular oxygen. The anaerobic treatment process occurs without air (molecular/free oxygen) to degrade or assimilate organic pollutants. Wastewater with higher BOD (>500 mg L^{-1}) is subjected to anaerobic treatment. Wastewater treatment encompasses various phases, which sequentially eliminate the pollutant load, facilitating improved treatment effluent quality. It involves primary, secondary, and tertiary treatment. Using bar screens and grit chambers at the initial phase, primary treatment is a physical technique that primarily intends to remove heavy particles and non-degradable items. In wastewater treatment, secondary treatment involves a biological process. Table 1 enlists a few biological treatments and their efficacy in eliminating the variety of pollutants present in wastewater.

Table 1. Different treatment technologies and their efficiency in wastewater treatment.

S. No.	Treatment Technology	Characteristic Features	Wastewater	Efficiency (%)	Refs.
1.	Aerobic granular sludge	• Self-immobilized cells • Extracellular polymeric substances maintain the stability of sludge bed • Adhesion and aggregation favour the DO gradient	Piggery wastewater	• COD - 98%; NH_3--97%; Total N-92.4% • Kanamycin, Tetracycline, ciprofloxacin, ampicillin, and erythromycin – 88.4%	[10]
2.	Annamox bacteria *Candidatus brocadia*	• Anaerobic ammonium oxidation process, which combines directly ammonia and nitrite into dinitrogen gas • Eliminates two-stage process –nitrification and denitrification • Eliminates the necessity of organic carbon sources for notification	Piggery wastewater	• 58.5% N removal at DO < 0.5 mg L^{-1}	[11]
3.	Constructed wetland	• It mimics the natural wetland system, facilitates water treatment, and enhances pollutant removal efficiency. • Entails reeds and other aquatic macrophytes, native microorganisms with naturally existing filter beds with sand/soil • Cost-effective method of wastewater treatment	Potato Farm wastewater	• 96% BOD ; 99% TSS; 90% TP	[12]
			Aquaculture wastewater	• 73.2% TSS; 43.8% No_3-N; 14.3% NH_4-N	[13]
			Dairy wastewater	• 92% BOD; 17% NH_4-N; 40% TP	[14]
			Pig farm wastewater	• 86.3% COD; 64.5% TSS; 85% NH_4-N; 77.9%TP	[15]
4.	Sequencing Batch Biofilm reactor	• A hybrid biofilm reactor grows biomass on a tiny bio-carrier that circulates freely inside the reactor. • Reduced bulking Problem • No need for sludge recycling	Dairy wastewater	• 81.8% COD • 85.1% NH_4-N	[16]

(Table 1) cont.....

S. No.	Treatment Technology	Characteristic Features	Wastewater	Efficiency (%)	Refs.
5.	Integrated aerobic sequencing batch reactor and anaerobic pulsed bed filter	• Immobilized biomass in the anaerobic filter • Pulsed with bio-cord bed • Anaerobic process followed by aerobic process	Beverage effluent	• COD-97% • BOD-98% • TSS-80% • TN-82%	[17]

Aerobic Treatment

It mainly entails the oxidation of organic substances with the assistance of heterotrophic and aerobic organisms while oxygen is present. The bacteria use the dissolved oxygen in the wastewater or extra oxygen provided *via* various aeration facilities as an oxygen source. Aerobic and facultative bacteria have to use a fraction of their energy to synthesize new cells from the residual material, while the rest is metabolized to minimal energy molecules for endogenous respiration. A simple schema of the oxidation process is illustrated in Fig. (**2**).

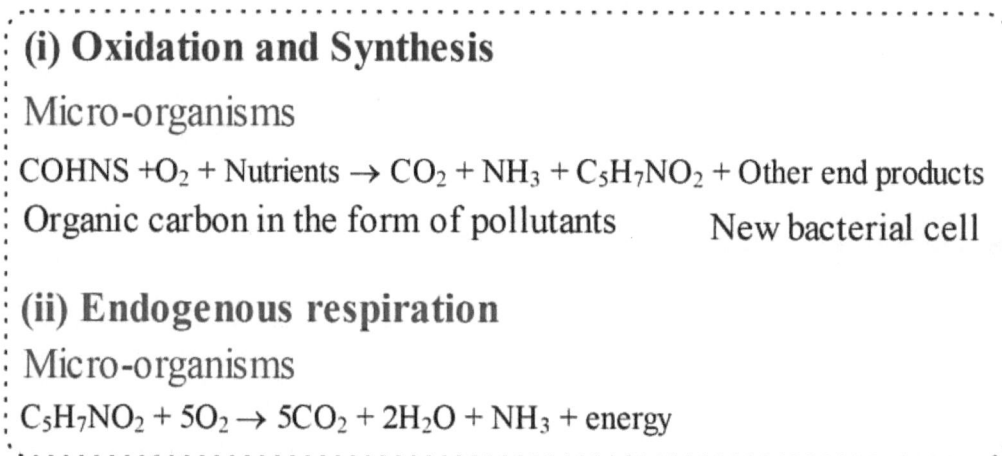

(i) Oxidation and Synthesis

Micro-organisms

$$COHNS + O_2 + Nutrients \rightarrow CO_2 + NH_3 + C_5H_7NO_2 + Other\ end\ products$$

Organic carbon in the form of pollutants   New bacterial cell

(ii) Endogenous respiration

Micro-organisms

$$C_5H_7NO_2 + 5O_2 \rightarrow 5CO_2 + 2H_2O + NH_3 + energy$$

Fig. (2). Schematic illustration of the oxidation process and endogenous respiration.

The aeration technique was mainly used to eliminate volatile organic compounds present in the wastewater. It includes aerated lagoons and oxidation ponds. An illustration of a situation where aeration comes from both the atmosphere and photosynthetic algae is an oxidation pond. Algae can only develop in areas where sunlight can penetrate. In the oxidation ponds, mutualism was known to occur between aerobes and photosynthetic algae. Algae utilize the inorganic compounds mineralized by the bacteria, and aerobes utilize the algae's oxygen release. In the case of aerated lagoons, aerators artificially give aeration, which keeps the microbial biomass in suspended nature, providing sufficient oxygen for the aerobes. The activated sludge process is an aerobic process in which maximum pollutant degradation can be achieved because of the high rate of microbial growth and their respiration rate due to the nutrients in the wastewater.

Anaerobic Treatment

Wastewater having higher BOD and is difficult to treat under aerobic treatment is subjected to anaerobic treatment. Even though anaerobic wastewater takes a lot of time, it has multiple advantages such as enhanced quality of treated wastewater, capacity to assimilate high load, less sludge generation, less energy requirement, and generation of biogas as the outcome. The four primary stages of anaerobic digestion—hydrolysis, acidogenesis, acetogenesis, and methanogenesis—are complicated and dynamic (Fig. (3).

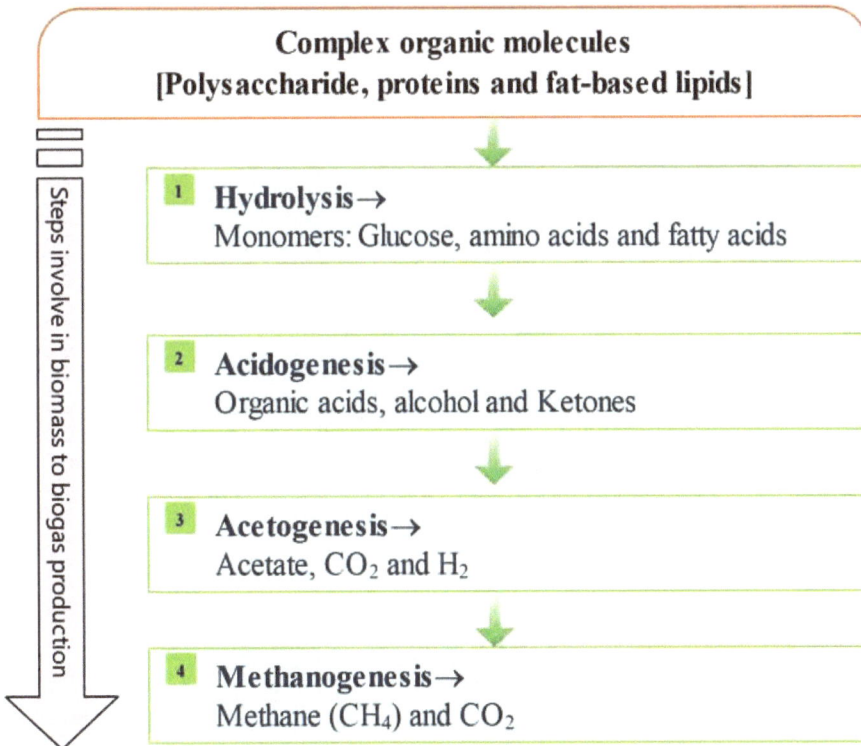

Complex organic molecules
[Polysaccharide, proteins and fat-based lipids]

Steps involve in biomass to biogas production

1. **Hydrolysis→**
   Monomers: Glucose, amino acids and fatty acids

2. **Acidogenesis→**
   Organic acids, alcohol and Ketones

3. **Acetogenesis→**
   Acetate, CO_2 and H_2

4. **Methanogenesis→**
   Methane (CH_4) and CO_2

Fig. (3). Schematic illustration of the chemistry involved in the anaerobic transformation of complex organic molecules to fuel.

- *(i). Hydrolysis*: Complex organic molecules are broken down by the process of hydrolysis, which includes the dissociation of chemical bonds by water. Proteins are broken down into peptides and amino acids from polysaccharides, fatty acids and glycerol from triglycerides, and heterocyclic nitrogen compounds, ribose, and inorganic phosphate from nucleic acids with the aid of hydrolytic bacteria such as *Shigella, Salmonella,* and *E. coli.*

- *(ii). Acidogenesis*: Acidogenesis is the process by which soluble components from the preceding reaction, such as sugars, amino acids, and fatty acids, are converted into organic acids (such as acetic, formic, propionic, lactic, butyric, and succinic acids), alcohols, and ketone (Ethyl alcohol, methyl alcohol, and acetone) with the help of obligate and facultative acidogenic anaerobes such as *Clostridium, Lactobacillus, Micrococcus, Peptococcus, Streptococcus, Desulfomonas,* and *Escherichia coli.*

- *(iii). Acetogenesis*: Next, it undergoes acetogenesis, where the volatile fatty acids are transformed into acetate with carbon dioxide and hydrogen with the help of acetogenic bacteria such as *Enterobacter wolinii* and *Synthrophomonas wolfie.*

- *(iv). Methanogenesis*: Hydrogenotrophic methanogens (Hydrogen-utilizing bacteria) and Acetotrophic methanogens (Acetate-splitting bacteria) are the two categories of anaerobes involved in the generation of gaseous fuel. *Methanosarcina* and *Methanothrix* are the two main genera involved in the thermophilic digestion of substrates. Methanogens such as *Methanobrevibacter sp.* and *Methanogenium sp.* favor methane generation from acetate [9]. Factor 420, Factor 430, coenzyme M, methanopterin, and methanofuran are some of the coenzymes of methanogens involved in methanogenesis.

BIOREACTORS

Bioreactors are "the engineered device or system that supports the encompassed biologically active environment" in which there will be microorganisms and biochemical substances derived from microbes that determine the rate of degradation. It is further assorted into two categories based on microbial growth: suspended and attached growth reactors. While attached growth reactors include trickling filters, rotating biological contractors, root zone reed beds, packed beds, fluidized beds, and integrated fixed film reactors, whose benefits and drawbacks are discussed in Table **2**, suspended growth includes activated sludge, extended aeration, aerated lagoons, waste stabilization ponds, and fermentors. Hybrid reactors are extensively employed to accomplish maximum pollutant load removal from wastewater.

MICROBIAL DYNAMICS IN WASTEWATER TREATMENT

Many microorganisms, including bacteria, fungi, viruses, and protozoans, are involved in the biological treatment. These organisms use the organic content in the wastewater to sustain their development and reproduction. The microbial community in the wastewater determines the rate and the type of pollutant degradation. That microbial community tends to change based on the type of wastewater, loading rate, operating parameters, *etc*. These microbial communities define the biomass species structure, which governs the metabolic pathways favoring pollutant removal, resulting in fine-grade treated effluent.

Table 2. Advantages and limitations of different bioreactors and the treatment process (Adapted from [2, 18].

S. No.	Bioreactor	Advantages	Limitations
1.	Stirred tank	Easy to establish and handle, utilize suspended growth of microbes and applicable to the aerobic and anaerobic process	Limited to low capacities
2.	Rotating biological contractor	Efficient pollutant removal needs less power input and is simple to handle, easy to maintain, and has a low space requirement, low sludge production	Odor problems may occur, requiring a skilled technical operator and a continuous electric supply. Must be shielded from the sun, wind, and rain.
3.	Trickle bed biofilm	Attached is the growth of microbes Lesser handling cost because of downflow Higher biomass escalates the bioconversion rate	Especially for BOD removal under aerobic condition Lesser capacity because of the reduced feed
4.	Moving bed biofilm	Heterogeneous edition of the stirred tank, higher biomass concentration in biofilm improves the bioconversion rate.	Less competent while compared with column reactors Higher agitation rate results in biofilm disturbance
5.	Fluidized bed biofilm	Function under higher capacities with a higher bioconversion rate. The extent of bioconversion is directly proportional to the flow rate of incoming feed.	Chance of shortfall of aggregates of biofilm Overhead cost is higher as compared to trickle bed
6.	Semi-fluidized bed biofilm reactor	Requires lesser volume reactor and higher bioconversion rate at high capacity, directly proportional to feed flow rate while maintaining a constant reactor volume.	Overhead cost is more than fluidized beds; the chance of continuous and circulating mode is impossible.
7.	Inverse fluidized biofilm reactor	Reduced overhead/operating cost because of the downflow, applicable to larger size particles, higher bioconversion rate	Lower capacity while evaluating against fluidized/semi-fluidized bed requires a reactor with more volume.

(Table 2) cont.....

S. No.	Bioreactor	Advantages	Limitations
8.	DSFF bioreactor	Easy to establish and handle, Need not require support particles, and reduced overhead cost due to downflow. Capacity can be increased with multiple tubes/columns	Currently limited to the anaerobic treatment process, it requires a reactor with greater volumes to treat higher capacities
9.	UASB reactor	Simple in construction, does not require support particles, higher bioconversion rate even at high capacity	Limited to the anaerobic treatment process, it employs diverse cultures of microbes, requires a higher bioconversion rate, and requires more startup time.
10.	Sequencing Batch Reactor	Flexible in cycle modification, reliable in cost; Fewer flow applications, tolerant to variations in wastewater strength, higher efficiency in the removal, flexible in variant operations, reduced chance of sludge bulking	It consumes More Energy, has limitations such as adjusting cycle times for small communities, and requires frequent sludge disposal.
11.	Membrane Bioreactor	Higher quality of effluent, ability to treat higher load, higher degradation capacity, minimum generation of sludge does not require more space, energy efficient, efficient pollutant removal.	Membrane fouling problem, limitations in aeration, Sludge in the external membrane, highly expensive

The biological treatment process boosts microbial diversity [19]. Under aerobic conditions, microbes tend to eliminate organic matter present in the wastewater, thereby reducing the BOD of the wastewater, whereas, in anaerobic conditions, the microbes and their metabolites actively transform and reduce the toxic contaminants present in the wastewater. Wastewater includes both favorable (aerobic – *Pseudomonas, Zooglea, Chromobacter, and Flavobacterium; anaerobic-bdellovibrio bacteriovorus* and *Brocadia annamoxidans*) and pathogenic organisms (*Escherichia coli, Vibrio cholera* and *Helicobacter pylori*).

Nitrifiers (*Nitrosomonas* and *Nitrobacter*), denitrifiers (*Thiobacillus*), and sulfate-reducing bacteria (*Desulfotomaculum* and *Desulfovibrio*) also facilitate nutrient removal. Filamentous bacteria in the wastewater facilitate the transfer of nitrate and nitrite nitrogen. Protozoans in the wastewater maintain a slimy layer and assist in removing bacteria and other protozoans. Fungi also play a defined role in the treatment process. Common fungi found in the treatment plants are *Subbaromyces splendens, Ascoidea rubescens, Fusarium aquaeductuum, Geotrichum candidum, Trichosporon cutaneum,* Yeast *(Saccharomyces)* and *Phanerochaete chyrosporium*. Another multicellular organism named rotifier tends to exist commonly in the activated sludge treatment of wastewater, which removes the non-flocculant bacteria and aids in floc formation and clarification of the effluent.

Bacteria

Bacteria degrade the organic matter in the wastewater mainly by adsorption and absorption. They fall into different categories based on oxygen requirements, such as obligate, facultative, obligate anaerobes, and micrographs. The bacterial communities in the wastewater are governed by various factors such as oxygen utilization, Dissolved oxygen, pH, Temperature, Nutrients, wastewater composition, retention time, and competition. These heterogeneous groups of bacteria treat the wastewater by forming flocs, secreting extracellular polymeric substances, biofilms, and aerobic granules. Table **2** lists some predominant microbes, their abundance, and wastewater treatment efficiency.

Alphaproteobacteria and *Beta proteobacteria* were found to be predominant in wastewater. Denitrifying bacteria such as *Rhodocyclus, Dechloromonas, Zoogloea,* and *Simplispira* are responsible for phosphorous removal bacteria [20]. Rhodoplanes, Lysobacter, and Leucobacter genera exhibited potential for xenobiotics degradation in wastewater treatment plants. *Zobellella denitrificans* strain A63 was identified as a novel denitrifying bacteria with 89.9 percent in treating saline wastewater under constructed wetland systems [21]. Table **3** represents a few microbes and their importance in pollutant removal. The initial population of archaea and bacteria varied from 48 and 52% to 0.005 and 99.99%. *Clostridia* belonging to firmicutes are responsible for converting simple sugars to organic acids. *Actinobacteria* and *Lactobacillus* are responsible for producing propionic acid and Lactic acid. *Bifidobacteriaceae* and *Coriobacteriaceae* convert lactic acid, resulting in acetic acid and propionic acid. Besides this, *Pseudoarmibacter* explains the buildup of volatile fatty acids by converting carbohydrates to organic acids [22]. In some instances, crude or purified enzymes were extracted from desired microbes used to treat the effluent. Dairy and residential sewage are treated using an alkaline protease enzyme obtained from *Bacillus pseudofirmus* SVB1, which is associated with reducing the overall cost of treatment by replacing the commercial enzymes [23].

Fungi

Mycoremediation refers to the involvement of fungi in wastewater treatment by removing contaminants by degradation or adsorption by the mycelium or the enzymes secreted by them. Fungi are reported to degrade biopolymers such as lignin, hemicellulose, and cellulose. It was reported that oyster mushrooms could remove 93 per of total aromatic hydrocarbons. Among the other fungi, white rot fungi were widely known to degrade aromatic and chlorinated compounds such as bisphenol A, bisphenol S, and nonyl phenol from wastewater [24]. Besides, Filamentous fungi in the wastewater can trap microalgae to form pellets, efficiently removing pollutants and enhancing biofuel production [25]. Arbuscular mycorrhizal fungi treat domestic wastewater that colonizes plants' roots, such as *Canna indica, Canna flaccid,* and *Watsonia borbonica.*

Table 3. Predominant microbes in wastewater treatment.

S. No.	Predominant Microorganism		Abundance	Treatment System/ Method	Pollutant Removal Efficiency	Refs.
1.	Phyla	Euryarchaeota	7.60-44.42% 4.65-37.98% 9.72-27.13% 8.39-22.69% 0.73-13.20%	Anaerobic pulsed bed filter with aerobic sequencing batch reactor	COD-97% BOD-98% TSS-80% TN-82%	[17]
		Proteobacteria				
		Bacteroidetes				
		Firmicutes				
		Synergistetes				
	Genus	*Acinetobacter*				
		Streptococcus				
		Cloacibacillus				
		Methanobacterium				
		Methanothrix				
		Methanospirillum				
2.	Phyla	Actinobacteria	49% 41% 6% 3%	Sequencing Batch Reactor	COD-23%	[22]
		Firmicutes				
		Bacteroidetes				
		Proteobacteria				
3.	*Dracena sanderiana* *Bacillus cereus*		-	Phytoremediation	Bisphenol 89.5%	[29]
4.	*Nannochloropsis oculata* *Bacillus polymyxa*		-	Co-immobilization	NH_4 59.85% PO_4^{3-} 90.5%	[30]
5.	*Chlorella vulgaris* *Bacillus Licheniformis*		-	Algae – Bacterial Consortium	TN – 88.82%; NH_4 84.98%; PO_4^{3-} 84.87% COD –82.25%	[31]
6.	*Bacillus flexus* RMWW II		-	Biodegradation of lignin - Rice mill wastewater Treatment	COD- 84% Lignin – 20%	[32]
7.	*Bacillus sp.*		-	Forward osmosis membrane bioreactor	Total nitrogen - 79%	[33]

(Table 3) cont.....

S. No.	Predominant Microorganism		Abundance	Treatment System/ Method	Pollutant Removal Efficiency	Refs.
8.	Phyla	Proteobacteria	26.7-48.9%	Activated sludge process	-	[34]
		Bacteroidetes	19.3-37.3%			
		Chloroflexi	2.9-17.1%			
		Acidobacteria	1.5-13.8%			

Phytoremediation

Phytoremediation involves the accumulation and degradation of various pollutants from contaminated soil and water by plants. In wastewater treatment, constructed wetlands were established to mimic the natural degradation process in which the root exudates, microflora, and toxic and anoxic conditions play a major role in eliminating wastewater pollutants. The nitrification and denitrification process facilitates nitrogen removal. Phosphorous can be removed from wastewater by plant uptake and maybe by the precipitation of insoluble phosphates.

Typha and Phragmites are commonly used reed species in constructed wetlands in which typha has the capacity even at various water depths. There are two types of surface and subsurface flow, in which surface flow is exclusively used for polishing the effluent. In the case of subsurface flow, they are further classified into two types: horizontal and vertical flow wetlands. This subsurface method is employed to treat different wastewater, likely domestic wastewater, paper mill wastewater, industrial wastewater, and mining runoff. Rhizospheric microbes play a significant role in mineralizing and translocating nutrients from soil to plant.

Plant-associated beneficial bacteria are crucial for plants' health and development under various circumstances. Plant growth-promoting rhizobacteria are bacteria that colonize the plant's roots and have favorable variables that determine the growth and development of plants [26]. Rhizospheric microbes attained 54.13 percent, 67.27 percent, and 47.34 percent removal in COD, TN, and TP, and *Trichococcus sp.* majorly influences the performance and resists the concentration variation [27]. It is also reported that the dominant microbial communities depend highly on oxygen, dissolved organic carbon, and pollutant concentration [28].

Phycoremediation

Wastewater treatment is essential for a sustainable earth. Since wastewater contains a huge amount of organic matter, it will disturb the homeostasis of the environment. Microalgae cultivation in wastewater has evolved to be environmentally friendly and cost-efficient, eliminating organic pollutants and generating valuable byproducts. Algae assimilate nutrients such as nitrogen, phosphorous, potassium, and other nutrients, releasing oxygen during photosynthesis. This oxygen will be utilized by the bacteria present in the wastewater, and microalgae fix carbon dioxide *via* respiration (Fig. **4**). Various technologies in wastewater treatment are significant contributors to carbon emissions.

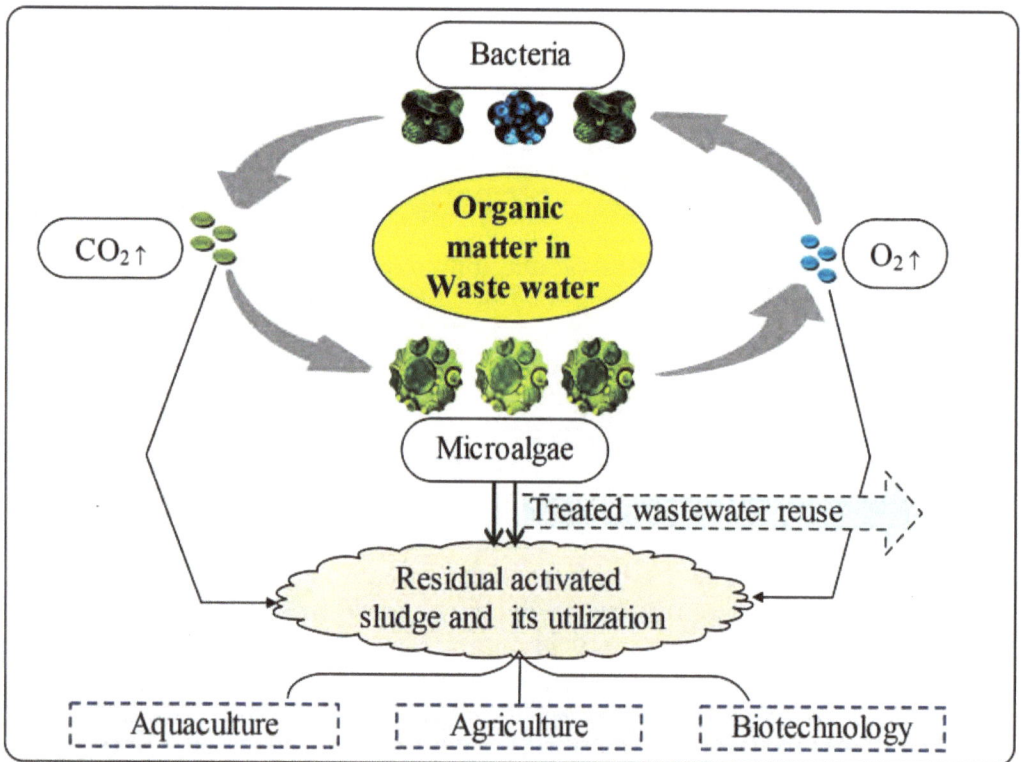

Fig. (4). Schematic illustration of microalgae in wastewater treatment.

The efficiency of microalgae in wastewater treatment is presented in Table **4**. A proposed strategy for carbon sequestration is microbiome-based carbon sequestration. Innovative strategies to increase sequestration and use include microalgae culture, microbial electrolysis, microbial electrosynthesis, microbial fuel cells, and engineered wetlands. Following the degradation of organic materials, electroactive bacteria break down carbon dioxide into carbonates and bicarbonates. Similarly, microbial electrosynthesis (MES) during wastewater treatment not only aids in carbon capture but also generates secondary products of economic significance (generation of polyhydroxy alkanoates by gram-negative rod *Aeromonas hydrophila* bacteria).

Microbial carbon capture cells (MCCs) have also been used recently to remediate wastewater whilst concurrently generating power and sequestering carbon. Additionally, it has been shown that the culture of microalgae rapidly absorbs CO_2 while simultaneously emitting O_2 as an outcome of photosynthesis. Considering carbon sequestration and byproduct generation are two benefits, microbe-based wastewater treatment has great potential [35]. Microalgae sequesters carbon and proves to be a cost-effective and sustainable method in biofuel production, which could be the best alternative to conventional fossil fuels [36]. Since wastewater is rich in various micro and macronutrients, it serves as a reliable growth substrate for micro-algae, resulting in protein build-up due to nitrogen assimilation in the wastewater. These microalgae are fed as feedstock for larvae and bivalves in aquaculture [37]. Microalgae has significant potential in generating bioactive compounds such as lipids, pigments, polysaccharides, terpenoids, phenols, vitamins, and minerals apart from biofertilizers and biofuel [38].

Table 4. The efficiency of microalgae in wastewater treatment.

S. No.	Culture	Wastewater	Objective	Refs.
1.	*Rhodosporidium toruloides Tetradesmus obliquus*	Brewery wastewater	Lipid production	[39]
2.	*Coelastrella sp*	Distillery wastewater	Nutrient removal and lipid production	[40]
3.	*Chlamydomonas reinhardtii, Chlorella vulgaris, and Scenedesmus obliquus*	Brewery wastewater	Nutrient removal and lipid production	[41]
4.	*Chlorella vulgaris, Spiruna platensis, and Haematococcus pluvialis*	Membrane-treated distillery wastewater	Biomass production and nutrient removal	[37]
5.	*Chlorella emersonii*	Municipal wastewater Dairy and brewery wastewater	Biomass production and nutrient removal	[42]
6.	*Chlorella vulgaris*	Textile wastewater	Biodiesel production and nutrient removal	[43]
7.	*Chlorella sorokiniana*	Municipal wastewater	Biofuel and nutrient removal	[44]
8.	*Chlorella sorokiniana*	Winery wastewater	Nutrient removal	[45]
9.	*Auxenochlorella protothecoides*	Winery wastewater	Nutrient removal	[45]
10.	*Scenedesmus dimorphus*	Brewery wastewater	Nutrient removal	[46]
11.	*Desmodesmus sp.*	Piggery wastewater	Nutrient removal	[47]
12.	*Scenedesmus quadracauda*	Dairy wastewater	Nutrient removal	[48]

THE WAY FORWARD

Even though biological treatment is inexpensive and eco-friendly, it struggles to remove the micropollutants in nanogram levels, which insists on a different approach to treat these pollutants. Biological treatment can be coupled with other engineered systems to eliminate trace pollutants. Another challenge faced in the anaerobic process is the generation of hydrogen sulfide, which can be addressed by biotrickling filtration coupled with microaerobic desulfurization, which exhibits a high potential in the case of pilot and commercial scale treatment plants in removing hydrogen sulfide. The current treatment technology should change to transform wastewater remediation into wastewater mining so that the optimum extraction of the resources can be achieved.

CONCLUSION

Microorganism growth and efficiency were primarily influenced by the type, composition, characteristics, loading rate of wastewater, and its treatment process. Besides this, microbes can degrade a wide array of environmental pollutants. Consequently, biological methods are a feasible solution for the complex wastewater treatment system in which anaerobic treatment methods stand ahead because of lesser sludge generation, low energy requirement, and energy generation. In recent decades, aerobic and anaerobic systems have been clustered together to degrade the pollutants in wastewater. These integrated biological systems are anticipated to treat wastewater with high organic pollutant rates. Extensive research on these integrated systems is essential so that the advantages of this system and the limitations that exist in other biological systems can be addressed through feasible options.

REFERENCES

[1]   Cai W, Long F, Wang Y, Liu H, Guo K. Enhancement of microbiome management by machine learning for biological wastewater treatment. Microb Biotechnol 2021; 14(1): 59-62.
        [http://dx.doi.org/10.1111/1751-7915.13707] [PMID: 33222377]

[2]   Narayanan CM, Narayan V. Biological wastewater treatment and bioreactor design: a review. Sustain Environ Res 2019; 29(1): 33.
        [http://dx.doi.org/10.1186/s42834-019-0036-1]

[3]   Zhou H, Lian L, Yan S, Song W. Insights into the photo-induced formation of reactive intermediates from effluent organic matter: The role of chemical constituents. Water Res 2017; 112: 120-8.
        [http://dx.doi.org/10.1016/j.watres.2017.01.048] [PMID: 28153698]

[4]   Ward CP, Cory RM. Complete and partial photo-oxidation of dissolved organic matter draining permafrost soils. Environ Sci Technol 2016; 50(7): 3545-53.
        [http://dx.doi.org/10.1021/acs.est.5b05354] [PMID: 26910810]

[5]   Bradley PM, Battaglin WA, Clark JM, *et al.* Widespread occurrence and potential for biodegradation of bioactive contaminants in Congaree National Park, USA. Environ Toxicol Chem 2017; 36(11): 3045-56.
        [http://dx.doi.org/10.1002/ *etc.*3873] [PMID: 28636199]

[6] Tran NH, Reinhard M, Gin KYH. Occurrence and fate of emerging contaminants in municipal wastewater treatment plants from different geographical regions-a review. Water Res 2018; 133: 182-207.
[http://dx.doi.org/10.1016/j.watres.2017.12.029] [PMID: 29407700]

[7] Verlicchi P, Al Aukidy M, Zambello E. Occurrence of pharmaceutical compounds in urban wastewater: Removal, mass load and environmental risk after a secondary treatment—A review. Sci Total Environ 2012; 429: 123-55.
[http://dx.doi.org/10.1016/j.scitotenv.2012.04.028] [PMID: 22583809]

[8] Samer M. Biological and chemical wastewater treatment processes. Wastewater Treat Eng 2015; 150: 212.

[9] Anukam A, Mohammadi A, Naqvi M, Granström K. A review of the chemistry of anaerobic digestion: Methods of accelerating and optimizing process efficiency. Processes (Basel) 2019; 7(8): 504.
[http://dx.doi.org/10.3390/pr7080504]

[10] Wang S, Ma X, Wang Y, Du G, Tay JH, Li J. Piggery wastewater treatment by aerobic granular sludge: Granulation process and antibiotics and antibiotic-resistant bacteria removal and transport. Bioresour Technol 2019; 273: 350-7.
[http://dx.doi.org/10.1016/j.biortech.2018.11.023] [PMID: 30448688]

[11] Meng J, Li J, Li J, *et al.* Enhanced nitrogen removal from piggery wastewater with high NH_4^+ and low COD/TN ratio in a novel upflow microaerobic biofilm reactor. Bioresour Technol 2018; 249: 935-42.
[http://dx.doi.org/10.1016/j.biortech.2017.10.108] [PMID: 29145120]

[12] Bosak V, VanderZaag A, Crolla A, Kinsley C, Gordon R. Performance of a constructed wetland and pretreatment system receiving potato farm wash water. Water 2016; 8(5): 183.
[http://dx.doi.org/10.3390/w8050183]

[13] Mahmood T, Zhang J, Zhang G. Assessment of constructed wetland in nutrient reduction, in the commercial scale experiment ponds of freshwater prawn Macrobrachium rosenbergii. Bull Environ Contam Toxicol 2016; 96(3): 361-8.
[http://dx.doi.org/10.1007/s00128-015-1713-3] [PMID: 26679323]

[14] Gorra R, Freppaz M, Ambrosoli R, Zanini E. Seasonal performance of a constructed wetland for wastewater treatment in alpine environment 2007.

[15] De La Mora-Orozco C, González-Acuña IJ, Saucedo-Terán RA, Flores-López HE, Rubio-Arias HO, Ochoa-Rivero JM. Removing organic matter and nutrients from pig farm wastewater with a constructed wetland system. Int J Environ Res Public Health 2018; 15(5): 1031.
[http://dx.doi.org/10.3390/ijerph15051031] [PMID: 29883370]

[16] Ozturk A, Aygun A, Nas B. Application of sequencing batch biofilm reactor (SBBR) in dairy wastewater treatment. Korean J Chem Eng 2019; 36(2): 248-54.
[http://dx.doi.org/10.1007/s11814-018-0198-2]

[17] Tyagi VK, Liu J, Poh LS, Ng WJ. Anaerobic–aerobic system for beverage effluent treatment: Performance evaluation and microbial community dynamics. Bioresour Technol Rep 2019; 7: 100309.
[http://dx.doi.org/10.1016/j.biteb.2019.100309]

[18] Talaiekhozani A. A review on different aerobic and anaerobic treatment methods in dairy industry wastewater. Goli A, Shamiri A, Khosroyar S, Talaiekhozani A, Sanaye R, Azizi K A Rev Differ Aerob Anaerob Treat Methods Dairy Ind Wastewater. J Environ Treat Tech 2019; 7: 113-41.

[19] Zhu S, Wu H, Wu C, Qiu G, Feng C, Wei C. Structure and function of microbial community involved in a novel full-scale prefix oxic coking wastewater treatment O/H/O system. Water Res 2019; 164: 114963.
[http://dx.doi.org/10.1016/j.watres.2019.114963] [PMID: 31421512]

[20] Liu H, Huang Y, Duan W, Qiao C, Shen Q, Li R. Microbial community composition turnover and function in the mesophilic phase predetermine chicken manure composting efficiency. Bioresour

Technol 2020; 313: 123658.
[http://dx.doi.org/10.1016/j.biortech.2020.123658] [PMID: 32540690]

[21]   Fu G, Zhao L, Huangshen L, Wu J. Isolation and identification of a salt-tolerant aerobic denitrifying bacterial strain and its application to saline wastewater treatment in constructed wetlands. Bioresour Technol 2019; 290: 121725.
[http://dx.doi.org/10.1016/j.biortech.2019.121725] [PMID: 31301568]

[22]   Lim JX, Zhou Y, Vadivelu VM. Enhanced volatile fatty acid production and microbial population analysis in anaerobic treatment of high strength wastewater. J Water Process Eng 2020; 33: 101058.
[http://dx.doi.org/10.1016/j.jwpe.2019.101058]

[23]   Sen S, Dasu VV, Mahajan D. Potential application of Bacillus pseudofirmus SVB1 extract in effluent treatment. Biocatal Agric Biotechnol 2019; 20: 101250.
[http://dx.doi.org/10.1016/j.bcab.2019.101250]

[24]   Grelska A, Noszczyńska M. White rot fungi can be a promising tool for removal of bisphenol A, bisphenol S, and nonylphenol from wastewater. Environ Sci Pollut Res Int 2020; 27(32): 39958-76.
[http://dx.doi.org/10.1007/s11356-020-10382-2] [PMID: 32803603]

[25]   Chu R, Li S, Zhu L, *et al.* A review on co-cultivation of microalgae with filamentous fungi: Efficient harvesting, wastewater treatment and biofuel production. Renew Sustain Energy Rev 2021; 139: 110689.
[http://dx.doi.org/10.1016/j.rser.2020.110689]

[26]   Abedinzadeh M, Etesami H, Alikhani HA. Characterization of rhizosphere and endophytic bacteria from roots of maize (*Zea mays* L.) plant irrigated with wastewater with biotechnological potential in agriculture. Biotechnol Rep (Amst) 2019; 21: e00305.
[http://dx.doi.org/10.1016/j.btre.2019.e00305] [PMID: 30705833]

[27]   Zhao X, Guo M, Chen J, *et al.* Successional dynamics of microbial communities in response to concentration perturbation in constructed wetland system. Bioresour Technol 2022; 361: 127733.
[http://dx.doi.org/10.1016/j.biortech.2022.127733] [PMID: 35932946]

[28]   Wang J, Man Y, Ruan W, *et al.* The effect of rhizosphere and the plant species on the degradation of sulfonamides in model constructed wetlands treating synthetic domestic wastewater. Chemosphere 2022; 288(Pt 2): 132487.
[http://dx.doi.org/10.1016/j.chemosphere.2021.132487] [PMID: 34626651]

[29]   Suyamud B, Thiravetyan P, Gadd GM, Panyapinyopol B, Inthorn D. Bisphenol A removal from a plastic industry wastewater by *Dracaena sanderiana* endophytic bacteria and *Bacillus cereus* NI. Int J Phytoremediation 2020; 22(2): 167-75.
[http://dx.doi.org/10.1080/15226514.2019.1652563] [PMID: 31468977]

[30]   Wang S, Liu J, Li C, Chung BM. Efficiency of *Nannochloropsis oculata* and *Bacillus polymyxa* symbiotic composite at ammonium and phosphate removal from synthetic wastewater. Environ Technol 2019; 40(19): 2494-503.
[http://dx.doi.org/10.1080/09593330.2018.1444103] [PMID: 29466933]

[31]   Ji X, Li H, Zhang J, Saiyin H, Zheng Z. The collaborative effect of Chlorella vulgaris-Bacillus licheniformis consortia on the treatment of municipal water. J Hazard Mater 2019; 365: 483-93.
[http://dx.doi.org/10.1016/j.jhazmat.2018.11.039] [PMID: 30458425]

[32]   Kumar A, Priyadarshinee R, Singha S, *et al.* Biodegradation of alkali lignin by Bacillus flexus RMWW II: analyzing performance for abatement of rice mill wastewater. Water Sci Technol 2019; 80(9): 1623-32.
[http://dx.doi.org/10.2166/wst.2020.005] [PMID: 32039894]

[33]   Song H, Liu J. Forward osmosis membrane bioreactor using *Bacillus* and membrane distillation hybrid system for treating dairy wastewater. Environ Technol 2021; 42(12): 1943-54.
[http://dx.doi.org/10.1080/09593330.2019.1684568] [PMID: 31647375]

[34]   Xu S, Yao J, Ainiwaer M, Hong Y, Zhang Y. Analysis of bacterial community structure of activated sludge from wastewater treatment plants in winter 2018.
[http://dx.doi.org/10.1155/2018/8278970]

[35]   Zhu X, Lei C, Qi J, *et al.* The role of microbiome in carbon sequestration and environment security during wastewater treatment. Sci Total Environ 2022; 837: 155793.
[http://dx.doi.org/10.1016/j.scitotenv.2022.155793] [PMID: 35550899]

[36]   Liu X, Hong Y. Microalgae-based wastewater treatment and recovery with biomass and value-added products: a brief review. Curr Pollut Rep 2021; 7(2): 227-45.
[http://dx.doi.org/10.1007/s40726-021-00184-6]

[37]   Amenorfenyo DK, Li F, Zhang Y, Li C, Zhang N, Huang X. Effects of Microalgae Grown in Membrane Treated Distillery Wastewater as Diet on Growth and Survival Rate of Juvenile Pearl Oyster (Pinctada fucata martensii). Water 2022; 14(17): 2702.
[http://dx.doi.org/10.3390/w14172702]

[38]   Amenorfenyo DK, Huang X, Zhang Y, *et al.* Microalgae brewery wastewater treatment: potentials, benefits and the challenges. Int J Environ Res Public Health 2019; 16(11): 1910.
[http://dx.doi.org/10.3390/ijerph16111910] [PMID: 31151156]

[39]   Dias C, Gouveia L, Santos JAL, Reis A, da Silva TL. Rhodosporidium toruloides and Tetradesmus obliquus populations dynamics in symbiotic cultures, developed in brewery wastewater, for lipid production. Curr Microbiol 2022; 79(2): 40.
[http://dx.doi.org/10.1007/s00284-021-02683-7] [PMID: 34982231]

[40]   Vasistha S, Balakrishnan D, Manivannan A, Rai MP. Microalgae on distillery wastewater treatment for improved biodiesel production and cellulose nanofiber synthesis: A sustainable biorefinery approach. Chemosphere 2023; 315: 137666.
[http://dx.doi.org/10.1016/j.chemosphere.2022.137666] [PMID: 36586450]

[41]   Su Y, Zhu X, Zou R, Zhang Y. The interactions between microalgae and wastewater indigenous bacteria for treatment and valorization of brewery wastewater. Resour Conserv Recycling 2022; 182: 106341.
[http://dx.doi.org/10.1016/j.resconrec.2022.106341]

[42]   Schagerl M, Ludwig I, El-Sheekh M, Kornaros M, Ali SS. The efficiency of microalgae-based remediation as a green process for industrial wastewater treatment. Algal Res 2022; 66: 102775.
[http://dx.doi.org/10.1016/j.algal.2022.102775]

[43]   Javed F, Rehman F, Khan AU, Fazal T, Hafeez A, Rashid N. Real textile industrial wastewater treatment and biodiesel production using microalgae. Biomass Bioenergy 2022; 165: 106559.
[http://dx.doi.org/10.1016/j.biombioe.2022.106559]

[44]   Paddock MB, Fernández-Bayo JD, VanderGheynst JS. The effect of the microalgae-bacteria microbiome on wastewater treatment and biomass production. Appl Microbiol Biotechnol 2020; 104(2): 893-905.
[http://dx.doi.org/10.1007/s00253-019-10246-x] [PMID: 31828407]

[45]   Higgins BT, Gennity I, Fitzgerald PS, Ceballos SJ, Fiehn O, VanderGheynst JS. Algal–bacterial synergy in treatment of winery wastewater. npj Clean Water 2018; 1(1): 6.
[http://dx.doi.org/10.1038/s41545-018-0005-y]

[46]   Lutzu GA, Zhang W, Liu T. Feasibility of using brewery wastewater for biodiesel production and nutrient removal by *Scenedesmus dimorphus*. Environ Technol 2016; 37(12): 1568-81.
[http://dx.doi.org/10.1080/09593330.2015.1121292] [PMID: 26714635]

[47]   Luo L, Shao Y, Luo S, Zeng F, Tian G. Nutrient removal from piggery wastewater by Desmodesmus sp. CHX1 and its cultivation conditions optimization. Environ Technol 2018.
[PMID: 29513087]

[48]   Daneshvar E, Zarrinmehr MJ, Koutra E, Kornaros M, Farhadian O, Bhatnagar A. Sequential

cultivation of microalgae in raw and recycled dairy wastewater: Microalgal growth, wastewater treatment and biochemical composition. Bioresour Technol 2019; 273: 556-64.
[http://dx.doi.org/10.1016/j.biortech.2018.11.059] [PMID: 30476864]

CHAPTER 7

Role of Microbes and Microbiomes in Solid Waste Management for Ecosystem Restoration

Sathya Velusamy[1,*], Murugaiyan Sinduja[2], R. Vinothini[3] and Govindaraj Kamalam Dinesh[4,5,6]

[1] *Environmental Scientist, Tamil Nadu Pollution Control Board, Chennai, Tamil Nadu, India*

[2] *National Agro Foundation, Taramani, Chennai, Tamil Nadu, India*

[3] *MIT College of Agricultural and Technology, Musiri, Tamil Nadu, India*

[4] *Division of Environment Science, ICAR-Indian Agricultural Research Institute (IARI), New Delhi, India*

[5] *Division of Environmental Sciences, Department of Soil Science and Agricultural Chemistry, SRM College of Agricultural Sciences, SRM Institute of Science and Technology, Baburayanpettai-603201, Chengalpattu, Tamil Nadu, India*

[6] *INTI International University, Persiaran Perdana BBN, Putra Nilai, 71800 Negeri Sembilan, Malaysia*

Abstract: Solid waste disposal is a major issue that is getting worse every day as more people move into cities. Solid waste disposal in India and other developing nations frequently involves open dumping and incineration. This practice increases both health risks and already existing pollution issues. To solve this issue with the least amount of environmental impact possible, it is urgently necessary to employ sustainable techniques. There are many environmentally friendly ways to manage solid waste, including composting, vermicomposting, and anaerobic digestion, with additional benefits like generating byproducts. Among them, biological agents play a significant role in solid waste management. This chapter overviews solid waste management and a long-term solution using biological agents and microbes. The distinctive characteristics of microorganisms can be effectively used to revive the environment. Microorganisms can be used as "miracle cures" for biodegradation and remediation of contaminated sites. Today, microorganisms and nanotechnology are used in nano-bioremediation to clean up radioactive waste effectively. Additionally, using genetically modified organisms (GMOs) in severely polluted areas makes the microorganisms beneficial for human welfare and ecosystem restoration. Numerous environmental phenomena, both natural and man-made, depend on microorganisms for maintenance. They perform beneficial roles that improve and optimize human life. Waste management is one of

---

* **Corresponding author Sathya Velusamy:** Tamil Nadu Pollution Control Board, Chennai, Tamil Nadu, India; E-mail: sathyavelu1987@gmail.com

Shiv Prasad, Govindaraj Kamalam Dinesh, Murugaiyan Sinduja, Velusamy Sathya, Ramesh Poornima & Sangilidurai Karthika (Eds.)

these areas where microorganisms are being used. The proper disposal of the vast amounts of waste that people produce throughout the course of a day presents a significant challenge, one that the government and environmental organizations are constantly looking for new solutions to. Utilizing microorganisms is a crucial component of effectively combating this threat. It discusses the numerous functions of microorganisms in the environment, such as the management of solid waste, and it concludes by highlighting some recent developments in microbiological solid waste management.

Keywords: Contamination, Environment, Microbes, Solid waste management.

INTRODUCTION

India is facing a challenge in waste management due to the tremendous quantity of waste products produced due to urbanization and industrialization. The commercial and industrial sectors benefit the country's GDP while it questions waste disposal and management. Around 377 million people in 7,935 towns and cities generate 62 million tonnes of municipal solid waste (MSW) per year. Out of this amount, only 43 million tonnes (MT) is collected. Out of which, only 11.9 MT is treated, and the untreated 31 MT is dumped in landfill sites [1]. One of the essential services provided by municipal authorities to keep the nation clean is solid waste management (SWM). Appropriate waste segregation at the source is a prerequisite for effective waste management. Depending on the type of waste materials, the segregated waste should be taken further for recycling or resource recovery. The final residue should be appropriately managed and disposed of in sanitary landfills [2]. After proper treatment and disposal of other waste products that cannot be recycled or reused, sanitary landfills are the only remaining option for disposing of municipal solid waste. This waste management option is severely constrained by the cost of transportation [3]. The Environment Protection Act of 1986 established all statutory guidelines for the disposal of generated waste and other management standards as waste generation increased day by day.

Special wastes are subject to different regulations and compliances in areas like the type of authorization, record-keeping, and disposal methods, among others. Although the legal protections are sufficient, all municipalities indiscriminately dump their waste at locations on the edge or inside cities. Therefore, according to experts, India's waste management and disposal system is ineffective [3]. Microorganisms have solved numerous issues that humanity has faced in maintaining environmental quality. They have been successfully applied in several fields, including treating municipal and industrial waste, genetic engineering, environmental protection, and human and animal health. Using microorganisms, feasible and affordable responses that were previously impossible to achieve using chemical or physical engineering techniques are now possible [4]. Microbial technologies have also been successfully used to address a variety of environmental problems, particularly those related to waste management [5].

SOLID WASTE MANAGEMENT SCENARIO IN INDIA

The waste generation rate has exponentially increased in the last ten years due to the accelerated economy and rapid population growth. In urban India, 68.8 million tonnes of municipal solid waste are produced annually, or about 1,88,500 tonnes per day [6]. However, only 24% of this massive waste is processed correctly, treated, and disposed of. Even though there are scientifically sound techniques such as composting, incineration, and landfilling, open disposal is a popular method since it is the least expensive and easy to adopt [7]. Open dumping is followed by waste landfilling as the primary method of waste treatment and disposal, but the need for larger areas restricts solid waste disposal, particularly in larger cities [8]. Approximately the combined area of the three most populous Indian cities, 1,400 sq. km. of land, will be needed for landfilling solid waste in India by 2047.

Furthermore, the high amounts of hazardous secondary pollutants they generate, such as smells, leachate, and greenhouse gases, restrict the use of landfills for waste treatment. This circumstance points to the need for a practical procedure, such as MSW pretreatment, which is carried out to homogenize the trash and make it easier for waste to be treated using particular biological technologies before being dumped. Despite having access to a multitude of knowledge and resources, municipalities and local governments are nevertheless under considerable pressure to develop a waste treatment technology that is both inexpensive and sustainable. The most comprehensive process that can be used to manage these wastes is composting, especially in the case of Indian genera, where 50–60% of the MSW (C/N ratio 23) collected is biodegradable. Additionally, composting has several environmental advantages over landfilling organic waste. In addition to decreased greenhouse gas emissions from landfills, soil properties like texture, porosity, organic matter, and NPK availability for agricultural applications have improved. In order to improve the waste recycling system, we must comprehend the technical details of this natural, wet waste recycling process and the effects of any operational changes.

ROLE OF MICROORGANISMS IN THE ENVIRONMENT

To transform environmental natural resources into more palatable forms for consumption, there are several anthropogenic activities, including chemical synthesis. Man also causes pollution issues because of the production of goods. The better way to manage the waste generated and discarded in the environment is to integrate it back into it. That strategy makes use of microorganisms, most frequently yeast, bacteria, or fungi. These microbes or their byproducts are added to the substrates that result in the production of the desired industrial products, such as biocatalysts, biomass fuel production, biomonitoring, bioleaching (biomining), bio-detergent, biotreatment of pulp, biotreatment of wastes (bioremediation), biofiltration, aquaculture treatments, textile biotreatment, and so on [9]. As participants in the carbon and nitrogen cycles and performing other crucial tasks like recycling the remains of dead organisms and waste materials through decomposition, microorganisms are also essential to humans and the environment [10]. In addition, microorganisms play a crucial role as symbionts in most higher-order multicellular organisms.

Microbes in Solid Waste Management

Utilizing cutting-edge scientific methods and equipment, microbial biotechnology in solid waste utilizes a range of microorganisms under regulated circumstances without disrupting the ecology. The most popular and successful WM techniques are composting, biodegradation, bioremediation, and biotransformation. Numerous microorganisms, including *Bacillus sp., Corynebacterium sp., Staphylococcus sp., Streptococcus sp., Scenedesmus platydiscus, S. quadricauda, S. capricornutum, Chlorella vulgaris, etc.*, have been successfully used for solid waste [11].

SOLID WASTES

Solid waste refers to all discarded waste, including sectors such as household, commercial, non-hazardous institutions and industries, street sweeping, construction, agriculture, and other non-hazardous and non-toxic solid waste.

Sources and Types of Wastes

Waste is unwanted substances that are discarded by human society. These include urban, industrial, and agricultural wastes, which contain a fair amount of organic waste that can be used to enhance microbial processes to recover valuable resources. Wastes generated through human activities include:

Municipal wastes: Municipal trash is the most common and substantial contributor. It is the garbage that people and households generate as they go about their daily lives. Any society can be a significant source, although modern urban cultures provide significantly more than rural societies. The output of urban waste is immense, and the collection and disposal procedure is costly and time-consuming. Municipal garbage is made up of both liquid and solid waste. Solid waste includes both homogeneous and heterogeneous wastes generated in urban and periurban areas. Paper, glass, and metal can all be recycled, but they must be separated before or after collection. After decomposing, the other organic leftovers can efficiently enhance soil nutrients. However, there will always be some parts that cannot be recycled or decomposed, which can be used to generate energy but will require special disposal or treatment methods.

Industrial wastes: Sludges, product residues, kiln dust, slags, and ashes are examples of industrial waste. The three industries that generate the greatest industrial waste are metallurgy, non-metallurgy, and food processing. Waste can be classified into three types based on the raw materials used, the manufacturing processes, and the product outlets. Not all trash is the same; it may contain inorganic fractions, organic fractions, biodegradable fractions, and nonbiodegradable compounds, which may be recyclable, among other characteristics. The primary problem for industrial waste treatment is the proper handling of the liquid waste produced. Liquid waste can be alkaline or acidic, with organic and inorganic portions that are dissolved, suspended, or inseparable.

Agricultural wastes: Agricultural wastes are plant materials that are left over after crops have been collected or processed. They include a high concentration of cellulose (35%-50%), hemicelluloses (15%-35%), and lignin (15%-25%). They are thought to be inexpensive, plentiful, and widely available. The most promising carbon source for microbial activities is lignocellulosic biomass. Agricultural waste may be converted into sustainable energy using renewable fuels. Animal manure and other agricultural wastes must not create new pathogen and disease transmission pathways between animals, humans, and the environment.

Mining and quarrying wastes: Mining and quarrying can be extremely damaging to the environment. They have a direct influence on the countryside by leaving pits and heaps of garbage. Sulfur dioxide and other pollutants from the extraction procedures can also contaminate the air and water, putting wildlife and local inhabitants at risk. Trash from mining and quarrying accounts for 15% of total trash in Western Europe and 31% in Eastern Europe. Mining and quarrying wastes are a source of pollution in the environment. To avoid this, this waste must be put to use. Mining and quarrying can be extremely harmful to the environment. They have a direct impact on the countryside by leaving pits and heaps of waste debris. The extraction operations can also pollute the air and water with sulfur dioxide and other pollutants, putting wildlife and local populations in dangerA variety of factors influence the waste treatment and disposal issues that emerging and developed countries face. The scale of urbanization and industrialization, intensive agriculture, community economic position, and other factors have a significant impact on the type and volume of trash generated, as well as waste treatment alternatives [12].

SYNOPSIS OF SOLID WASTE MANAGEMENT

Public opinion on solid waste management will likely be either burning or burying in landfills. Though these landfills and incineration are a vital part of the waste management processes, certain other elements are involved in the integrated solid waste management (ISWM) approach, which aims to reduce the quantity and presence of toxic elements in solid waste [13]. Solid waste treatment and disposal methods will be selected based on waste volume, type, and nature. The important treatment methods and disposal options for the management of solid waste are as follows:

Thermal Treatment

Thermal treatment is when heat is used to degrade or treat solid waste materials. Some of the standard thermal treatment techniques are discussed below [14].

- *Incineration* involves combusting waste material with oxygen, producing energy recovery for electricity or heating. The significant advantages are the volume reduction of waste materials and the decrease in greenhouse gas production.
- *Gasification and pyrolysis* are similar processes in which biodegradable waste has been decomposed in the presence of low oxygen and high temperature. Pyrolysis is the process of complete absence of oxygen, whereas gasification is carried out under low oxygen.
- *Open Burning* is a commonly followed method worldwide since it is a less expensive solution. However, it creates severe environmental hazards by releasing toxic elements such as hexachlorobenzene, carbon monoxide, particulate matter, polycyclic aromatic compounds, and ash.

Dumps and Landfills

Different types of dumps and landfill methods have been followed for dumping and managing solid waste; some are discussed below [15].

Sanitary landfills are a common solution for the solid waste disposal problem. The residuals or unutilized or immobile materials obtained after the treatments, such as composting and energy conversion, are dumped in the engineered landfills to avoid environmental pollution.

Controlled dumps are relatively the same as sanitary landfills but may lack one or two criteria of sanitary landfills. Possibly, there is no or incomplete gas management, lack of record maintenance, or irregular cover.

Bioreactor landfills are the outcome of recent research. Microbes are used in this method to speed up the degradation of the organic portion of the waste, but the limiting factor is the requirement of moisture for microbial sustenance and digestion.

BIOLOGICAL TREATMENT

Composting

is one of the commonly followed methods in which the organic waste is decomposed aerobically by the action of microorganisms and earthworms. The essential types of composting are static pile composting, windrow composting, in-vessel composting, vermicomposting, *etc.*

Role of Microbes in Composting of Solid Waste

Composting is a biological method that decomposes organic matter and converts it into a stable humus-like compound. Compost is a dark brown, crumby material produced from decomposing organic waste from plants and animals. The final product will be delicate organic matter and humus [16]. The biodegradable waste produced in the kitchen and agriculture can be composted and used as organic manure for agriculture farms. This compost will help raise the level of soil organic matter, which is beneficial for the environment. The compost will have good fertilizer value, and it is an excellent soil conditioner that improves organic content, soil texture, permeability, and moisture-holding capacity of the soil. The municipal solid waste produced in India generally consists of 35%–40% organic matter. The waste can be recycled through composting, which is the traditional method of waste disposal [17]. It is also a process in which the decomposition of organic waste occurs naturally and finally yields nutrient-rich organic manure. It is a biological method mainly mediated by microorganisms such as fungi and bacteria, which convert degradable organic waste into humus-like substances. Like the finished product, the soil is rich in carbon and nitrogen and is an excellent soil conditioning and growing medium for plants [18].

Composting Process

There are two essential types of composting: aerobic and anaerobic. Aerobic composting is the process in which organic waste is degraded in the presence of oxygen, and the end products are CO_2, NH_3, water, and heat. This process can be used to manage any organic waste, but it requires the right substrate and circumstances, such as a moisture content of 60-70% and a carbon-to-nitrogen ratio (C/N) of 30/1. If there are any major deviations, the degrading process will be halted [19]. Generally, wood and paper act as significant sources of carbon, while sewage sludge and food waste supply nitrogen. Proper aeration and adequate oxygen supply are the most critical factors to be considered throughout the composting process. Aerobic microorganisms administrate aerobic composting under oxygen during the composting period [20]. Aerobic composting is commonly illustrated with essential conditions such as high temperature, no foul odor, and a short decomposition period [21].

Anaerobic composting is generally characterized by the lack of oxygen, and the end products are methane, carbon dioxide, ammonia, a trace amount of other gases and organic acids. It is mainly used to decompose animal and human waste but is now popularly used for municipal solid waste (SWM) and green waste [22].

Stages of the Composting Process

There are three main stages to composting, and *Phase 1* is the mesophilic growth stage, characterized by temperatures between 25 and 40 °C and bacterial growth. In phase *2*, the thermophilic stage, cellulose, lignin, and other resistant materials are broken down by bacteria, fungi, and actinomycetes (first-level consumers) when temperatures are between 50 and 60°C. To ensure pathogens and contaminants are destroyed, it is crucial to maintain the temperature at this level for at least one day during the thermophilic stage, where the upper limit can reach 70°C. The particulate matter will be disintegrated into carbohydrates, lipids, and proteins and then hydrolyzed into short-chained carbohydrates, long-chain fatty acids, and amino acids by enzymes such as protease, lipases, cellulase, and amylase produced during the process [23]. Chemical decomposition during the thermophilic phase is shown in Fig. (**1**).

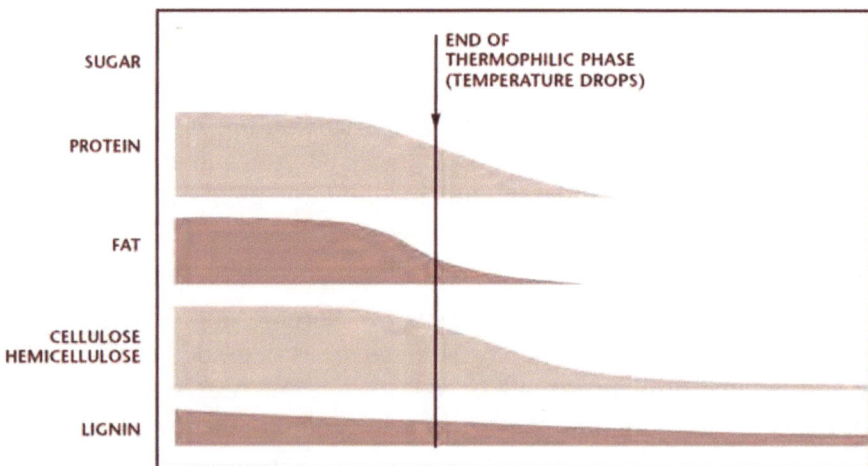

Fig. (1). Chemical decomposition during the thermophilic phase [24].

Phase 3 is the maturation stage, during which temperatures stabilize [25] and some fermentation happens. Nitrifying reactions turn the material into humus (Fig. 1). The goal is to create a stable material, which can be determined by the carbon-to-nitrogen (C/N) ratio; a well-composed material has a low C/N ratio. For instance, windrow material has a C/N of 15, whereas untreated new organic waste has a C/N of 30. The temperature range for different phases of composting is shown in Table 1.

Table 1. Temperatures range for different phases of composting.

Phases	Temperature Range	Process
Mesophilic phase	40°C	Decomposition of soluble sugars and production of organic acid
Thermophilic	>40°C	Degradation of complex organic matter such as cellulose and lignin.
Mesophilic Phase II	10-40°C	Degradation of polymers
Maturation Phase	20-30°C	Formulation of fulvic and humic acids

Aerobic vs Anaerobic Microorganisms

Aerobic organisms flourish in an environment with 5% or greater oxygen levels (The atmosphere contains about 21% oxygen). Generally, aerobic microbes are the most desired option since they decompose the materials more effectively and rapidly. If the oxygen is deficient, anaerobic microbes will flourish in the compost pile. However, the anaerobic condition is not preferred due to the lousy odor during decomposition. Aerobic microbes are an essential originator of decomposition, raising the compost pile's temperature. A variety of microorganisms flourished between the temperatures of 55°-155°F. Initially, the temperature of the compost pile depends on the atmospheric temperature. If the temperature drops below 70°F, no beneficial microbes will flourish. If the temperature goes above 140°F, most of the pathogens and weeds will be killed, and the most effective microbes will flourish at the temperature range of 70°-100°F. The variations in the temperature mainly depend on the substrate materials, method of composting, moisture, *etc.* If the compost pile temperature is between 90°-140°F, the composting process will be rapid in nature. In order to avoid pathogenic organisms and weed seeds, the best management option should be followed during the composting process [26]. The composition of microbes in municipal solid waste compost is shown in Fig. (**2**).

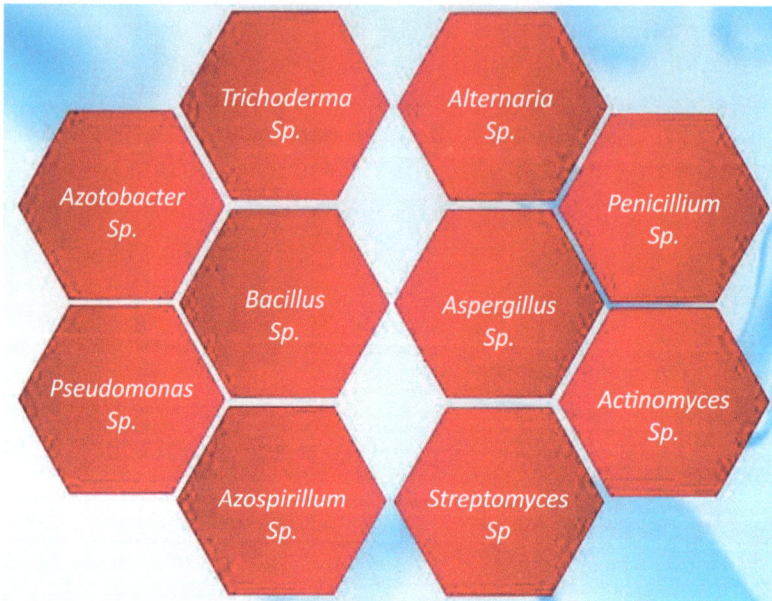

Fig. (2). Composition of microbes in municipal solid waste compost [27].

The significant challenges in the composting process are high variation in waste composition, protracted residence time, temperature sensitivity, and odor. The lignocellulose composition in agricultural waste is high and hard to decompose/degrade. Therefore, it may require pretreatment with acid and heat, but the energy consumption and its impact on the environment are the drawbacks of such pretreatment methods [28]. There is no universal method for composting since the composition, physio-chemical properties, and substrate waste differ from region to region, and these factors significantly determine the composting process. Co-composting is an integrated sustainable process that offers some advantages over composting. Recent studies have aimed at co-composting using different types of agricultural waste. Fig. (**3**) shows the microorganisms associated with the composting process.

Advantages of Composting

- Enhances soil health and nutrient status
- Improves water holding capacity of soil
- Less expensive and low capital cost
- Appropriate for the degradation of organic fractions and lignocellulose materials.
- Technically proven for larger scale
- Creates fertilizer value and saves landfill area

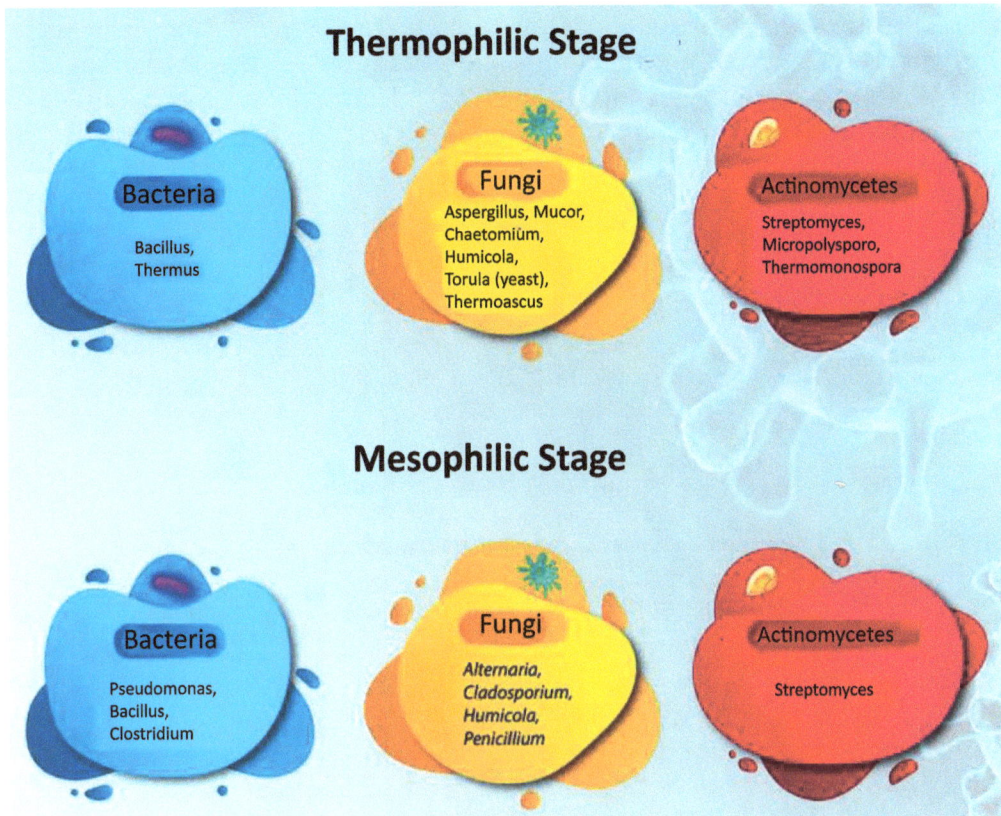

Fig. (3). Microorganisms associated with the composting process [29].

Limitations of Composting

- Odor nuisances during the composting process
- Chances for the spread of pathogenic and allergenic microorganisms
- Possibility for heavy metal pollution of soil if the compost materials contain high amount of heavy metals
- Groundwater pollution from runoff water is possible if impermeable materials do not cover the surface.

Anaerobic Digestion

Anaerobic digestion is a biological process that decomposes organic materials under an oxygen-free environment where aerobic composting should have oxygen for decomposition (Fig. **4**). It has major advantages such as the decrease in the emission of greenhouse gases and renewable energy generation in the form of biogas. The organic waste is degraded by a microbial consortium, resulting in methane-rich biogas, the alternative to natural gas. The liquid waste after this process can be used as organic manure based on the nutrient value. The energy conversion efficiency of anaerobic digestion is low if waste material has more lignocellulosic composition, which is hard to biodegrade by microbes [30]. At the same time, the potential of lignocellulosic biomass in yielding biogas is high compared to other feedstocks.

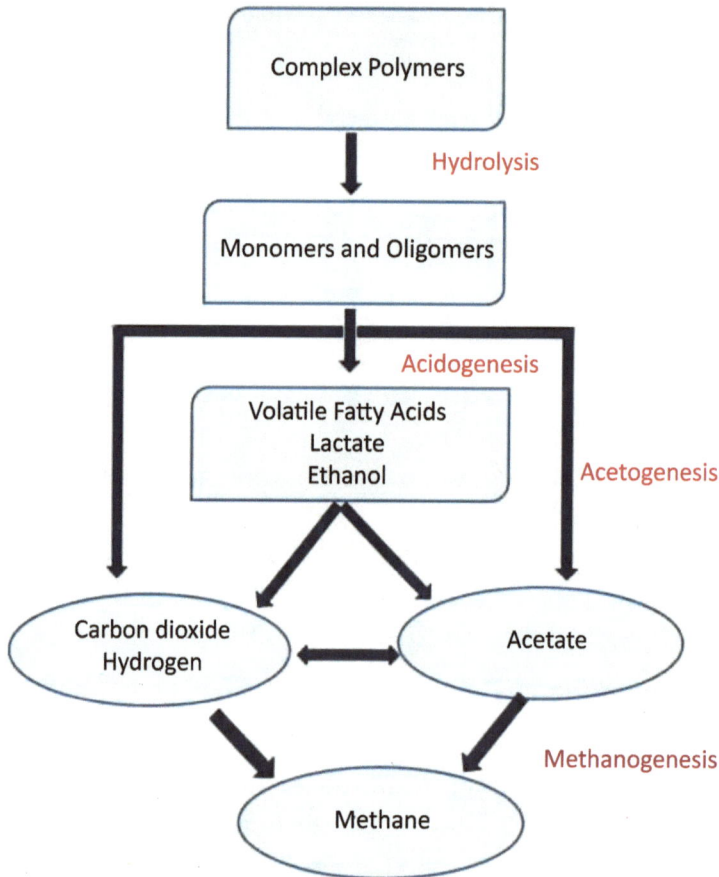

Fig. (4). Stages of anaerobic digestion.

Anaerobic digestion includes hydrolysis, acidogenesis, acetogenesis, and methanogenesis. Researchers mentioned that out of these four steps, hydrolysis of lignocelluloses is a rate-limiting step for anaerobic digestion [31]. The effectiveness of biogas production is low when there is no additional substrate during anaerobic digestion. There should be an improvement in the implementation of anaerobic digestion to improve the hydrolysis of lignocelluloses [32]. The use of microorganisms in anaerobic digestion would substantially increase enzyme production, and further, the degradation of lignocellulosic materials will be improved. White rot fungi are commonly used microbes for biofuel production [33]. Generally, fungi efficiently degrade lignin, whereas bacteria act on hemicellulose degradation.

However, the economic feasibility of fungal use is low because of the loss of polysaccharide components during the growth of fungi. Instead of using a single bacterial strain, a consortium is more efficient in biogas production. It allows a higher rate of hydrolysis in the mesophilic stage and enhances energy use efficiency [34]. It also adapts to pH and temperature changes while showing a higher resistance to toxic elements and contaminants. The decomposition of lignocelluloses is taken up by various microorganisms, and cellulolytic and hemicellulolytic enzymes are produced by them. Therefore, mixed strains are a better option than single strains for anaerobic digestion. Regular bioaugmentation of cellulolytic and hemicellulolytic microbes may gear up the process and increase biogas production by 15% or more [35]. Microbial activity during the anaerobic digestion of municipal solid waste is given in Table **2**.

Table 2. Microbial activity during the anaerobic digestion of municipal solid waste.

Acetate-oxidizing Bacteria	Microbial Description	Hydrogenotrophic Methanogens
AOR	Anaerobic, rod-shaped, gram-positive, and thermophilic (60 °C)	*Methanobacterium sp.* strain
Clostridium ultunense	Mesophilic (37 °C), spore-forming, gram-negative, and anaerobic	*Methanoculleus sp.* strain
Thermacetogenium phaeum	Thermophilic (between 55 and 58 °C), anaerobic, rod-shaped, gram-negative but with gram-positive bacteria	*Methanothermobacter thermoautotrophicus*
Thermotoga lettingae	Rod-shaped, gram-negative, nonspore, anaerobic, and thermophilic (65 °C)	*Methanothermobacter thermoautotrophicus or Thermodesulfovibrio yellowstonii*
Syntrophaceticusschinkii	spore-forming, mesophilic (between 25 and 40 °C), anaerobic, and having a variety of cell shapes	*Methanoculleus sp.* strain

Advantages and Disadvantages of Anaerobic Digestion (AD)

AD process is a practical way to treat various organic waste to bioenergy. Sewage sludge, animal dung, and solid organic waste from municipalities and industrial sources could be digested into valuable resources through biological reactions under anaerobic conditions [36]. The advantages and disadvantages of the AD process are listed in Table **3**.

Table 3. Advantages and disadvantages of anaerobic digestion.

Advantages	Disadvantages
Generation of the stable biogas process	Great care and expertise are required for the design and construction if it is operated on a large scale.
The emission of greenhouse gas reduced by means of methane recovery	The reuse of the energy recovered from anaerobic digestion should be established in proper scientific ways
Different organic waste and wastewater can be treated in a combined manner.	Methanogens are highly sensitive to a variety of chemical compounds.
The volume of solid portions of the waste may be reduced	Odor problems due to the production of sulfur compounds
Removal of pathogens from waste	Less biogas production during winter

EFFECTIVE MANAGEMENT OF SOLID WASTE

Pollution of the environment is a global problem. The relationship between people and the environment is not something new. It is common knowledge that urbanization and industrialization have caused a steady decline in the state of nature. Ambient air and water had been contaminated by urbanization and industrial growth. Urban life and industry both produce much unsafe waste. Researchers have studied how changes in four crucial areas, population, energy, industrialization, and urbanization, have affected the planet. Most researchers supported a sustained education and awareness program as the best way to ensure engaged citizen participation [37]. According to The Down to Earth, the institutional flaws, the underproductive management staff, and the poor financial standing of local authorities are to blame for India's waste clearance system's shortcomings. The inevitable results of ineffective waste treatment and management techniques are waste and its environmental effects. Given the increasing global importance of environmental concerns, waste treatment methods, systems, and processes must be environmentally sustainable. Different practices, strategies, objectives, controls, monitoring, and regulations of the financial and marketing aspects of production, environmental assessments of various treatments, evaluations, and policies can all be used to define waste management. To address the problems brought on by trash, its management, and various treatments, these variables should be integrated into a holistic strategy [38].

Current Management Practices of Urban Solid Wastes

Rising populations and rapid urbanization in developing countries led to the dramatic release of enormous amounts of urban solid wastes. A significant source of biomass is urban solid waste. More solid waste is produced in countries with higher financial success and metropolitan populations [39]. Planning, organizing, implementing, and weighing economic, esthetic, power, and preservation for future generations are all included in urban solid waste management [13]. This includes the processes involved in waste management, including creation, development, storage, collection, transportation, handling, and environmental disposal. The biological process of using microorganisms to degrade the biodegradable disposal components is the more sustainable way of disposal. Anaerobic and aerobic biological processes are the two types that exist. Waste management using heat process technology includes pyrolysis, gasification, combustion, and incineration. Waste may be converted into heat and electricity, making it a more sustainable energy source. In certain populated nations worldwide, energy recovery from MSW is gaining momentum.

Role of the Microbes and Microbiomes in Solid Waste Management

The oxidation or decomposition of different microorganisms converts organic chemicals (OC) into simpler and more stable end products. A paper claims that some microorganisms, including mesophilic bacteria, actinomycetes, fungi, and protozoa, may colonize a mound of biodegradable solid waste [40]. These microbes can flourish between 10 and 45°C and efficiently break down biodegradable substances. The thermophilic or active phase of composting can extend for many weeks. The thermophilic phase is when the majority of the OM is broken down [41]. Microorganisms use pollutants for growth, feeding, and reproduction. This is the main reason why different organic pollutants are transformed by microbes. Microbes use OC to produce C. Because it serves as the building block for new cells, C is essential for bacteria. Another type of energy used by microbes is C [42].

Recent Advances in Microbial Waste Management

Understanding the effects of soil erosion, unwanted sediment migration, chemical fertilizers and pesticides, and improper handling of human and animal wastes on the environment has an impact on the review of recent scientific advances in the practical use of microbes in environmental management and biotechnology. Around the world, these negative phenomena have led to significant environmental and social issues, for which we must look outside of traditional physical and chemical technologies for solutions [43]. Notably, developments in biological and biotechnological tools have made sustainable environmental cleanup procedures more applicable to waste management systems. Examples of such biotechnological biodegradation methods include bioremediation, biostimulation, bioaugmentation, phytoremediation, and other related methods.

Organic material undergoes biodegradation, converting it into nutrients other organisms can utilize. In the presence of ideal environmental conditions, sufficient nutrients, and microorganisms and their products, bioremediation uses contaminants, including hazardous substances, to be deconstructed. Utilizing microorganisms, bioremediation technology reduces, eliminates, contains, or transforms harmful contaminants found in soils, sediments, water, and air into benign ones. By altering the environment, existing bacteria that can perform bioremediation are prompted. Adding living, degradable microbial cells to the environment to supplement the native populations is known as bioaugmentation [5]. The microbes employed are referred to as bioremediators. Microbes may take months or years to clean up a site, depending on a number of factors, including high contaminant concentrations, pollutant-trapped areas, or the contaminated location.

Mycoremediation is a different kind of bioremediation. To decontaminate a space, fungi are used [44]. Its mycelium secretes an extracellular enzyme as well as acids that break down lignin and cellulose. When there is not enough leachate, liquid waste, such as sewage sludge, is used. Anaerobic, aerobic, or hybrid bioreactors are those that combine both aerobic and anaerobic processes. All three methods entail reintroducing collected leachate and water to keep the landfill moist. To reduce harmful emissions, the microorganisms responsible for decomposition are thus encouraged to decompose more quickly. Vertical or horizontal systems of pipes are used to pump air into the landfill in aerobic bioreactors. The amount of volatile organic compounds, leachate toxicity, and methane are reduced while decomposition is accelerated in an aerobic environment. Compared to conventional landfills, anaerobic bioreactors with leachate circulation produce methane at a higher and faster rate [45].

FUTURE PERSPECTIVE

As microbes can survive in harsh environments, there are many opportunities for their identification and classification, which may help find solutions to various human problems. Microbes produce unidentified products in large quantities. Therefore, determining beneficial microbial strains from different sources, such as municipal waste, is extremely important. It is well known that waste materials and soil are full of microorganisms that can produce antibiotics and enzymes, which can improve human health and make life easier.

- The management of municipal waste streams can be done in a variety of ways. These include biological, chemical, and physical techniques. Many times, conventional waste management techniques have some unfavorable effects.
- Scientific advancements in the use of microorganisms for environmental

management have lately been accomplished using hybrid applications, which integrate microbial techniques with physical and chemical ones. Vermiculture technologies, anaerobic digestion, and bioreactor landfills are a few examples of these.

- The following are practical recommendations from this discussion, given the advantages of using microorganisms in waste management in terms of cost and the environment, as well as the biodegradable nature of the vast amounts of waste produced by nations.
- Waste should be separated at the source to enable more effective and efficient waste management.
- Techniques for managing microbiological waste should be created and applied for both environmental remediation and their value-added benefits.
- Lastly, garbage collection systems need to be improved for more hygienic and sustainable environmental conditions, particularly in and around inhabited regions of the communities.

CONCLUSION

Microorganisms and their role in the degradation of waste materials have been proven through several research studies. Therefore, applying microbial consortia in composting and anaerobic digestion would be better for solid waste management than chemical and thermal methods. Besides efficiency, chemical and thermal methods involve high costs and energy consumption. Co-composting also provides other benefits, such as enhanced degradation and minimized loss of nutrients through valorization during the composting process. In anaerobic digestion, microorganisms and their activity also improve the biogas yield potential and efficiency. Hence, using microbes for municipal solid waste management is promising and paves the way for a sustainable solution to the environmental pollution problem arising from municipal solid waste.

ACKNOWLEDGEMENTS

The authors thank the Division of Environmental Science, ICAR-*Indian Agricultural Research Institute, New Delhi, India*, for their continual encouragement and unflinching support.

REFERENCES

[1]   Jurado MM, Camelo-Castillo AJ, Suárez-Estrella F, *et al.* Integral approach using bacterial microbiome to stabilize municipal solid waste. J Environ Manage 2020; 265: 110528.
[http://dx.doi.org/10.1016/j.jenvman.2020.110528] [PMID: 32421558]

[2]   Chen W, Jiao S, Li Q, Du N. Dispersal limitation relative to environmental filtering governs the vertical small□scale assembly of soil microbiomes during restoration. J Appl Ecol 2020; 57(2): 402-12.
[http://dx.doi.org/10.1111/1365-2664.13533]

[3]   Liu Q, Zhang Q, Jarvie S, *et al.* Ecosystem restoration through aerial seeding: Interacting plant–soil microbiome effects on soil multifunctionality. Land Degrad Dev 2021; 32(18): 5334-47.
[http://dx.doi.org/10.1002/ldr.4112]

[4]   Franco-Duarte R, Černáková L, Kadam S, *et al.* Advances in chemical and biological methods to identify microorganisms—from past to present. Microorganisms 2019; 7(5): 130.
[http://dx.doi.org/10.3390/microorganisms7050130] [PMID: 31086084]

[5]   Rastogi A, Zang X, Sunkara S, Gupta R, Khaitan P. Towards scalable multi-domain conversational agents: The schema-guided dialogue dataset. Proceedings of the AAAI Conference on Artificial Intelligence. 8689-96.
[http://dx.doi.org/10.1609/aaai.v34i05.6394]

[6]   Nandan A, Yadav BP, Baksi S, Bose D. Recent scenario of solid waste management in India. World Sci News 2017; (66): 56-74.

[7]   Khandelwal H, Thalla AK, Kumar S, Kumar R. Life cycle assessment of municipal solid waste management options for India. Bioresour Technol 2019; 288: 121515.
[http://dx.doi.org/10.1016/j.biortech.2019.121515] [PMID: 31125936]

[8]   Sharma BK, Chandel MK. Life cycle cost analysis of municipal solid waste management scenarios for Mumbai, India. Waste Manag 2021; 124: 293-302.
[http://dx.doi.org/10.1016/j.wasman.2021.02.002] [PMID: 33640669]

[9]   Bhat RA, Dar SA, Dar DA, Dar GH. Municipal solid waste generation and current scenario of its management in India. Int J Adv Res Sci Eng 2018; 7(2): 419-31.

[10]   Joshi R, Ahmed S. Status and challenges of municipal solid waste management in India: A review. Cogent Environ Sci 2016; 2(1): 1139434.
[http://dx.doi.org/10.1080/23311843.2016.1139434]

[11]   Mondal S, Wijewardena KP, Karuppuswami S, Kriti N, Kumar D, Chahal P. Blockchain inspired RFID-based information architecture for food supply chain. IEEE Internet Things J 2019; 6(3): 5803-13.
[http://dx.doi.org/10.1109/JIOT.2019.2907658]

[12]   Ganguly RK, Chakraborty SK. Integrated approach in municipal solid waste management in COVID-19 pandemic: Perspectives of a developing country like India in a global scenario. Case Studies in Chemical and Environmental Engineering 2021; 3: 100087.
[http://dx.doi.org/10.1016/j.cscee.2021.100087]

[13]   Abdel-Shafy HI, Mansour MSM. Solid waste issue: Sources, composition, disposal, recycling, and valorization. Egyptian Journal of Petroleum 2018; 27(4): 1275-90.
[http://dx.doi.org/10.1016/j.ejpe.2018.07.003]

[14]   Moustakas K, Loizidou M. Solid waste management through the application of thermal methods. In: Kumar S (Ed) Waste Management InTech Open. 2010; pp. 89-124.

[15]   Dixit A, Singh D, Shukla SK. Changing scenario of municipal solid waste management in Kanpur city, India. J Mater Cycles Waste Manag 2022; 24(5): 1648-62.
[http://dx.doi.org/10.1007/s10163-022-01427-4]

[16]   Meena AL, Karwal M, Dutta D, Mishra RP. Composting: phases and factors responsible for efficient and improved composting. Agric Food e-Newsletter. 2021; 1: 85–90.

[17]   Ayilara M, Olanrewaju O, Babalola O, Odeyemi O. Waste management through composting: Challenges and potentials. Sustainability (Basel) 2020; 12(11): 4456.
[http://dx.doi.org/10.3390/su12114456]

[18]   Brady NC, Weil RR, Weil RR. The nature and properties of soils. NJ: Prentice Hall Upper Saddle River 2008; Vol. 13.

[19]   Pujara Y, Pathak P, Sharma A, Govani J. Review on Indian Municipal Solid Waste Management

practices for reduction of environmental impacts to achieve sustainable development goals. J Environ Manage 2019; 248: 109238.
[http://dx.doi.org/10.1016/j.jenvman.2019.07.009] [PMID: 31319199]

[20]   Krishna RS, Mishra J, Meher S, Das SK, Mustakim SM, Singh SK. Industrial solid waste management through sustainable green technology: Case study insights from steel and mining industry in Keonjhar, India. Mater Today Proc 2020; 33: 5243-9.
[http://dx.doi.org/10.1016/j.matpr.2020.02.949]

[21]   Prajapati KK, Yadav M, Singh RM, Parikh P, Pareek N, Vivekanand V. An overview of municipal solid waste management in Jaipur city, India - Current status, challenges and recommendations. Renew Sustain Energy Rev 2021; 152: 111703.
[http://dx.doi.org/10.1016/j.rser.2021.111703]

[22]   Ugwu K, Herrera A, Gómez M. Microplastics in marine biota: A review. Mar Pollut Bull 2021; 169: 112540.
[http://dx.doi.org/10.1016/j.marpolbul.2021.112540] [PMID: 34087664]

[23]   Vargas-García MC, Suárez-Estrella F, López MJ, Moreno J. Microbial population dynamics and enzyme activities in composting processes with different starting materials. Waste Manag 2010; 30(5): 771-8.
[http://dx.doi.org/10.1016/j.wasman.2009.12.019] [PMID: 20096556]

[24]   Steger MF, Frazier P. Meaning in life: One link in the chain from religiousness to well-being. J Couns Psychol 2005; 52(4): 574-82.
[http://dx.doi.org/10.1037/0022-0167.52.4.574]

[25]   Ullah N, Mansha M, Khan I, Qurashi A. Nanomaterial-based optical chemical sensors for the detection of heavy metals in water: Recent advances and challenges. Trends Analyt Chem 2018; 100: 155-66.
[http://dx.doi.org/10.1016/j.trac.2018.01.002]

[26]   Yadav P, Samadder SR. Environmental impact assessment of municipal solid waste management options using life cycle assessment: a case study. Environ Sci Pollut Res Int 2018; 25(1): 838-54.
[http://dx.doi.org/10.1007/s11356-017-0439-7] [PMID: 29063409]

[27]   Rebollido R, Martinez J, Aguilera Y, Melchor K, Koerner I, Stegmann R. Microbial populations during composting process of organic fraction of municipal solid waste. Appl Ecol Environ Res 2008; 6(3): 61-7.
[http://dx.doi.org/10.15666/aeer/0603_061067]

[28]   Rouches E, Herpoël-Gimbert I, Steyer JP, Carrere H. Improvement of anaerobic degradation by white-rot fungi pretreatment of lignocellulosic biomass: A review. Renew Sustain Energy Rev 2016; 59: 179-98.
[http://dx.doi.org/10.1016/j.rser.2015.12.317]

[29]   Coccia AM, Gucci PMB, Lacchetti I, Paradiso R, Scaini F. Airborne microorganisms associated with waste management and recovery: biomonitoring methodologies. Ann Ist Super Sanita 2010; 46(3): 288-92.
[PMID: 20847463]

[30]   Kiyasudeen S K, Ibrahim MH, Quaik S, Ahmed Ismail S, Ibrahim MH, Quaik S, *et al.* An introduction to anaerobic digestion of organic wastes. Prospect Org waste Manag significance earthworms. 2016;23–44.

[31]   Merlin Christy P, Gopinath LR, Divya D. A review on anaerobic decomposition and enhancement of biogas production through enzymes and microorganisms. Renew Sustain Energy Rev 2014; 34: 167-73.
[http://dx.doi.org/10.1016/j.rser.2014.03.010]

[32]   Hashemi B, Afkhami H, Khaledi M, *et al.* Frequency of Metalo beta Lactamase genes, bla IMP1, INT 1 in Acinetobacter baumanii isolated from burn patients North of Iran. Gene Rep 2020; 21: 100800.
[http://dx.doi.org/10.1016/j.genrep.2020.100800]

[33]   Poszytek K, Ciężkowska M, Skłodowska A, Drewniak Ł. Microbial consortium with high cellulolytic activity (MCHCA) for enhanced biogas production. Front Microbiol 2016; 7: 324.
[http://dx.doi.org/10.3389/fmicb.2016.00324] [PMID: 27014244]

[34]   Kumar R, Qureshi M, Vishwakarma DK, *et al.* A review on emerging water contaminants and the application of sustainable removal technologies. Case Studies in Chemical and Environmental Engineering 2022; 6: 100219.
[http://dx.doi.org/10.1016/j.cscee.2022.100219]

[35]   Martin-Ryals A, Schideman L, Li P, Wilkinson H, Wagner R. Improving anaerobic digestion of a cellulosic waste *via* routine bioaugmentation with cellulolytic microorganisms. Bioresour Technol 2015; 189: 62-70.
[http://dx.doi.org/10.1016/j.biortech.2015.03.069] [PMID: 25864032]

[36]   Christy D, Egawa T, Yano Y, *et al.* Uniform growth of AlGaN/GaN high electron mobility transistors on 200 mm silicon (111) substrate. Appl Phys Express 2013; 6(2): 026501.
[http://dx.doi.org/10.7567/APEX.6.026501]

[37]   Nanda S, Berruti F. Municipal solid waste management and landfilling technologies: a review. Environ Chem Lett 2021; 19(2): 1433-56.
[http://dx.doi.org/10.1007/s10311-020-01100-y]

[38]   Sharholy M, Ahmad K, Mahmood G, Trivedi RC. Municipal solid waste management in Indian cities – A review. Waste Manag 2008; 28(2): 459-67.
[http://dx.doi.org/10.1016/j.wasman.2007.02.008] [PMID: 17433664]

[39]   Pal MS, Bhatia M. Current status, topographical constraints, and implementation strategy of municipal solid waste in India: a review. Arab J Geosci 2022; 15(12): 1176.
[http://dx.doi.org/10.1007/s12517-022-10414-w]

[40]   Gajalakshmi S, Abbasi SA. Solid waste management by composting: state of the art. Crit Rev Environ Sci Technol 2008; 38(5): 311-400.
[http://dx.doi.org/10.1080/10643380701413633]

[41]   Meena BL, Fagodiya RK, Prajapat K, *et al.* Legume green manuring: an option for soil sustainability. Legum soil Heal Sustain Manag. 2018;387–408.
[http://dx.doi.org/10.1007/978-981-13-0253-4_12]

[42]   Graboski AM, Martinazzo J, Ballen SC, Steffens J, Steffens C. Nanosensors for water quality control. Nanotechnology in the Beverage Industry. Elsevier 2020; pp. 115-28.
[http://dx.doi.org/10.1016/B978-0-12-819941-1.00004-3]

[43]   Ferronato N, Torretta V. Waste mismanagement in developing countries: A review of global issues. Int J Environ Res Public Health 2019; 16(6): 1060.
[http://dx.doi.org/10.3390/ijerph16061060] [PMID: 30909625]

[44]   Akpasi SO, Anekwe IMS, Tetteh EK, *et al.* Mycoremediation as a Potentially Promising Technology: Current Status and Prospects—A Review. Appl Sci (Basel) 2023; 13(8): 4978.
[http://dx.doi.org/10.3390/app13084978]

[45]   Azizi M, Schmieder RE, Mahfoud F, *et al.* Endovascular ultrasound renal denervation to treat hypertension (RADIANCE-HTN SOLO): a multicentre, international, single-blind, randomised, sham-controlled trial. Lancet 2018; 391(10137): 2335-45.
[http://dx.doi.org/10.1016/S0140-6736(18)31082-1] [PMID: 29803590]

Part 5: Traditional and Advanced Techniques for Ecosystem Restoration

CHAPTER 8

Current State and Future Prospects of Microbial Genomics in Ecosystem Restoration

Saraswathy Nagendran[1,*] and **Pooja Mehta**[1]

[1] SVKM`s Mithibai College of Arts, Chauhan Institute of Science & Amrutben Jivanlal College of Commerce and Economics Vile Parle (W), Mumbai, Maharashtra 400056. India

Abstract: Ecosystem degradation through human actions is a global phenomenon. The international society has established goals to stop and reverse these trends, and the restoration industry faces the vital but difficult challenge of putting these goals into practice. Microbial communities are integral to all ecosystems because they perform critical roles like nutrient cycling and other geochemical processes. They are the indicators of the success of ecological restoration, including plantation forests, post-mining areas, oil and gas activities, invasive species management, and soil stabilization. Since the last 2 decades, advancements in microbial genomics have allowed researchers to focus on microbial ecology and dynamics of environmentally balanced vis-a-vis damaged ecosystems. Advancements have significantly improved our capacity to define diversity in microbial ecology and its putative functions in meta-omics methods brought about by developments in high-throughput sequencing (HTS) and bioinformatics. These tools may boost the likelihood that damaged ecosystems will be restored. The current article focuses on using meta-omics techniques to monitor and assess the outcomes of ecological restoration projects and to monitor and evaluate interactions between the various organisms that make up these networks, such as metabolic network mapping. We provide an overview of functional gene editing with the CRISPR/Cas technology to improve microbial bioremediation. The existing understanding will be strengthened by creating more efficient bioinformatics and analysis processes.

Keywords: CRISPR-Cas, Ecosystem restoration, Gene editing, Microbial genomics, Metagenomics.

** **Corresponding author Saraswathy Nagendran:** SVKM`s Mithibai College of Arts, Chauhan Institute of Science & Amrutben Jivanlal College of Commerce and Economics Vile Parle (W), Mumbai, Maharashtra 400056. India; E-mail: saraswathynagendran@gmail.com*

Shiv Prasad, Govindaraj Kamalam Dinesh, Murugaiyan Sinduja, Velusamy Sathya, Ramesh Poornima & Sangilidurai Karthika (Eds.)

INTRODUCTION

The process of recovering a degraded ecosystem in order to restore biodiversity and ecological functioning of species equivalent to that of a reference environment is ecological restoration [1 - 3]. Restoration is a tool of global significance to stop biodiversity loss on a global scale [4, 5], and restoration of degraded natural ecosystems has gained recognition [2, 3, 6]. A large portion of the restoration work is predicated on the idea that all animal and microbial species would develop organically when plant species colonize the area, leading to proper ecosystem functioning [7]. However, the significance of the soil microbial community (SMC) to the plant community's expansion, diversity, and abundance cannot be overstated. Due to its diverse impacts on various species, the SMC substantially affects plant performance [8]. It regulates the variety and quantity of plant species [9, 10]. Microbial communities have returned in some restoration projects [11] but not in others, where decades have passed, and microbial community development and nutrient cycling have not occurred [10, 12].

The effectiveness of restoration initiatives must, therefore, be assessed, and without monitoring, it is impossible to determine if restoration efforts have been made or not [13]. Researchers and practitioners can improve future restoration results by evaluating the success of restoration and comparing it to the efficacy of different remediation processes and adjustments [1, 6, 14]. This is crucial for optimizing the restoration program's effectiveness, achieving the anticipated results, and getting the most out of the financial investment. Monitoring offers the chance for adaptive management by showing whether additional interventions may be required to meet goals [6]. Adaptive management can aid in reducing monitoring periods by highlighting situations where it is doubtful that completion conditions will be reached and additional remediation work is thought to be required. As a result, sensitive indications are required to track the degree of restoration [15]. Microbes are essential for constructing soil composition, nutrient cycling, growth and development of plants, the emergence of biodiversity, and ecosystem operations, making them key drivers of ecological restoration [10, 16]. A stable ecosystem is different from a degraded environment in that it has a variety of microbial populations. Thus, these SMCs can serve as reliable indicators to determine how much ecological restoration has been accomplished [10].

In order to count, visualize, and sequence bacterial DNA, microbiologists are increasingly skipping the cultivation process altogether [17]. Several reasons justify this shift in the use of techniques for assessing microbes and microbial communities. First, recent technological developments, cost, and scalability have made DNA sequencing a viable option for biological monitoring, especially for hard-to-survey populations like microorganisms [14]. Second, with these technologies, it is possible to quickly, accurately, non-destructively, and reliably provide biodiversity data on a range of organisms in an ecosystem. Third, SMCs are dynamic, and little is known about the ecological services they can provide. This intertwine module can be clarified by meta-omics approach to understanding how complex bacteria function in the ecosystem [18]. In this chapter, we delve into the significance of microbial genomics and meta-omics-based tools in ecological restoration

META-OMICS FOR TRACKING AND RESTORATION

Assessing and Tracking Cryptic and Functional Ecosystem Components

Employing field-based monitoring and restoration assessment techniques, surveying microbes or microbial populations is challenging or impossible. In contrast, it is now possible to precisely and quickly describe and quantify these diverse and functional taxonomic groups by sequencing large quantities of environmental DNA or RNA utilizing genomic and, in particular, meta-omic technologies. Eukaryotes like fungi can be tracked using meta-omic methods [19]. However, it is most beneficial for prokaryotes monitoring. On average, 5% of taxa were amenable to the culture-dependent techniques employed to classify prokaryotes. The use of meta-omic has shown that Earth's microbiota is much more diverse than previously believed and that only around 1% of it is culturable. Metabarcoding, metagenomics, metatranscriptomics, and metaproteomics are the meta-omics methods that are most frequently employed.

Metabarcoding

It is a technique that identifies multiple taxa by amplifying DNA barcodes from bulk samples like water or soil environmental DNA (eDNA) using universal primers followed by DNA sequencing and bioinformatics analysis. In a metabarcoding process, DNA is extracted from materials (such as soil, aquatic systems, *etc.*) and then amplified using polymerase chain reaction (PCR). After the amplified product is sequenced and identified using a reference database, sequencing data can be evaluated using a taxonomy-independent methodology to evaluate changes in community profiles. Metabarcoding-based systems can more quickly and affordably generate comprehensive information than traditional monitoring techniques, depending on the monitoring goal [20]. Furthermore, it enables data collection on populations that would otherwise be impossible to see, such as soil microorganisms. This strategy potential has been observed in varied systematic groups, including plants, invertebrates, and vertebrates, as well as in a wide range of diverse environmental contexts, including the tropics, the boreal forest, and alpine meadows [21, 22]. However, it has only recently been used in a restoration context [6, 23].

The bacterial and fungal communities at three mining site restoration sites in Western Australia were examined across chronosequences to see if there were any steady changes in SMC diversity, community structure, and functionality. Although the community's composition underwent directional changes, an indication of microbial recovery, it was discovered that these variations were dependent on the microbe's location and type (bacteria or fungi). The study of functional diversity provided more information about changes in site environments and microbial retrieval than a community structure analysis alone. These results validate that the HTS of eDNA is an effective method for keeping track of the intricate modifications to the SMC after restoration. Future eDNA monitoring of mine site rehabilitation can involve sample archiving to better understand community alterations over time. The inclusion of additional biological categories and substrates, such as soil fauna, would also help to provide a more comprehensive knowledge of biodiversity recovery [14].

On South Australia's Mount. Bold, a water catchment reserve of the Mt. Lofty Ranges (35.07°S, 138.42°E), is an active restoration site where metabarcoding for ecological restoration monitoring was carried out. The open eucalypt woodland that once covered this catchment has been removed and grazed since the turn of the 20th century. The local Eucalyptus leucoxylon grassy woodland community was intended to be recreated when grazing ended in 2003, and restoration work started in 2005. In order to investigate the fungal reactions to restoration, the soil was collected throughout a 10-year chronosequence using eDNA metabarcoding of fungal internal transcribed spacer (ITS) barcodes. After only ten years of aggressive intrinsic plant revegetation, a significant transition toward the indigenous fungal population was seen in the fungus community.

In older recovered environments, there was an increase in the rarefied sequence abundance of Agricomycetes and other Basidiomycota. Although Ascomycota dominated the microbial community, their rarefied sequence abundance declined over the course of the restoration chronosequence. It validated that eDNA metabarcoding is a powerful restoration tracking method that permits the measurement of variation in significant fungal indicator groups associated with functional recovery and is often disregarded in restoration monitoring since they are subterranean [19]. Because eDNA metabarcoding can swiftly, precisely, non-destructively, and reliably yield biodiversity data on various species, from soil microorganisms to animals, it is increasingly employed as a restoration monitoring tool [6].

Metagenomics

Understanding microbial communities in environmental systems has been frequently done by analyzing a microbial community's entire genomic content [24]. The two most popular metagenomic approaches are targeted metagenomics and shotgun metagenomics. Targeted metagenomics examines the diversity of a particular gene to find all of that gene sequence in the environment. Targeted metagenomics is most frequently used to analyze the phylogenetic diversity and comparative abundance of a particular gene in a sample. This method is frequently used to evaluate a sample's diversity of small subunit ribosomal RNA (SSU rRNA) sequences (16S/18S rRNA). SSU rRNA sequencing is frequently used by microbial ecologists to analyze the taxonomic richness of an ecosystem. It may also be used to study how environmental contaminants alter the makeup of microbial populations.

Targeted metagenomics requires isolating environmental DNA and amplifying the target gene using broad-range PCR. Next-generation sequencing is then used to sequence the amplified PCR product. Hundreds of samples may be examined simultaneously using next-gen sequencing, which generates thousands of SSU rRNA reads for each sample. The research's use of universal PCR primers places restrictions on targeted metagenomics, although it does capture the variation of a specific gene of interest [25, 26]. Additionally, different bioinformatics analyses may bias the estimates of total diversity [27].

Targeted metagenomics gives a wide-ranging list of the microbial species present in varied samples and allows for a detailed evaluation of changes in the microbial community before disturbance [24]. For example, the 5000 acres of wetlands habitat in the New Jersey Pine Barrens at the Franklin Parker Preserve include cranberry bogs, old-growth swamps of Atlantic white cedar (commonly known as AWC; *Chamaecyparis thyoides*), and former bogs that are being transformed into AWC forests. The findings of this investigation showed that the carbon use efficiency increased with the number of years of restoration. However, carbon use efficiency is far from matching that of old-growth AWC soils. The comparative abundance of DNA sequences from a variety of copiotrophic bacterial groups (Bacteroidetes and Proteobacteria), complex carbon-decomposing fungal groups (Sordiomycetes, Mortierellales, and Thelephorales), and collembolan and formicid invertebrates, were found to be increasing, which was associated with the carbon-cycle trends as revealed by metagenomic examination of eDNA taken from these soils. These characteristics are all crucial for the functions of the soil carbon cycle and as indicators of soils that are further along in the soil succession. These results imply that the restoration efforts under evaluation are improving the soil's capacity to utilize carbon, improving key guilds of soil biota, and constructing evaluation models for the soil reclamation of restored wetlands using metagenomic analysis of soil eDNA [28].

Another example is the spatiotemporal meta-analysis that demonstrates the ecological regeneration of bacterial populations in the Godavari River during the COVID-19 pandemic. The Kumbh Mela and other mass bathing events (MBE) are known to upset the ecology, which impacts the healthy bacterial communities in the river water in India. Lockdowns and travel limitations allowed for the assessment of the impacts of modest anthropogenic activity on the ecology of the river water and variation in bacterial communities, including antibiotic-resistant strains. Targeted metagenomics discovered that the variety of bacteria increased by 0.87 times during the lockdown's restricted activity. *Verrucomicrobia* (1.8%), *Actinobacteria* (1.2%), *Proteobacteria* (70.6%), *Bacteroidetes* (22.5%), and *Cyanobacteria* (1.1%) all demonstrated a significant increase. There were few allochthonous bacterial populations of human origin. According to alleged metagenomics functional profiling, genes associated with virulence and antibiotic resistance were significantly down [29].

Shotgun metagenomics uses genomic sequencing to examine the total genetic diversity of an environmental population. This process includes obtaining eDNA, which is then broken apart to construct sequencing libraries. Sequencing these libraries then determines the entire genetic makeup of the sample. A powerful technique for assessing the functional potential of a microbial community is shotgun metagenomics. Depth of sequencing or the major drawback of shotgun metagenomics. Very deep sequencing is often necessary to obtain a wide-ranging list of the genes included in an eDNA sample. A thorough examination of the functional potential requires complete genomic coverage of each organism in the community.

Shotgun metagenomics usually only examines the genetic makeup of a community's less populated members while sparingly sampling its dominating bacterial species. Additionally, accurate annotation of diverse gene sequences is necessary for interpreting metagenomic sequencing data, which can be challenging since many of these sequences lack homologs in the current sequence databases [30]. In a chronosequence that included reference soils that had not been mined and mine soils that had been mined for 6-31 years since reforestation, Sun and Badgley (2019) analyzed 15 soil metagenomes. Changes in taxonomic and functional patterns suggested a transition from copiotrophic to oligotrophic species, an increase in the breakdown of refractory carbon sources, and the impact on flora and fauna. As the chronosequence age advanced, there may have been more synergistic and antagonistic interactions among microorganisms due to increases in the genes for transposable elements, virulence, defense, and stress response. Ammonia and nitrite-oxidizing bacteria considerably increased among nitrogen cycling groups upon restoration, but only minor alterations in essential nitrogen cycle functional genes were observed. The scarce number of methanotrophs and methane monooxygenase genes in all restored soils were used to explain earlier results that showed methane consumption at these sites did not rise more than three decades after reforestation. It highlighted the roles of soil microbes in restoring ecosystem services [31].

Without deep sequencing and reads that can be appropriately assembled into sufficiently long contigs, it can be exceedingly challenging to link a functioning gene with a specific species in a population using a phylogenetic anchor. Assembling metagenomic sequences into entire genomes has been the goal of several computer algorithms to better understand the possible functionality of different species within a community [24, 32, 33]. Numerous microarray-based methods have been developed [34]. Microarray technologies that are most often used are PhyloChip and GeoChip. Ten thousand nine hundred and ninety three subfamilies from 147 phyla can be examined using the 16S rRNA-based microarray PhyloChip [35]. The functional gene microarray GeoChip allows the analysis of 152,414 genes from 410 gene categories [36]. Without in-depth sequencing, microarray methods can give extensive insights into microbial ecology [34].

Moreover, they provide thorough annotation for all the different taxa and genes found on the chip, reducing the requirement for high-quality homologs in the database for precise classification. The limitation of only being able to detect the genes covered on the chip with microarray-based approaches prevents the discovery of novel genes or pathways in a sample. Microarray-based approaches are typically a beneficial adjunct to sequencing-based techniques [24]. It is feasible to use metagenomics to assess the functional gene abundance of SMCs along a gradient of nitrogen to determine whether soil fertilization alters the organization of the SMCs such that microbes that are found in soil containing high organic matter are now predominant [37]. This monitoring strategy may be beneficial when a thorough knowledge of changes in nutrients is required during restoration, such as in post-agricultural contexts where the objective is to diminish the impact of agricultural fertilizer usage on soil fertility [38].

For instance, the meta-genomic GeoChip sequencing technology and 16 S rRNA gene and ITS sequencing performed by Illumina were used to evaluate the changes in SMC and functional genes during grassland restoration. While oligotrophs like *Acidobacteria*, *Chloroflexi*, and *Planctomycetes* fell in relative abundance during grassland restoration, copiotrophs like *Actinobacteria*, *Proteobacteria*, and *Bacteroidetes* grew. The meta-genomic GeoChip sequencing results also corroborated the changes in microbial eco-strategies. Early in the restoration process, resistant carbon degradation (*pgu, glx, lig, mnp*), carbon fixation (*accA, aclB, acsA, rbcL*), nitrogen fixation (*nifH*), and nitrification (*amoA, hao*) related genes predominated among the microbial functional genes. In the later years of restoration, the denitrification-related genes (*nosZ, nirS, narG, napA*) and labile carbon degradation (*amyA, amyX, apu, sga, abfA*) predominated. The biotic components in the soil were primarily responsible for the modifications in the functional genes of microbes. A framework illustrating how soil nutrient cycling processes and microbial eco-strategies have evolved over time was developed [39].

Another study used metagenomic sequencing to examine 18 SMCs in phosphorous-deficient deteriorated mining sites in southern China. The ecological restoration used eight varieties of plants along with two plant growth-promoting organisms. According to the findings, soil phosphorus (P) cycling had improved as a result of restoration because the restored site had higher relative abundances of functional genes for P-cycling than the unrestored location. All soil samples included the *gcd* gene, which codes for an enzyme that aids in the solubilization of inorganic P, and it was the most critical contributor to the amount of bioavailable soil P. 39 bacterial genomes were rebuilt for the most part that contained *gcd* and represented a variety of diverse species of phosphate-solubilizing microbial species. Significant relationships were identified between these genomes' relative abundance and the bioavailable soil P, pointing to their possible role in promoting soil P cycling. Moreover, the scaffolds of the 39 genomes contained 84 mobile genetic elements that included *gcd*, demonstrating the importance of phage-associated horizontal gene transfer in soil microorganisms' development of distinct metabolic capabilities related to P cycling [40].

Metagenomics can also be used to look into relationships between various soil microorganisms and plants. On the Loess Plateau in Northwestern China, a 30-year chronosequence of restored grasslands was examined for such a relationship. The findings demonstrated that the richness of plant, bacterial, and fungal species all grew as the number of years following the grassland restoration rose and as their community structure changed across six distinct habitats. An examination of co-occurrence networks using data on the development of microbial communities revealed that network complexity increased as the restoration process went on (25 and 30 years). The bacterial and fungal communities' alpha and beta diversities were significantly and favorably connected to plant communities. While deterministic processes and soil edaphic variables largely governed the formation of the bacterial community, the composition of the plant community and both deterministic and stochastic processes largely governed the structure of the fungus community. This information strongly showed that numerous ecological mechanisms altered the bacterial and fungal populations throughout the ecological restoration of the grasslands. These findings shed light on how restored habitats are connected above and below ground, which might have an impact on ecological restoration techniques and the maintenance of biodiversity in dry and semiarid grassland ecosystems [41].

Metatranscriptomics and Metaproteomics

Environmental systems are increasingly using metatranscriptomics and metaproteomics. These techniques offer crucial new insights into actively expressed genes in a microbial community, and as a result, they are accurate predictors of the microbial activities that are expressed at the time of sampling under the conditions present. RNA is isolated from an environmental sample for Metatranscriptomics. cDNA is produced from the RNA and sequenced, just like in metagenomics. The genes that are actively expressed in a sample are listed using this method. Metatranscriptomic study of environmental RNA has revealed the expression of functional genes and the role of uncommon bacteria in nitrogen-fixing and sulfur oxidation in acid mine drainage [42]. By assessing the recovery of such microorganisms, this method holds the potential for determining whether entire ecosystems could return during the rehabilitation of polluted land, such as post-mining areas.

For instance, genome-centric metagenomics and metatranscriptomics were used to study the diversity, metabolic capability, and gene expression profile of sulfate-reducing bacteria in a restored acidic mine wasteland with aerobic or microaerophilic conditions. Sixteen metagenome-assembled genomes (MAGs) with reductive dsrAB were found to be of medium to high quality. Twelve of them belonged to Acidobacteria and Deltaproteobacteria groups and contained three novel sulfate species-reducing microbes. An analysis of the genomes of three supplementary cultured model organisms and seven high-quality MAGs revealed that the genes encoding glycoside hydrolases, oxygen-tolerant hydrogenases, and cytochrome c oxidases were more abundant in Acidobacteria-related sulfate-reducing microorganisms than in Deltaproteobacteria-related microorganisms. The opposite pattern was seen for the genes producing superoxide reductases and thioredoxin peroxidases. Viral genome sequences were discovered using VirSorter in all three cultured model species and five of the 16 MAGs. In their hosts, these prophages carried enzymes catalyzing glycoside hydrolysis and antioxidation. Furthermore, 15 of the 16 sulfate-reducing microorganisms identified here were active *in situ*, according to the metatranscriptomic study. The sulfate-reducing microorganism transcripts were dominated by an acidobacterial MAG harboring a prophage, which expressed many genes related to its reaction to oxidative stress and competition for organic materials [43].

Using metatranscriptomics, dynamic variations in mycorrhizal cooperation were discovered in boreal woods. Metatranscriptomic examination of samples from the boreal forest, the biggest terrestrial carbon store, revealed the seasonally resolved transcriptomes of Norway spruce roots and more than 350 species of root-associated fungus. It was discovered that there were considerable changes in the fungal population connected with the host plants' functional reaction to increasing nutrient availability. Specialist ectomycorrhizal species, which are essential for the enzymatic degradation of resistant carbon, and metabolically adaptable organisms with strong melanin production in cell walls were distinguished by differences in their occurrence and host-coordination. This study combines functional data from the boreal root microbiome with taxonomic data from all the kingdoms, offering a perspective that was missing from models of global carbon cycling [44].

Another sample from a boreal lake was used to assess the coverage of the HTS approach. The composition of the phytoplankton community, as determined by targeted Metatranscriptomics was compared to results from conventional light microscopy. Using the directed primer-independent high-throughput sequencing (HTS) technique, the community topologies of the 16S and 18S rRNA genes were investigated. A study of 83 samples from boreal lakes revealed that a large number of prokaryotic cyanobacteria and eukaryotic phytoplankton were comparable across HTS and microscopy data. Because there weren't enough sequences in the reference library, the results demonstrated poor concordance at the genus level. HTS is proven to be more effective when looking for species that are well-represented in reference databases but have poor morphological traits. Targeted metatranscriptomics was found to be a reliable technique to support microscopy analysis. Metatranscriptomics can also be used without relating the sequences to any particular taxon. The sequences are directly indexed to their environmental indicator values. However, this necessitates the collection of larger sample sets due to the fact that the coverage of eukaryotic species' molecular barcodes is yet insufficient [45].

Metaproteomics, a new field of study, aims to capture a snapshot of all the proteins that are present in a defined environment or community at a specific time. Profiling microbial enzymes may be a sensitive indicator of the soil ecosystem because it links the phylogeny and function of SMC, showing the level of the individual dominating organism and the community level [46]. High-resolution mass spectrometry is paired with enzymatic protein digestion, liquid chromatography, and other techniques for metaproteomics. It sheds light on the variety of proteins present in a sample of environmental matter, including posttranslational modifications in proteins that might affect their activity [47].

The abundance of bacterial proteins and peptides at different mine land restoration phases was compared to a non-restored site and a native environment (a dense evergreen forest) in the eastern Amazon. The abundance of soil proteins was much higher in the restoring and native soils when compared to the non-restored sites. In terms of SMC, the transitional and advanced areas were discovered to be the most comparable to native soil. The aggregation of many hydrolases and oxidoreductases is proof that the three steps of soil revegetation produced more comparable enzyme abundances and functions to those found in native soil when compared to non-restored areas. Additionally, after the start of restoration activities, essential ecological processes, including the cycling of carbon and nitrogen, appeared to return to the soil swiftly (*i.e.,* 4 years). The enzymes are effective as indicators of the environmental monitors of mineland restoration programs. Metaproteomics allows for the monitoring of the biochemical processes that occur underground throughout the restoration stages [48].

Assessing and Monitoring Interactions in Restored Ecological Networks

Not only species diversity but also the interactions between organisms are essential for successfully restoring healthy ecosystems. Reconstructing these interactions with accuracy allows the measurement of network features and the development of distinctive objectives for restoration. The nitrogen cycle, for instance, controls the fluxes of both organic and inorganic nutrients and is essential for a wide variety of organisms utilized in restoration, such as plants [49]. Nutrient and energy changes indicate the metabolic activity of the microbes from the ecosystem. By estimating and mapping these intricate metabolic interconnections using meta-omics, a metabolic network might be created that permits the evaluation and targeting of crucial metabolic processes and connections by restoration efforts.

Meta-omics can also be utilized to assess the biological processes that are engaged in nutrient cycling before, during, and after restoration activities to quantify the impact of the restoration project. Meta-omics, for instance, can pinpoint the species in charge of nutrient transfer [50]. The shape and accessibility of some essential minerals, including nitrogen, are crucial for the health of ecosystems. Given that inorganic forms of nitrogen are inefficient for species to consume, it is essential to ascertain the abundance of nitrogen fixers and how they react to restoration. A quantitative approach that estimates metabolomic turnover (Predictive Relative Metabolomic Turnover) and describes microbial metabolic networks may also be used to integrate metagenomic or metatranscriptomic data to create a metabolic activity map [51].

The rebuilding of oyster reefs in coastal locations to reduce levels of organic nitrogen in coastal water illustrates this principle. The abundance of organic nitrogen contributed by agricultural fertilizer runoff may lead to eutrophication, which may have serious ecological effects. Oyster reefs have the ability to produce and secrete biosolids that act as a source of carbon for naturally occurring microbial denitrifiers, leading to an increased rate of denitrification [52, 53]. Such restoration programs can be tracked *via* metagenomics and metatranscriptomics by detecting denitrifier amount, activity, and response to perturbation [54]. However, it is challenging to comprehend vast ecological networks because of the need for wide-spread sampling to accurately capture their temporal flux. However, if these networks are identified, the data can drive restoration practitioners to evaluate ecosystem function more comprehensively (*i.e.,* at the level of both macro- and microorganisms) [38].

Amplicon sequencing and co-occurrence network analyses were utilized to examine the interactions between soil bacterial and fungal populations in reaction to slash-and-burn practice and spontaneous revegetation of tropical forests in Papua New Guinea (covering around 60 years after the practice). Despite no discernible discrepancies between the networks of bacteria and fungi at different phases of succession, fungal networks were shown to be substantially more complex and variable than bacterial ones. In the majority of successional stages, bacterial core co-occurrence that is steadily present across all sub-networks in a stage was more frequent than fungi-related ones, showing better stability of bacterial interconnections along succession. The frequent prevalence of long-lasting connections between ectomycorrhizal fungi (such as *Boletaceae* and *Russulaochroleuca*) and bacterial taxa (such as *Sporosarcina, Acidimicrobiale,* and *Bacillaceae*) suggests that these taxa play significant ecological roles in the restoration of the environment. The combined results revealed new information on microbial links in reaction to slash-and-burn agriculture and the succeeding ecosystem restoration, which improved the understanding of microbial involvement in ecological restoration [55].

GENE EDITING FOR RESTORATION

The restoration of contaminated soil, surface water, and groundwater is now possible naturally, affordably, and sustainably through utilizing biological agents such as bacteria, fungi, and other organisms or their enzymes [56]. Recent developments in high-throughput genomic profiling have uncovered numerous genes that react to various biological processes, including developments in bioremediation. Through gene editing, scientists can regulate gene expression at specific locations, gaining new insights into the development of microbial bioremediation [57]. An outstanding approach to gene editing allows for DNA manipulation using molecular scissors, a designed nuclease [58]. Editing entails targeting a desired sequence region by inserting or deleting a customized guide sequence complementary to the target gene sequence [59]. The application of gene editing techniques can considerably enhance the bioremediation processes, including removing xenobiotics, transforming toxic compounds, and breaking pesticides into easily metabolizable compounds [60, 61].

The aforementioned expectations can be met by using the three main gene editing methods: CRISPR-Cas, Zinc Finger Nucleases (ZFN), and Transcription Activator-Like Effector Nucleases (TALEN) [62 - 64]. These gene editing techniques precisely cut the DNA, which is subsequently repaired by the error-prone repair mechanism [65, 66]. Synthetic restriction enzymes are used by ZFNs and TALEN to cleave a precise DNA sequence using the Zinc finger DNA binding domains and TAL effector DNA binding domains, respectively [58]. By introducing more complex genes and creating bacteria with the greatest possible quality, CRISPR gene editing technology aims to enhance microbes [60, 67]. It primarily modifies the wild-type genome to produce the intended modified bacteria with the required functionality [68].

Crispr-Cas

CRISPR-CAS describes the most effective and innovative gene editing types [66]. The CRISPR-Cas systems appear in three categories: Types I, II, and III, as well as countless subtypes [69]. Each system has a different Cas depending on how it is functioning, *i.e.*, model organisms. A DNA endonuclease called Cas9 is directed by RNA to target foreign DNA and inhibit it. The complementary foreign sequence is a 30–40 bp direct repeat sequence called CRISPR, which is separated from it by a spacer region. crRNA is produced following transcription and processing. The gRNA is then obtained by using CRISPRs (guide RNA). crRNP (Ribonucleoprotein), which is created by the combination of crRNA and Cas protein, breaks the DNA/RNA of the intrusive organism. The specificity and functionality of CRISPR are due to the exact binding of gRNA to the target DNA region. CRISPR/Cas9 can be utilized to change (add or remove) the target gene by nicking the double-strand at the specific site [70]. The ideal promoters for sgRNA and Cas9 transcription, the codon that optimizes the variation of Cas9, and the appropriate expression system are all necessary for achieving the CRISPR-Cas sgRNA sequence. Because it is well adapted to bacterial and archaeal systems, the CRISPR technology is drawing the attention of molecular biologists [58].

Transcription Activator-Like Effector Nucleases

The term "TALENs" refers to transcription activator-like effector nucleases. It is a state-of-the-art tool for gene editing. TALENs contain TAL proteins. The harmful bacterium species Xanthomonas first released these proteins. TAL proteins have the ability to bind to sequences as small as 1-2 nucleotides due to their immense strength. Additionally, the associated nucleases are particularly good at binding because of the presence of 34 amino acid tandem repeats. Currently, gene knock-in (homology-directed repair) and gene knockout (non-homologous end joining) of the gene of interest are the two most popular ways to use TALENs. Due to two protein domains—first for breaking apart sequences and the second for locating and binding a precise location—TALENs are a potent gene editing method [58].

Zinc Finger Nucleases

Zinc Finger Nucleases (ZFNs) are the most frequently used endonucleases. They are synthetic restriction enzymes. Zinc Finger Nucleases have ZFPs (Zinc Finger Proteins), which are eukaryotic transcription factors and thus have DNA-binding domains. The Folk1 (nucleotide cleavage domain) from *Flavobacterium okeanokoites* is also present in ZFNs. Based on the target site, several ZFPs (typically four to six) surround the cleavage domain. These ZFPs' 18 bp selectivity might make precise gene editing possible. ZFPs are 30 amino acids long and have an alpha helix as opposed to two antiparallel sheets. This gene-editing tool is described along with a gene knockout and knock-in for effective gene editing.

TALENs and ZFNs are more expensive, complicated, and difficult for researchers to utilize than CRISPR-Cas [71]. It has the benefit of making it possible to analyze gene interactions, the genetic and phenotypic links between them, and gene knockout systems using different genes. Off-target mutation, which can result in death, genomic disintegration, and limited applicability, is a disadvantage of the CRISPR-Cas system [72]. The use of the gene, as mentioned above, editing methods must be restricted due to the possibility of unintended or intentional discharge of the changed organism into the environment [73].

Previously, many plant genomes were modified for phytoremediation of potential hazardous pollution reduction [74 - 77]. CRISPR technology could improve organic bioremediation without posing significant risks or costs [66, 78]. In a review, Basharat *et al.* provided an exciting idea for a potential phytoremediation scenario based on plant genome reprogramming utilizing CRISPR (2018). Additionally, genome-edited microbes are being used in various fields, such as bioremediation, agriculture, and human health, to create probiotic, edible microorganisms [58, 66].

Restoration has historically been a crucial strategy for reversing habitat loss and restoring functioning, but it is questionable to what degree this will be adequate in light of changing climatic conditions. With the help of cutting-edge genetic technologies like CRISPR, restoration may now proactively match the adaptation of target species to expected future climatic circumstances, potentially increasing resilience to stress in vulnerable and degraded ecosystems. As a result, restoring to historical baselines or being ready for the future is a crucial choice that will decide the sustainability of restoration amidst environmental and climate change [79].

For instance, a recent review by Coleman *et al.* (2020) provided an overview of the various justifications for restoration, such as restoring lost or damaged ecosystems to their original historical states or strengthening or altering the course of the future. With synthetic biology and CRISPR, it is now possible to completely redefine genetic baselines and population resilience to introduce engineered advantageous genetic elements within restored or vulnerable populations and to enable the development of customized restoration or assisted adaptation programs for particular stressors of interest [80]. That involves modifying a species' genome at a single site to alter the quantity of metabolites [81], engineering speciation through the use of designer karyotypes [82], or engineering the microbiome to produce sequence-specific anti-microbials [83]. These restoration activities, emphasizing declining kelp forests and other critical maritime habitats, also apply to broader restoration efforts [84-87].

CONCLUSION AND FUTURE PERSPECTIVES

Microbial genetics can command transformation in ecological restoration, the way the human microbiome has been done in healthcare and medicine. In the field of microbial genomics, meta-omics methods, in particular, aid in a better understanding of the microbial biodiversity that is crucial to the ecosystem's health. As a result, it can provide ingenuity to help with some of the fundamental problems facing ecological restoration. However, other factors, such as site accessibility and property ownership that are outside the purview of microbial genomics, will be needed for successful and economically advantageous restoration procedures in addition to the use of microbial genomics. Microorganisms need to be more widely recognized as indicators for tracking and evaluating ecological restoration. Governments, local authorities, participants in bio-monitoring programs, restoration ecologists, *etc.*, must work together extensively on this, and outreach, education, and training must also be provided. Bioinformatics is particularly important for processing and interpreting massive data produced by these high-throughput sequencing techniques, and multidisciplinary scientific collaboration is essential for successfully deploying meta-omics methodologies for restoration initiatives.

Additionally, as with any innovation, proof of concept is required to show that the suggested procedures are viable. Therefore, case studies are crucial to substantiate the advantages and applications of meta-omics. Additionally, it is vital to create risk frameworks and involve the appropriate parties (such as governmental, academic, and community organizations) in weighing potential risks with potential advantages of using genomic techniques. Considering risks proactively can help find potential solutions to reduce the unintended effects of emerging technology, including genome editing. Furthermore, ensuring funding for the research and developing a robust set of accessory elements, such as biotechnology and computing infrastructure, is essential to unleashing the full potential of microbial genomics.

ACKNOWLEDGEMENTS

The authors thank the management of SVKM's Mithibai College of Arts, Chauhan Institute of Science & Amruthben Jivanlal College of Commerce and Economics for their continual encouragement and unflinching support.

REFERENCES

[1]   Gann GD, McDonald T, Walder B, *et al.* International principles and standards for the practice of ecological restoration. Res Eco. 2019; 27: pp. (S1)S1-S46.

[2]   Hobbs RJ, Harris JA. Restoration Ecology: Repairing the Earth's Ecosystems in the New Millennium. Restor Ecol 2001; 9(2): 239-46.
[http://dx.doi.org/10.1046/j.1526-100x.2001.009002239.x]

[3]   Suding KN. Toward an Era of Restoration in Ecology: Successes, Failures, and Opportunities Ahead. Annu Rev Ecol Evol Syst 2011; 42(1): 465-87.
[http://dx.doi.org/10.1146/annurev-ecolsys-102710-145115]

[4]   Aronson J, Alexander S. Ecosystem Restoration is Now a Global Priority: Time to Roll up our Sleeves. Restor Ecol 2013; 21(3): 293-6.
[http://dx.doi.org/10.1111/rec.12011]

[5]   Benayas JMR, Newton AC, Diaz A, Bullock JM. Enhancement of biodiversity and ecosystem services by ecological restoration: a meta-analysis. Science 2009; 325(5944): 1121-4.
[http://dx.doi.org/10.1126/science.1172460] [PMID: 19644076]

[6]   van der Heyde M, Bunce M, Nevill P. Key factors to consider in the use of environmental DNA metabarcoding to monitor terrestrial ecological restoration. Sci Total Environ 2022; 848: 157617.
[http://dx.doi.org/10.1016/j.scitotenv.2022.157617] [PMID: 35901901]

[7]   Holl KD. Restoration of Tropical Forests. In: van Andel J, Aronson J, Eds. Restoration Ecology. 1st ed. Wiley 2012; pp. 103-14.https://onlinelibrary.wiley.com/doi/10.1002/9781118223130.ch9 [Internet]
[http://dx.doi.org/10.1002/9781118223130.ch9]

[8]   Bever JD, Dickie IA, Facelli E, *et al.* Rooting theories of plant community ecology in microbial interactions. Trends Ecol Evol 2010; 25(8): 468-78.
[http://dx.doi.org/10.1016/j.tree.2010.05.004] [PMID: 20557974]

[9]   Hodge A, Fitter AH. Microbial mediation of plant competition and community structure. Funct Ecol 2013; 27(4): 865-75.
[http://dx.doi.org/10.1111/1365-2435.12002]

[10]   Singh Rawat V, Kaur J, Bhagwat S, Arora Pandit M, Dogra Rawat C. Deploying microbes as drivers and   indicators   in   ecological   restoration.   Res   Eco   2023;   31(1):   e13688. https://onlinelibrary.wiley.com/doi/10.1111/rec.13688

[11]   Scott DA, Baer SG, Blair JM. Recovery and Relative Influence of Root, Microbial, and Structural Properties of Soil on Physically Sequestered Carbon Stocks in Restored Grassland. Soil Sci Soc Am J 2017; 81(1): 50-60.
[http://dx.doi.org/10.2136/sssaj2016.05.0158]

[12]   Bonner MTL, Herbohn J, Gregorio N, *et al.* Soil organic carbon recovery in tropical tree plantations may depend on restoration of soil microbial composition and function. Geoderma 2019; 353: 70-80.
[http://dx.doi.org/10.1016/j.geoderma.2019.06.017]

[13]   Thompson SA, Thompson GG. Adequacy of rehabilitation monitoring practices in the Western Australian mining industry. Ecol Manage Restor 2004; 5(1): 30-3.
[http://dx.doi.org/10.1111/j.1442-8903.2004.00172.x]

[14]   van der Heyde M, Bunce M, Dixon K, Wardell-Johnson G, White NE, Nevill P. Changes in soil microbial communities in post mine ecological restoration: Implications for monitoring using high throughput DNA sequencing. Sci Total Environ 2020; 749: 142262.
[http://dx.doi.org/10.1016/j.scitotenv.2020.142262] [PMID: 33370926]

[15]   Glasl B, Bourne DG, Frade PR, Thomas T, Schaffelke B, Webster NS. Microbial indicators of environmental perturbations in coral reef ecosystems. Microbiome 2019; 7(1): 94.
[http://dx.doi.org/10.1186/s40168-019-0705-7] [PMID: 31227022]

[16]   Harris J. Soil microbial communities and restoration ecology: facilitators or followers? Science 2009; 325(5940): 573-4.
[http://dx.doi.org/10.1126/science.1172975] [PMID: 19644111]

[17]   Christine M. Microbial Diversity Unbound: What DNA-based techniques are revealing about the planet's hidden biodiversity. Bioscience 2004; 54(12): 1064-8.
[http://dx.doi.org/10.1641/0006-3568(2004)054[1064:MDU]2.0.CO;2]

[18]   Michealsamy A, Thangamani L, Manivel G, Kumar P, Sundar S, Piramanayagam S, *et al.* Current Research and Applications of Meta-Omics Stratagems in Bioremediation: A Bird's-Eye View. J Apple Biotechnol Rep [Internet] 2021; 8(2).
[http://dx.doi.org/10.30491/jabr.2020.237662.1248]

[19]   Yan D, Mills JG, Gellie NJC, Bissett A, Lowe AJ, Breed MF. High-throughput eDNA monitoring of fungi to track functional recovery in ecological restoration. Biol Conserv 2018; 217: 113-20.
[http://dx.doi.org/10.1016/j.biocon.2017.10.035]

[20]   Ji Y, Ashton L, Pedley SM, *et al.* Reliable, verifiable and efficient monitoring of biodiversity *via* metabarcoding. Ecol Lett 2013; 16(10): 1245-57.
[http://dx.doi.org/10.1111/ele.12162] [PMID: 23910579]

[21]   Creer S, Deiner K, Frey S, *et al.* The ecologist's field guide to sequence-based identification of biodiversity. Methods Ecol Evol 2016; 7(9): 1008-18.
[http://dx.doi.org/10.1111/2041-210X.12574]

[22]   Deiner K, Bik HM, Mächler E, *et al.* Environmental DNA metabarcoding: Transforming how we survey animal and plant communities. Mol Ecol 2017; 26(21): 5872-95.
[http://dx.doi.org/10.1111/mec.14350] [PMID: 28921802]

[23]   Gellie NJC, Mills JG, Breed MF, Lowe AJ. Revegetation rewilds the soil bacterial microbiome of an old field. Mol Ecol 2017; 26(11): 2895-904.
[http://dx.doi.org/10.1111/mec.14081] [PMID: 28261928]

[24]   Techtmann SM, Hazen TC. Metagenomic applications in environmental monitoring and bioremediation. J Ind Microbiol Biotechnol 2016; 43(10): 1345-54.
[http://dx.doi.org/10.1007/s10295-016-1809-8] [PMID: 27558781]

[25]   Klindworth A, Pruesse E, Schweer T, *et al.* Evaluation of general 16S ribosomal RNA gene PCR primers for classical and next-generation sequencing-based diversity studies. Nucleic Acids Res 2013; 41(1): e1.
[http://dx.doi.org/10.1093/nar/gks808] [PMID: 22933715]

[26]   Parada AE, Needham DM, Fuhrman JA. Every base matters: assessing small subunit RRNA primers for marine microbiomes with mock communities, time series and global field samples. Environ Microbiol 2016; 18(5): 1403-14.
[http://dx.doi.org/10.1111/1462-2920.13023] [PMID: 26271760]

[27]   Rinke C, Schwientek P, Sczyrba A, *et al.* Insights into the phylogeny and coding potential of microbial dark matter. Nature 2013; 499(7459): 431-7.
[http://dx.doi.org/10.1038/nature12352] [PMID: 23851394]

[28]   Eaton WD, Shokralla S, McGee KM, Hajibabaei M. Using metagenomics to show the efficacy of forest restoration in the New Jersey Pine Barrens. Genome. 2017; 60: pp. (10)825-36.
[http://dx.doi.org/10.1139/gen-2015-0199]

[29]   Jani K, Bandal J, Shouche Y, *et al.* Extended Ecological Restoration of Bacterial Communities in the Godavari River During the COVID-19 Lockdown Period: a Spatiotemporal Meta-analysis. Microb Ecol 2021; 82(2): 365-76.
[http://dx.doi.org/10.1007/s00248-021-01781-0] [PMID: 34219185]

[30]   Delmont TO, Simonet P, Vogel TM. Describing microbial communities and performing global comparisons in the 'omic era. ISME J 2012; 6(9): 1625-8.
[http://dx.doi.org/10.1038/ismej.2012.55] [PMID: 22717882]

[31]   Sun S, Badgley BD. Changes in microbial functional genes within the soil metagenome during forest ecosystem restoration. Soil Biol Biochem 2019; 135: 163-72.
[http://dx.doi.org/10.1016/j.soilbio.2019.05.004]

[32]   Imelfort M, Parks D, Woodcroft BJ, Dennis P, Hugenholtz P, Tyson GW. GroopM: an automated tool for the recovery of population genomes from related metagenomes. PeerJ 2014; 2: e603.
[http://dx.doi.org/10.7717/peerj.603] [PMID: 25289188]

[33]   Sharon I, Banfield JF. Genomes from Metagenomics. Science 2013; 342(6162): 1057-8.
[http://dx.doi.org/10.1126/science.1247023] [PMID: 24288324]

[34]   Hazen TC, Rocha AM, Techtmann SM. Advances in monitoring environmental microbes. Curr Opin Biotechnol 2013; 24(3): 526-33.
[http://dx.doi.org/10.1016/j.copbio.2012.10.020] [PMID: 23183250]

[35]   Hazen TC, Dubinsky EA, DeSantis TZ, *et al.* Deep-sea oil plume enriches indigenous oil-degrading bacteria. Science 2010; 330(6001): 204-8.
[http://dx.doi.org/10.1126/science.1195979] [PMID: 20736401]

[36]   Zhou A, He Z, Qin Y, *et al.* StressChip as a high-throughput tool for assessing microbial community responses to environmental stresses. Environ Sci Technol 2013; 47(17): 9841-9.
[http://dx.doi.org/10.1021/es4018656] [PMID: 23889170]

[37]   Fierer N, Lauber CL, Ramirez KS, Zaneveld J, Bradford MA, Knight R. Comparative metagenomic, phylogenetic and physiological analyses of soil microbial communities across nitrogen gradients. ISME J 2012; 6(5): 1007-17.
[http://dx.doi.org/10.1038/ismej.2011.159] [PMID: 22134642]

[38]   Breed MF, Harrison PA, Blyth C, *et al.* The potential of genomics for restoring ecosystems and biodiversity. Nat Rev Genet 2019; 20(10): 615-28.
[http://dx.doi.org/10.1038/s41576-019-0152-0] [PMID: 31300751]

[39]   Liao J, Dou Y, Yang X, An S. Soil microbial community and their functional genes during grassland restoration. J Environ Manage 2023; 325(Pt A): 116488.
[http://dx.doi.org/10.1016/j.jenvman.2022.116488] [PMID: 36419280]

[40]   Liang JL, Liu J, Jia P, *et al.* Novel phosphate-solubilizing bacteria enhance soil phosphorus cycling following ecological restoration of land degraded by mining. ISME J 2020; 14(6): 1600-13.
[http://dx.doi.org/10.1038/s41396-020-0632-4] [PMID: 32203124]

[41]   Guo Y, Hou L, Zhang Z, *et al.* Soil microbial diversity during 30 years of grassland restoration on the Loess Plateau, China: Tight linkages with plant diversity. Land Degrad Dev 2019; 30(10): 1172-82.
[http://dx.doi.org/10.1002/ldr.3300]

[42]   Hua ZS, Han YJ, Chen LX, *et al.* Ecological roles of dominant and rare prokaryotes in acid mine drainage revealed by metagenomics and metatranscriptomics. ISME J 2015; 9(6): 1280-94.
[http://dx.doi.org/10.1038/ismej.2014.212] [PMID: 25361395]

[43]   Li JT, Jia P, Wang XJ, *et al.*. Metagenomic and metatranscriptomic insights into sulfate-reducing bacteria in a revegetated acidic mine wasteland. NPJ Biofilms Microbiomes 2022; 8(1): 71.

[44]   Law SR, Serrano AR, Daguerre Y, *et al.* Metatranscriptomics captures dynamic shifts in mycorrhizal coordination in boreal forests. Proc Natl Acad Sci USA 2022; 119(26): e2118852119.
[http://dx.doi.org/10.1073/pnas.2118852119] [PMID: 35727987]

[45]   Vuorio K, Mäki A, Salmi P, Aalto SL, Tiirola M. Consistency of Targeted Metatranscriptomics and Morphological Characterization of Phytoplankton Communities. Front Microbiol 2020; 11: 96.
https://www.frontiersin.org/articles/10.3389/fmicb.2020.00096
[http://dx.doi.org/10.3389/fmicb.2020.00096] [PMID: 32117126]

[46]   Abiraami TV, Singh S, Nain L. Soil metaproteomics as a tool for monitoring functional microbial communities: promises and challenges. Rev Environ Sci Biotechnol 2020; 19(1): 73-102.
[http://dx.doi.org/10.1007/s11157-019-09519-8]

[47]   Hettich RL, Pan C, Chourey K, Giannone RJ. Metaproteomics: harnessing the power of high performance mass spectrometry to identify the suite of proteins that control metabolic activities in microbial communities. Anal Chem 2013; 85(9): 4203-14.
[http://dx.doi.org/10.1021/ac303053e] [PMID: 23469896]

[48]   Trindade FC, Gastauer M, Ramos SJ, *et al.* Soil Metaproteomics as a Tool for Environmental Monitoring of Minelands. Forests 2021; 12(9): 1158.
[http://dx.doi.org/10.3390/f12091158]

[49]   Jacoby R, Peukert M, Succurro A, Koprivova A, Kopriva S. The Role of Soil Microorganisms in Plant Mineral Nutrition—Current Knowledge and Future Directions. Front Plant Sci 2017; 8: 1617.
https://www.frontiersin.org/articles/10.3389/fpls.2017.01617
[http://dx.doi.org/10.3389/fpls.2017.01617] [PMID: 28974956]

[50]   Faust K, Raes J. Microbial interactions: from networks to models. Nat Rev Microbiol 2012; 10(8): 538-50.
[http://dx.doi.org/10.1038/nrmicro2832] [PMID: 22796884]

[51]   Larsen PE, Collart FR, Field D, *et al.* Predicted Relative Metabolomic Turnover (PRMT): determining metabolic turnover from a coastal marine metagenomic dataset. Microb Inform Exp 2011; 1(1): 4.
[http://dx.doi.org/10.1186/2042-5783-1-4] [PMID: 22587810]

[52]   Cerco CF, Noel MR. Can oyster restoration reverse cultural eutrophication in Chesapeake Bay? Estuaries Coasts 2007; 30(2): 331-43.
[http://dx.doi.org/10.1007/BF02700175]

[53]   Higgins CB, Tobias C, Piehler MF, *et al.* Effect of aquacultured oyster biodeposition on sediment N$_2$ production in Chesapeake Bay. Mar Ecol Prog Ser 2013; 473: 7-27.
[http://dx.doi.org/10.3354/meps10062]

[54]   Dandie CE, Wertz S, Leclair CL, *et al.* Abundance, diversity and functional gene expression of denitrifier communities in adjacent riparian and agricultural zones. FEMS Microbiol Ecol 2011; 77(1): 69-82.
[http://dx.doi.org/10.1111/j.1574-6941.2011.01084.x] [PMID: 21385191]

[55]   Lin Q, Dini-Andreote F, Li L, *et al.* Soil microbial interconnections along ecological restoration gradients of lowland forests after slash-and-burn agriculture. FEMS Microbiol Ecol 2021; 97(5): fiab063.
[http://dx.doi.org/10.1093/femsec/fiab063] [PMID: 33899919]

[56]   Kumar V, Shahi SK, Singh S. Bioremediation: An Eco-sustainable Approach for Restoration of Contaminated Sites. In: Singh J, Sharma D, Kumar G, Sharma NR, Eds. Microbial Bioprospecting for Sustainable Development. Singapore: Springer 2018; pp. 115-36. [Internet]
[http://dx.doi.org/10.1007/978-981-13-0053-0_6]

[57]   Sahoo S, Routray SP, Lenka S, Bhuyan R, Mohanty JN. CRISPR/Cas-Mediated Functional Gene Editing for Improvement in Bioremediation: An Emerging Strategy. In: Kumar V, Thakur IS, Eds. Omics Insights in Environmental Bioremediation. Singapore: Springer Nature 2022; pp. 635-64. [Internet]
[http://dx.doi.org/10.1007/978-981-19-4320-1_27]

[58]   Jaiswal S, Singh DK, Shukla P. Gene Editing and Systems Biology Tools for Pesticide Bioremediation: A Review. Front Microbiol 2019; 10: 87. https://www.frontiersin.org/articles/10.3389/fmicb.2019.00087
[http://dx.doi.org/10.3389/fmicb.2019.00087] [PMID: 30853940]

[59]   Bier E, Harrison MM, O'Connor-Giles KM, Wildonger J. Advances in Engineering the Fly Genome with the CRISPR-Cas System. Genetics 2018; 208(1): 1-18.
[http://dx.doi.org/10.1534/genetics.117.1113] [PMID: 29301946]

[60]   Basu S, Rabara RC, Negi S, Shukla P. Engineering PGPMOs through Gene Editing and Systems Biology: A Solution for Phytoremediation? Trends Biotechnol 2018; 36(5): 499-510.
[http://dx.doi.org/10.1016/j.tibtech.2018.01.011] [PMID: 29455935]

[61]   Hussain I, Aleti G, Naidu R, *et al.* Microbe and plant assisted-remediation of organic xenobiotics and its enhancement by genetically modified organisms and recombinant technology: A review. Sci Total Environ 2018; 628-629: 1582-99.
[http://dx.doi.org/10.1016/j.scitotenv.2018.02.037] [PMID: 30045575]

[62]   Singh V, Gohil N, Ramírez García R, Braddick D, Fofié C. Fofié CK. Recent advances in CRISPR-Cas9 genome editing technology for biological and biomedical investigations. J Cell Biochem 2018; 119(1): 81-94.

[63]   Waryah CB, Moses C, Arooj M, Blancafort P. Zinc Fingers, TALEs, and CRISPR Systems: A Comparison of Tools for Epigenome Editing. In: Jeltsch A, Rots MG, editors. Epigenome Editing: Methods and Protocols [Internet]. New York, NY: Springer; 2018. p. 19–63.
[http://dx.doi.org/10.1007/978-1-4939-7774-1_2]

[64]   Wong DWS. Gene Targeting and Genome Editing. In: Wong DWS, Ed. The ABCs of Gene Cloning. Cham: Springer International Publishing 2018; pp. 187-97.
[http://dx.doi.org/10.1007/978-3-319-77982-9_20]

[65]   Arazoe T, Kondo A, Nishida K. Targeted Nucleotide Editing Technologies for Microbial Metabolic Engineering. Biotechnol J 2018; 13(9): 1700596.
[http://dx.doi.org/10.1002/biot.201700596] [PMID: 29862665]

[66]   Yadav R, Kumar V, Baweja M, Shukla P. Gene editing and genetic engineering approaches for advanced probiotics: A review. Crit Rev Food Sci Nutr 2018; 58(10): 1735-46.
[http://dx.doi.org/10.1080/10408398.2016.1274877] [PMID: 28071925]

[67]   Dangi AK, Sharma B, Hill RT, Shukla P. Bioremediation through microbes: systems biology and

metabolic engineering approach. Crit Rev Biotechnol 2019; 39(1): 79-98.
[http://dx.doi.org/10.1080/07388551.2018.1500997] [PMID: 30198342]

[68]   Dai Z, Zhang S, Yang Q, *et al.* Genetic tool development and systemic regulation in biosynthetic technology. Biotechnol Biofuels 2018; 11(1): 152.
[http://dx.doi.org/10.1186/s13068-018-1153-5] [PMID: 29881457]

[69]   Zhu Y, Klompe SE, Vlot M, van der Oost J, Staals RHJ. Shooting the messenger: RNA-targetting CRISPR-Cas systems. Biosci Rep 2018; 38(3): BSR20170788.
[http://dx.doi.org/10.1042/BSR20170788] [PMID: 29748239]

[70]   Shapiro RS, Chavez A, Collins JJ. CRISPR-based genomic tools for the manipulation of genetically intractable microorganisms. Nat Rev Microbiol 2018; 16(6): 333-9.
[http://dx.doi.org/10.1038/s41579-018-0002-7] [PMID: 29599458]

[71]   Ju XD, Xu J, Sun Z. CRISPR editing in biological and biomedical investigation. J Cell Biochem 2018; 119: 52-61.

[72]   Sun J, Wang Q, Jiang Y, *et al.* Genome editing and transcriptional repression in Pseudomonas putida KT2440 *via* the type II CRISPR system. Microb Cell Fact 2018; 17(1): 41.
[http://dx.doi.org/10.1186/s12934-018-0887-x] [PMID: 29534717]

[73]   Canver MC, Joung JK, Pinello L. Impact of Genetic Variation on CRISPR-Cas Targeting. CRISPR J 2018; 1(2): 159-70.
[http://dx.doi.org/10.1089/crispr.2017.0016] [PMID: 31021199]

[74]   Basharat Z, Novo LAB, Yasmin A. Genome Editing Weds CRISPR: What Is in It for Phytoremediation? Plants 2018; 7(3): 51.
[http://dx.doi.org/10.3390/plants7030051] [PMID: 30720787]

[75]   Bortesi L, Fischer R. The CRISPR/Cas9 system for plant genome editing and beyond. Biotechnol Adv 2015; 33(1): 41-52.
[http://dx.doi.org/10.1016/j.biotechadv.2014.12.006] [PMID: 25536441]

[76]   Zaidi SSA, Mahfouz MM, Mansoor S. CRISPR-Cpf1: A New Tool for Plant Genome Editing. Trends Plant Sci 2017; 22(7): 550-3.
[http://dx.doi.org/10.1016/j.tplants.2017.05.001] [PMID: 28532598]

[77]   Yin K, Gao C, Qiu JL. Progress and prospects in plant genome editing. Nat Plants 2017; 3(8): 17107.
[http://dx.doi.org/10.1038/nplants.2017.107] [PMID: 28758991]

[78]   Khorsandi H, Ghochlavi N, Aghapour AA. Biological Degradation of 2,4,6-Trichlorophenol by a Sequencing Batch Reactor. Environ Process 2018; 5(4): 907-17.
[http://dx.doi.org/10.1007/s40710-018-0333-4]

[79]   Coleman MA, Goold HD. Harnessing synthetic biology for kelp forest conservation. J Phycol 2019; 55(4): 745-51.
[http://dx.doi.org/10.1111/jpy.12888] [PMID: 31152453]

[80]   Coleman MA, Wood G, Filbee-Dexter K, *et al.* Restore or Redefine: Future Trajectories for Restoration. Front Mar Sci 2020; 7: 237. https://www.frontiersin.org/articles/10.3389/fmars.2020.00237 [Internet].
[http://dx.doi.org/10.3389/fmars.2020.00237]

[81]   Shih PM, Vuu K, Mansoori N, *et al.* A robust gene-stacking method utilizing yeast assembly for plant synthetic biology. Nat Commun 2016; 7(1): 13215.
[http://dx.doi.org/10.1038/ncomms13215] [PMID: 27782150]

[82]   Luo J, Sun X, Cormack BP, Boeke JD. Karyotype engineering by chromosome fusion leads to reproductive isolation in yeast. Nature 2018; 560(7718): 392-6.
[http://dx.doi.org/10.1038/s41586-018-0374-x] [PMID: 30069047]

[83]   Bikard D, Euler CW, Jiang W, *et al.* Exploiting CRISPR-Cas nucleases to produce sequence-specific

antimicrobials. Nat Biotechnol 2014; 32(11): 1146-50.
[http://dx.doi.org/10.1038/nbt.3043] [PMID: 25282355]

[84]   Webber BL, Raghu S, Edwards OR. Is CRISPR-based gene drive a biocontrol silver bullet or global conservation threat? Proc Natl Acad Sci USA 2015; 112(34): 10565-7.
[http://dx.doi.org/10.1073/pnas.1514258112] [PMID: 26272924]

[85]   Cristescu ME, Hebert PDN. Uses and Misuses of Environmental DNA in Biodiversity Science and Conservation. Annu Rev Ecol Evol Syst 2018; 49(1): 209-30.
[http://dx.doi.org/10.1146/annurev-ecolsys-110617-062306]

[86]   Piaggio AJ, Segelbacher G, Seddon PJ, *et al.* Is It Time for Synthetic Biodiversity Conservation? Trends Ecol Evol 2017; 32(2): 97-107.
[http://dx.doi.org/10.1016/j.tree.2016.10.016] [PMID: 27871673]

[87]   Stirling A, Hayes KR, Delborne J. Towards inclusive social appraisal: risk, participation and democracy in governance of synthetic biology. BMC Proc 2018; 12(S8) (Suppl. 8): 15.
[http://dx.doi.org/10.1186/s12919-018-0111-3] [PMID: 30079106]

CHAPTER 9

Role of Metagenomics and Microbial Diversity in the Restoration of Tropic and Temperate Ecosystems

Anushka Satpathy[1], Koel Mukherjee[1] and **Vinod Kumar Nigam[1,*]**

[1] *Department of Bioengineering & Biotechnology, Birla Institute of Technology, Mesra, Ranchi-835215, Jharkhand, India*

Abstract: The geographical area where all the abiotic and biotic factors interact with each other to make the bubble of life is known as the ecosystem. While many natural and artificial calamities occur to destroy the ecosystem, microbial diversity plays a vital role in maintaining and functioning it. The microbes constitute one-third of the earth's biomass and are composed of enormous genetic diversity from extremely hot (thermophilic) and moderate (mesophilic) to extreme cold (psychrophilic) climatic conditions. Therefore, the principal objective of microbiome research is to elucidate the relationship between microbial diversity and its function in maintaining or restoring the ecosystem. Recent advances in microbial ecology and metagenomic approaches have enabled detailed assessment of the highly complex communities, allowing the establishment of the link between diversity and the function performed by microbes. In this chapter, we will explore some advanced bioinformatic tools for metagenomic studies that can provide quantitative insights into the functional ecology of microbial communities. The detailed study will help us understand the complex microbial diversity in tropical and temperate ecosystems and their functional aspects in ecosystem restoration.

Keywords: Ecosystem restoration, *In silico* approaches, Microbial diversity, Metagenomics, Mesophilic, Psychrophilic, Thermophilic, Tropical ecosystem, Temperate ecosystem.

INTRODUCTION

For ages, nature has always provided us with all the necessities for carrying all forms of life in a wieldy and subtle way. Among all the life forms that live and interact in the ecosystem, humans are known to explore and exploit the ecosystem to meet the rising demands for resources and environmental services. Over the

* **Corresponding author Vinod Kumar Nigam:** Department of Bioengineering & Biotechnology, Birla Institute of Technology, Mesra, Ranchi-835215, Jharkhand, India; E-mail: vknigam@bitmesra.ac.in

Shiv Prasad, Govindaraj Kamalam Dinesh, Murugaiyan Sinduja, Velusamy Sathya, Ramesh Poornima & Sangilidurai Karthika (Eds.)

past 50 years, humankind has changed the ecosystem in a more hasty and extensive way than any dated time in human history [1]. The result of these changes, no doubt contributed to substantial net gains in economic development and the well-being of humans. However, it is achieved at the rising price of degrading the ecosystem services, exacerbating poverty in some groups of society, and, most importantly, increasing the risk of nonlinear changes, leading to the irreversible loss in the momentum and diversity of life on Earth [1]. These inevitable consequences impact the biotic and biophysical conditions of the ecosystem. They will lead to the competing demand for finite space and ecosystem services, ultimately resulting in a dearth of infinite resources, confining the recourse to abandon degraded areas, and shifting toward the exploitation of non-degraded ones [1 - 4].

It is quantified in many theories that the ecosystem had the capacity of self-replenishment from the perturbations, but keeping in mind the present scenario, it always becomes questionable whether the ecosystem is able to recover from the present perturbation. And if so, how long will it take? The self-replenishment theory always states that the ecosystem could gradually be replenished from the perturbation at a rate directly proportionate to the rate at which the perturbation is abated [5]. According to the current human impact, it can be easily speculated that such recovery will take centuries [6]. But, of course, while replenishing the ecosystem, the demands of humankind can neither be neglected nor can be kept to a halt.

So, conservation efforts must create a portfolio of imminent opportunities that keep an equilibrium between environmental protection and environmental services for the future generation of the burgeoning human population. One such approach is to combine natural and modern techniques. Microbes are known from eternities for upholding and reinstating the ecosystem and are found in all forms of the environment while managing their survival along with maintaining the ecosystem. Combining the contemporary science techniques like metagenomic or *in silico* approaches to clear our understating of microbial diversity and protagonist in ecosystem restoration in a broader view will provide great assistance in more progressive, innovative, and prompt repossession of bionetwork to achieve our goal of "meeting the needs of the present and future without compromising our mother nature".

DAMAGING AGENTS OF THE ECOSYSTEM

How is the ecosystem damaged? It is one of the most underrated questions of the era that needs serious attention. The pace of ecosystem damage from anthropogenic and natural impacts is rapidly leading to a biodiversity crisis. The ecosystem services underpinning human livelihood and the effect of destructed ecosystems on human livelihood are constantly being discussed and highlighted internationally. The Intergovernmental Panel for Climate Change Fifth Assessment Report (2014) indicated that the adaptation capacity of the ecosystem is far more limited than that of humans in response to destruction and damage [7]. But our ecosystem is not just limited to humans; our discussion should also not be concise to humankind only. The primary causes of ecosystem damage that are most discussed and well-known are anthropogenic processes and natural hazards. The anthropogenic process is a premeditated human bustle that is non-malicious but may have a negative impact on society and may trigger or catalyze other hazardous or destructive processes [8]. The well-identified examples of such activities are groundwater abstraction, surface mining, vegetation amputation, chemical detonations, infrastructure loading for urbanization, *etc.*

On the other hand, natural hazards are phenomena that occur naturally and may have a negative or deteriorative effect on the ecosystem or biodiversity. Such as earthquakes, landslides, volcanic eruptions, droughts, floods, subsidence, tropical storms, wildfires, *etc* [8]. Natural hazards are a phenomenon that cannot be controlled or prevented. However, nowadays, the major concern is the "interaction," *i.e.*, the effect of one phenomenon on other phenomena (either anthropogenic or natural) [8] (Fig. **1**).

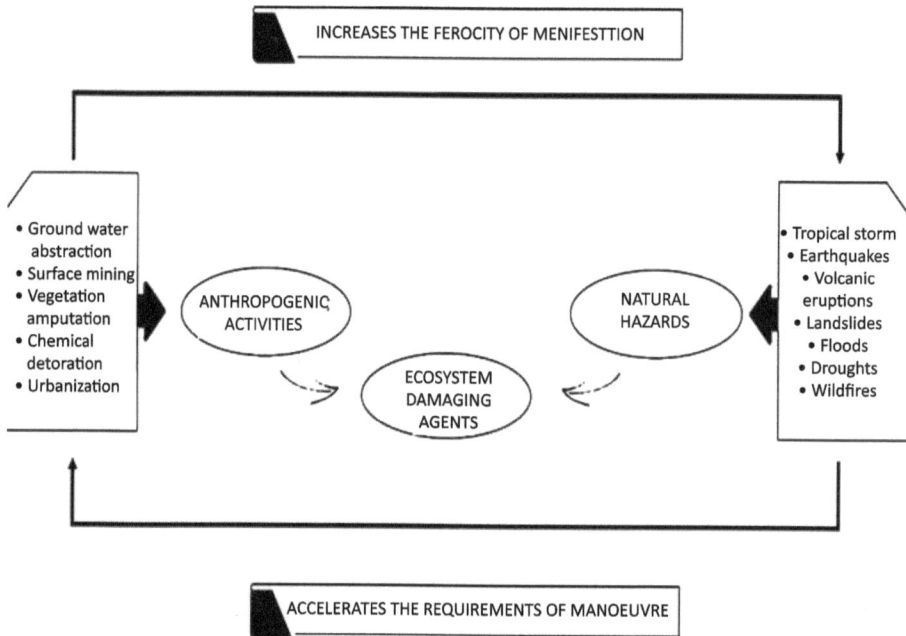

Fig. (1). >Systematic diagram showing the "interaction" of anthropogenic activities and natural hazards.

The occurrences, frequency, and severity of natural deathtraps, such as earthquakes, landslides, floods, collapsing, and sinkholes because of anthropogenic activities, increase the risk and difficulty in managing the calamities [9]. The anthropogenic activities act as a catalyst in ecosystem damage, such as removing groundwater or extracting solid material from the subsurface, leading to decreased pore pressure and a change in overall stress conditions. Mechanically removing water by pumping or intentionally adding water leads to increased erosive capacity and, hence, a shortage of fresh water. The unsystematic land use, such as the removal of trees for commercial or industrial purposes, deforestation to change the agricultural area, converting the forest to grassland for grazing purposes, excavation or removal of mass from the surface of the land for mining and infrastructure development or landfill formation, leads to the increased risk of earthquakes along with polluting the groundwater.

Changes in the flow of mother nature, like constructing reservoirs and dams and diverting the flow of water, can result in the eased surface holding pressure along with the change in surface hydrology, leading to the chances of floods as well droughts in the areas where the water flow has been diverted [8]. The consequences of this lead to the decrease or complete loss of biodiversity near these areas. Intentional detonation of conventional or nuclear explosives results in the generation and release of intense heat and energy, which can trigger high temperatures, leading to a greater extent of the greenhouse effect. The consequences are much harsher than a well-known catastrophe, *i.e.*, global warming, which plays a crucial role in initiating many natural holocausts like the melting of glaciers, depletion of ozone layers, and the most perilous forest fires that result in significant or sometimes complete loss of the forest ecosystem. The loss of the forest ecosystem is a massive forfeiture of human livelihood and the complete ecology. The different harmful chemicals excreted from different kinds of industries or sewage water are generally drained into the water bodies without proper treatment. They pollute the water bodies and cause natural disasters like volcano eruptions, which can blast a range of pollutants into the air, causing acid rain. They further get mixed with the soil and water sources and damage the aquatic and land ecosystems [10].

Modern days' livelihood has made our lifestyle so convenient, but the cost of this is over the necropolis of our ecosystem. The high-tech machinery and the advanced transport we use have saved our time and made our lives luxurious, but the waste they create, such as carbon dioxide, carbon monoxide, or nitric oxide emissions, has worsened the situation [11]. The nuclear waste, radioactive waste, or e-waste generated from all this advanced paraphernalia have created a newer issue as there is no technology to decompose or completely vanish them to date. The only option is to dump them in specific landfills that result in the underground and over-surface emission of radioactive radiation and the loss of a vegetative ecosystem. The area these landfills cover becomes completely useless and results in acres of barren land with an extinct or distorted ecosystem. It is a well-heard statement that "modern problems require modern solutions", but in the urge to find a modern solution, newer issues should not be created [1]. The reasons for ecosystem mutilation might be well known, but the core topic of discussion is to restore the injured ecosystem in a way that will not hinder or halt the development and will not harm our ecosystem any further.

MICROBIAL DIVERSITY

There is a lot of accumulating shreds of evidence indicating that microbes are the most vast and diverse organisms present on Earth. Microbial diversity is well known for its involvement in restoring temperate and tropic ecosystems. For example, anthropogenic activities elevate nitrogen deposition, exacerbating phosphate deficiency in the tropical ecosystem, which is uncommon in the temperate ecosystem [12]. Microbial communities respond to this disturbance in the tropical environment and restore the soil nitrogen by managing nitrogen cycling [13]. Similarly, in temperate grassy woodland, fungal saprotrophs and actinobacteria interact with each other and are involved in litter breakdown, and bacteria are intricated in soil nutrient cycling [14]. However, the critical understanding of microbial response to restoration activity is still obligatory [13, 14]. They carry a sundry array of metabolic activities to survive in the changing environment, and most of these activities are instrumental in fabricating the conditions favorable for the evolution of other forms of life. They are of great interest to the researchers to understand and explore their unique adaptions for survival. They can also be a virtuous source of vestige to reconnoiter more in-depth acquaintance of divergence (Table **1**).

We live on a planet that is predominated by "microbes" [15, 16]. They are the first form of cellular life that subsisted on the earth for more than 3.0 billion years before the macroscopical life form [17]. From that time to the present, microbes have gone through the origination of a spectacular array of diverse metabolic and physiological proficiencies, which help them exploit the horde of environments and microhabitats of the abiotic world. During this manipulating process, they continuously gain energy for their survival and reproduction and alter the geochemical conditions, expanding biotic or abiotic new habitats [17]. This new habitat provides both challenges and survival opportunities to the microbes and continuously leads to the evolution of distinct microbial types, which can endure any biogeochemical changes that take place in the ecosystem.

Table 1. List of industrially/biotechnologically imperative mesophiles, thermophiles, and psychrophiles.

Category	Microorganism	Industrial/ Biotechnological Importance	Refs.
Mesophile	*Acetivibriocellulolyticus*	Municipal waste and sewage sludge water treatment	[31]
Mesophile	*Halocellacellulosilytica*	Paper waste management	[32]
Mesophile	*A.albertis*	Bioleaching	[33]
Mesophile	*Leptosirillum sp.*	Bioleaching	[33]
Mesophile	*P.freudenreichii*	Used in dairy industries, Development of expression vector	[34]
Mesophile	*Lactobacillus sp.*	Dairy industries	[35]
Mesophile	*Candida kefir*	Dairy industries	[35]
Mesophile	*Kluyveromycesmarxianus*	Dairy industries	[35]
Mesophile	*Lactococcus lactis*	Production of bacteriocin to control contamination in cheese production	[36]
Mesophile	*Geotrichumcandidum*	Fermentation industries	[35]
Thermophile	*Pyrococcuswoesei*	Used in the sugar industry and starch processing	[37]
Thermophile	*P.furiosus*	Used in fruit industries, production of alcohol	[38]
Thermophile	*Humicola lanuginose*	Detergent industries	[39]
Thermophile	*Myceliophthorathermophila*	Animal feed industry	[40]
Thermophile	*Bacillus lichniformis*	Food industry, protein recovery, and protein fortification	[41]
Thermophile	*Thermus aquaticus*	Application in molecular biology and genetic engineering	[42]
Thermophile	*Bacillus brevis*	Baking, brewing, and leather industries	[43]
Thermophile	*T.litoralis*	Molecular biology, reverse transcription-PCR	[44]
Thermophile	*Aspergillus terreus*	Breast cancer treatment	[45]
Psychrophile	*Pseudoalteromonashaloplanktis*	Application in molecular biology, production of DNA ligase	[46]
Psychrophile	*Vibro sp.*	Molecular biology application, production of alkaline phosphatase	[47]
Psychrophile	*Sclerotinia borealis*	Cheese ripening, fruit juice, and wine industries	[48]
Psychrophile	*Pseudomonas fluorescens*	Antibacterial agent, food storage, production of alanine racemase	[49]
Psychrophile	*Candida humicola*	Pharmaceutical and cosmetic industries	[50]
Psychrophile	*Alteromonashaloplanktins*	Paper, bleach, and food industries	[51]

From passing the torch of preeminence in life to developing multicellular life to silently diminishing from the center stage of life, microbes are always the driving force and continue to diversify and dominate [17]. Microbes are always famous for their tough survival instincts and are present in every place on earth one can imagine or even in the places where one cannot imagine. Their diversity is always in the limelight, and the discussion can continue for ages, so we will restrict our discussion here mainly to their diversity based on temperature. They are broadly classified into three major groups, *i.e.*, mesophiles, thermophiles, and psychrophiles.

Mesophiles

The term "mesophiles" was first recorded in 1925-30 and is the combination of two words: "meso", which comes from the Greek word *"mésos"* meaning "middle" and "phile", which comes from the Latin word "-philus" meaning "enthusiast for" [18]. As this term specifically refers to microorganisms, it is used for microorganisms that love to live in the moderate or middle-temperature range. Mesophiles or mesophilic microorganisms are the organisms that grow best at a judicious temperature that is neither too hot nor too cold. The optimal temperature for the growth of mesophilic microbes lies in the range of 20°C to 45°C, *i.e.*, 68 to 113°F [18]. They have a diverse classification and belong to bacteria, archaea, and fungi.

In the case of habitat adoption, mesophiles can be aerobic or anaerobic and are the most common organisms in the human microbiome, as the human body temperature is 37°C. Most of them are also known as human pathogens, along with being in the category of essential fermentation microbes. Mesophiles are also the most geographically found microbes because of their optimum growth temperature, which is also the most optimum temperature range of any natural source, making them easily accessible for isolation from different sources under natural circumstances.

Thermophiles

Thermophiles are microorganisms that thrive at relatively high temperatures ranging from 41°C to 122°C, *i.e.*, 106 to 252°F [19]. Isn't it so astonishing to survive in such a high-temperature range being a unicellular life form? But this is the essence of microbial ecology, which never fails to surprise us. "Thermophile" comes from the Greek word "thermotita" and "philia", which mean "heat" and "love", respectively. Mainly, thermophiles are archaea, but bacteria and fungi are nowhere behind in this case. It has been suggested in the literature that thermophilic eubacteria are among the earliest bacteria present [20]. Based on the optimum growth temperature, thermophiles are sorted into three major categories, *i.e.*, facultative thermophiles, obligate thermophiles, and hyperthermophiles [21].

Facultative thermophiles are also called moderate or simple thermophiles, and this group can thrive in both high temperatures, such as 64 °C (147.2°F), as well as low temperatures below or up to 50 °C (122°F). On the other hand, obligate or extreme thermophiles require a temperature range as high as 65-79°C (149-174.2 °F) for survival and growth. At the same time, hyperthermophiles require temperatures above 80°C (176°F) for optimal growth. Many factors are responsible for deciding the optimal growth temperature of microbes, but the most evident explanation is the presence of specific genetic elements (alleles) that can amend the temperature-sensitive phenotype of the organisms [21]. Furthermore, there is a school of thought that thermophiles have more stable cellular components, and according to Gaughran and Allen, it is believed that rapid resynthesis of damaged cellular constituents because of the heat is the key to the stability of thermophiles in high temperatures as compared to mesophiles [22].

Thermophiles have a special arrangement to cope with high temperatures, such as their membrane constitutes saturated fatty acids that maintain cell unification at high temperatures. Their proteins have the litheness at high temperatures to evade deactivation or denaturation at elevated temperatures. They also have a special reverse gyrase that allows the coiling of the molecules into a more heat-resistant form [23]. Most thermophiles are anaerobes as they require sulfur instead of oxygen for their cellular respiration. They can oxidize sulfur to sulfuric acid and, thus, can be adapted to very low pH, *i.e.*, they are acidophiles and thermophiles. They are the potentates of warm/hot and sulfur-rich inhabitants and are usually allied with volcanisms or geothermally frenzied regions of the earth, such as hot springs like Yellowstone Nation Park, deep-sea hydrothermal vents, geysers, fumaroles as well as putrefying plant matter, for example, peat bogs and compost [21]. The enzymes from thermophiles are of great interest and in high demand both in industrial and biological science because of their thermal stability (Table **1**). Besides being commercially important, there is a presence of photosynthetic pigments, which give them characteristic eye-fascinating colours.

Psychrophiles

Psychrophiles come from the ancient Greek word *"psukhrós"*, which means 'cold/frozen loving'. Psychrophiles, called cryophiles, are basically extremophilic organisms and prefer to grow in cold temperatures in contrast to mesophiles and thermophiles. The optimum temperature range for the growth and reproduction of cryophiles is from -20°C to 20°C (-4°F to 68°F) [24]. Psychrophile microbes are generally bacteria or archaea, but some eukaryotes, such as snow algae, lichens, phytoplankton, fungi, and wingless midges, are also sorted as cryophiles. The habitation of psychrophiles is pervasive as a large fraction of our telluric surface experiences temperatures equal to or lower than 10°C [25]. They are found in places that are beyond our mind, such as permafrost, glaciers, snowfields, polar ice, and deep ocean waters.

They are also detected in pockets of sea ice with high salinity, and their microbial activity has also been restrained in soils frozen below -39°C [26]. Not only in cold temperatures, but they have also adapted to the environmental constraints of their habitats, such as high pressure in the deep sea and high salinity in some sea ice [27]. It is apparent that to survive in such freezing temperatures, cryophiles must have some unique adaption. They are secured from bittering and expansion of ice by ice-induced desiccation and vitrification between temperatures -10°C to 26°C as long as they cool slowly [24]. Overcoming the stiffing of their lipid membrane is essential to survive and maintain their functionality. Hence, to accomplish this, they acclimate to the membrane structure having short unsaturated fatty acids as they allow the lipid membrane to have a lesser melting point and intensify its fluidity than long saturated fatty acids [28]. Their membrane also constitutes carotenoids, which help modulate the membrane fluidity [29].

In addition to the above adaptation, their enzymes are hypothesized to have an activity-stability-flexibility type of relationship to adapt to the harsh cold, and the flexibility of the enzyme surges to compensate for the freezing upshot. They also synthesize antifreeze protein, which keeps the internal cellular space liquid to protect the DNA and other cellular components from freezing or crystallizing when the temperature drops below the water's freezing point [29]. Certain cryophiles can transition to the viable but nonculturable (VBNC) state, where they can respire and use substrate for metabolism. It is a reversible state in an optimum environment. Actinobacteria were found to survive in the VBNC state for about 500,000 years in permafrost conditions of Antarctica and Siberia [30].

APPROACHES OF METAGENOMICS

The study of microbial diversity is a crucial step toward understanding microbial ecology in a different ecosystem [52]. In the past two decades, many molecular and bioinformatic tools have advanced, are allied to classic microbiology, and provide new and different aspects of microbial ecology. For example, molecular biology tools have helped researchers to answer some important questions about microbial richness, and microbial abundance sequencing tools provide consistent information about the species of microbes present in the environment [53]. But knowing all this information is not enough. The biggest questions that arise are: what is the functionality of this species? Why are these microbes present in that particular environment? What is their liability? To overcome these challenges, the development and advancement of new technology, which can provide the key answer to all the questions, is required; this is when the metagenomic approaches come into the limelight.

Metagenomic is defined as an approach that provides direct genetic analysis of the genome contained in an environmental sample. In recent years, the advancement of metagenomic tools has opened new horizons for exploring microbial communities and refined our knowledge of microbiological parameters [53]. In addition, it has increased the opportunities for immense studies on microbial diversity for a subterranean comprehension of the composition and purposes that microorganisms play in a wide range of ecosystems. One of the well-known hurdles in understanding and exploring microbial diversity is that not all microbes are culturable in *in vitro* conditions. In the environment, almost 99% of the lineages in the denizen microbial community cannot be cultured [54].

Metagenomics is a powerful tool for studying unculturable microbes in ecological communities without being culture-bias and for understanding the real diversity using rapidly emerging DNA sequencing [54]. Throwing light on understanding the microbial community is necessary for many areas, such as discriminating taxonomic and functional profiles of microbes associated with humans, global ocean microbiomes, and especially different soil ecosystems. Furthermore, in order to facilitate better comparative analysis of results and interpretations of sequence data, the amalgamation of metagenomic data and associated metadata is requisite [55]. There are lots of metagenomic tools/software for the analysis of microbial diversity, but here we will concise our discussion to the preliminary knowledge of some majorly focused and actively used software/tools such as MLST, MOTHUR, EstimateS, QIIME, PHACCS.

Multi-Locus Sequence Typing (MLST)

Multi-locus sequence typing (MLST) is a metagenomic tool for decoding microbial diversity. As the name suggests, MLST is an unambiguous technique for characterizing isolated microbes by typing multiple loci of internal fragments of multiple housekeeping genes using DNA sequencing. MLST was first used for *Neisseria meningitidis*, a human pathogen, but as slowly the versatility of MLST was divulged, it has become one of the vital tools of metagenomics for studying microbial diversity [56]. It is based on the well-known principles of multi-locus enzyme electrophoresis but with a slight deviation. In this technique, alleles are assigned at multiple housekeeping loci directly by DNA sequencing rather than indirectly *via* their gene product electrophoretic mobility [57]. The basic workflow of MLST involves data collection, data analysis, and multi-locus sequence analysis [56] (Fig. **2**).

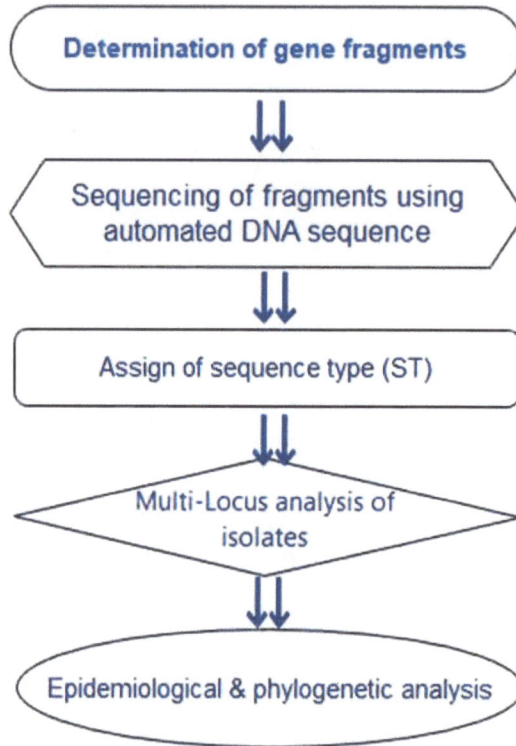

Fig. (2). Basic workflow of multi-locus sequence typing (MLST).

In the first step, gene fragments are determined using a nucleotide sequence. Then, irrefutable identification of variation is obtained. Next, in data analysis, approximately 450-500 bp internal fragments are precisely sequenced on both DNA strands using an automated DNA sequencer. All the unique sequences are given a specific allele number and then combined onto an allelic profile to assign a sequence type (ST).

After combining and assigning the STs, if a new allele is found, it is verified and deposited in the database. In the final multi-locus analysis step, the allelic profile of the isolates is compared to deduce the relatedness among them, and the epidemiological and phylogenetic studies are performed for different clonal complexes by comparing their STs. During this identifying and sequencing process, a huge set of data was generated, so bioinformatics tools are used to manage, arrange, compare, and merge all data in analyzable form. Finally, a dendrogram is constructed using a matrix of pairwise differences in allelic profiles, which shows the relatedness among the isolates. Through the dendrogram, the isolates having identical or very analogous allelic profiles are displayed, which are assumed to be derived from common ancestors. If the difference is more than three loci out of seven loci, then they are likely to be variable and should not be deduced for phylogenetic analysis. The advantage of MLST over other techniques is that the sequence data it uses to analyze are unambiguous, and the allelic profile of the isolates can be easily compared with the other available information in a large central database *via* digital mode [57]. In contrast to MLST, many typing procedures compare the sequence data by comparing DNA fragment size on gel, which is tedious and ambiguous.

MLST also has an advantage in the case of unculturable microbes as allelic profiles can be obtained by PCR amplifying the seven housekeeping loci directly from any sample source, and hence, unculturable isolates can also be precisely characterized [58]. The curb of MLST is that it appears to be best in population genetic studies, but it is expensive, and sometimes, it lacks discriminatory analysis to differentiate between strains because of the conserved sequence of housekeeping genes. So, to expand the discriminatory analysis, a multi-virulenc--locus sequence typing (MVLST) approach has been established, as virulence factors are also important factors for diversity analysis along with population genetics [59].

MOTHUR

MOTHUR was first released in 2009, and it is a package of open software for bioinformatics data processing. Many computational tools like ARB, DOTUR, SONS, LIBSHUFF, UniFrac, rRNA-specific database, AMOVA, HOMOVA, *etc.*, are used to address ecologically relevant questions [60]. Although these tools are widely used, they have some limitations when sequencing capacity increases and the studies become more complex. One such drawback is that all online rRNA-sequencing databases include aligners, classifiers, and analysis databases. However, these tools permit restricted generic analysis, and whether transferring enormous data sets across the internet for analysis is a justifiable process remains a big question.

These analyzing tools were designed to analyze the sequence of 10^2 to $10^{4,}$ but as the number of sequences expanded, the requirement for refactorization of existing software to use more efficient algorithms started coming forward. This boutique nature of existing tools has also restricted their amalgamation and further advancement. As a result of this limitation and as our sequencing capacity upsurges and our research questions become more erudite, the need for easily flexible and maintainable software increases, leading to MOTHUR software's development. It is a single software platform capable of analyzing the data generated from various sequencing methods such as 454 pyrosequencing, illumine HiSeq and MiSeq, sanger, PacBio, and Ion torrent.

MOTHUR can be very effortlessly used to analyze the DNA sequence of uncultured microbes. Moreover, MOTHUR had overcome the drawbacks of other tools by implementing the algorithms present in the preceding tools like SONS, DOTUR, LIBSHUFF, ʃ-LIBSHUFF, Tree Climber, and UniFrac along with the incorporation of additional eye-catching features such as (a) over twenty-five calculators for enumerating significant ecological constraints for measuring α and β diversity, (b) envisioning tools including heat maps, Venn diagrams, and dendrograms, (c) functions for selecting sequence assortments based on quality, (d) a NAST-based sequence aligner(5), (e) a pairwise sequence distance calculator, and (f) the aptitude to call individual command either from within MOTHUR, using files with lists of commands, or directly from the command line [60].

All these additional features in MOTHUR have provided greater flexibility in setting up analysis conduits. Along with these features, MOTHUR is designed in a way that is object-oriented, free, and platform-independent, *i.e.*, instead of realizing MOTHUR once a year with lots of changes, it is better to regularly update with minor changes to address bugs and add newer suggestions and features quickly. MOTHUR has also given leniency to the invaginators to modify the software according to their own analysis method. It also encourages the user to form pages that describe how they have used the software to analyze their data so that the other investigators can also get a varied way of analyzing and designing their experiment [60]. The field of microbial diversity has always faced a great revolution, and software like MOTHUR has opened new possibilities for designing our experiments in a more sophisticated manner.

Quantitative Insight into Microbial Ecology (QIIME)

QIIME is a metagenomic software that accomplishes microbial community analysis. It is used to interpret nucleic acid sequence data of different microbes and to process, analyze, and visualize microbiome data [61]. It is designed to analyze the data generated from the sample of amplicons of marker genes such as 16S rRNA and 18S rRNA, internal transcribes spacer (ITS), cytochrome oxidase I (COI), shotgun metagenomics, and untargeted metabolomics. QIIME performs in 4 basic protocols: (a) Acquiring an example study and demultiplexing DNA sequences, (b) Picking OTUs, assigning taxonomy, inferring phylogeny, and generating an OUT table, (c) Alpha diversity within samples and rarefaction curves, (d) Beta diversity between samples and beta diversity plots [61].

In the first step, analysis was performed on amplicons of marker genes (16S rRNA and 18S rRNA). Then, DNA sequence data from the microbiome and example data set were acquired, and the DNA sequence was assigned to the study of the microbial community. After the completion of acquiring and assigning the DNA sequence, the multiplexed reads are assigned to the samples based on their barcode and are filtered based on the characteristics of each sequence, and the poor-quality or abstruse reads are removed. The second step requires a demultiplexed sequence generated in the first step, and it entails picking operational taxonomic units (OTUs) based on the sequence similarity within the sequence reads and then selecting the representative sequence from each OTU. In this step, taxonomic identities were assigned using a reference database, the OTUs sequence was aligned, and the phylogenetic tree and OTU table were constructed. The phylogenetic tree and OTU table epitomize the copiousness of each OTU in each microbial sample. In the third step, the OTU table and phylogenetic tree constructed in the previous step were used to compute the alpha diversity.

Alpha diversity is basically the diversity that assed within the microbial community [61]. The alpha diversity was computed for each microbial community, and a rarefaction curve, *i.e.*, a graph of diversity v/s sequence depth, was generated. In this section, the phylogenetic tree and OTU table are used to compute the diversity between the community, *i.e.*, the beta diversity of all microbial communities [61]. The principal coordinates analysis (PCoA) plots and distance histogram plots were generated to deduce the relationship between the microbial community and gain knowledge of microbiome diversity [62]. QIIME is one of the promising metagenomic tools that perform straightforward analyses with minimum intervention of users and clears the default protocols performed (Fig. **3**).

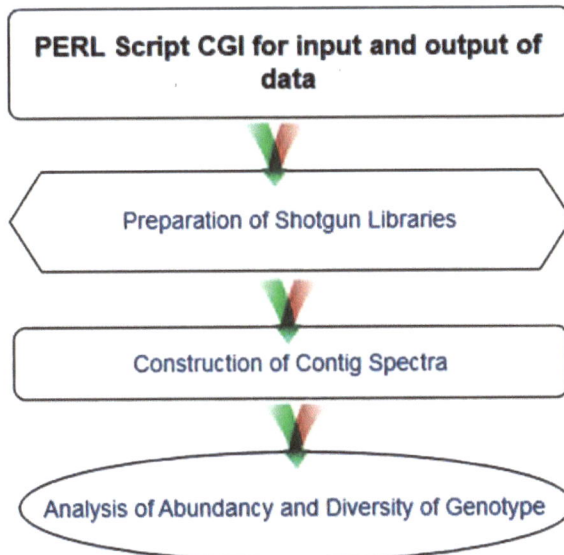

PERL Script CGI for input and output of data

Preparation of Shotgun Libraries

Construction of Contig Spectra

Analysis of Abundancy and Diversity of Genotype

Fig. (3). Flowchart of quantitative insight into microbial ecology (QIIME).

PHAge Communities from Contig Spectrum (PHACCS)

As we know, viruses are one of the important microbes in the ecosystem, and most environmental viruses are phases that infect prokaryotic cells. By blighting the prokaryotes, phages significantly impact the microbial community, but despite their importance, very limited information about phage miscellany is known. The most significant hurdle in learning about viruses or phages is the uncultivability in *in-vitro* conditions. Traditionally, prokaryotic cells were cultivated to study the phage assortment, dynamics, and ecology, and the plates were then infected with the phages [63]. However, through this technique, sufficient information was not possible to gather, as there was no proper method to justify their diversity by morphology or by just cultivating and observing the phages.

So, a new metagenomic approach for gathering more statistics about phage biodiversity is required. This leads to developing PHAge Communities from Contig Spectrum (PHACCS) [64]. To get the natural phage miscellany, PHACCS is an online computational tool that is based on the contig spectrum (Fig. **4**). The shotgun libraries are created from the environmental metagenomes, and the genetic data of each genotype is chronicled and analyzed both qualitatively and quantitatively. The metagenomic data is used in the contig spectrum, which is gritted by arbitrary shotgun DNA fragments. The contig spectrum is actually a vector that consists of assemblies of overlying sequences called contigs of size q, *i.e.*, the number of sequences in the group.

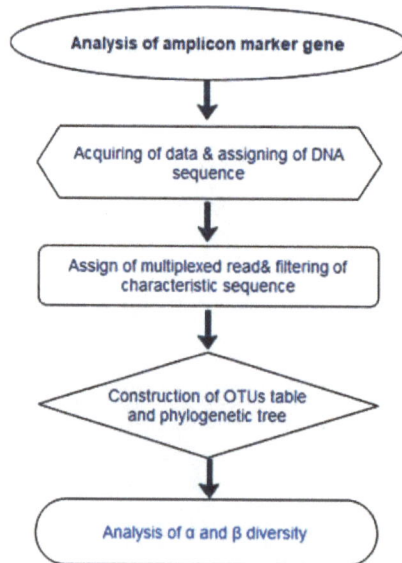

Fig. (4). Flowchart of PHAge Communities From Contig Spectrum (PHACCS).

The stringency in which the parameters are amassed can be speckled accordingly, so the sequence having the same genotype will only overlap. PHACCS, based on the contig spectrum, delivers information about the richness and miscellany of genotypes within a community [64]. The size of the contigs elects the abundance of the particular genotype, *i.e.*, for a single genotype, the larger the contigs in the contig spectrum, the higher the numeral of replicas of that particular genotype and the more copious the genotype. For deducing the diversity, the inputs PHACCS entails are (i) a viral group's contig spectrum, (ii) average genome size, (iii) average shotgun DNA sequence length, and (iv) the minutest overlap length used for mustering. PHACCS has an HTML interface for which it uses a common gateway interface (CGI) script written on PERL for the input and output of data [64].

It also generates rank-abundance curves, which give statistics on the abundance of a particular genotype based on rank. The genotype ranked foremost is the most abundant genotype. Indeed, there are some limitations of PHACCS, such as the contig spectrum obtained gives an approximation of viral diversity because the larger virus and RNA viruses are not characterized in the shotgun library, and the desire to eliminate the false contigs, *i.e.*, contigs between DNA fragments from different genotypes, the true contigs are also omitted, *i.e.*, DNA fragments that are selected as non-overlapping basically when they belong to the same genotype [64]. Despite all these limitations, PHACCS is a boon in the sea of metagenomic software, which has conveniently helped ecologists get the idea of the most dramatic microbes.

TEMPERATE AND TROPICAL ECOSYSTEM RESTORATION

Whenever we hear the word ecosystem restoration, the first question that comes to mind is, what is ecosystem restoration? Is it simply to restore the things back to their original form as they were, or should we modify the damaged ecosystem in a better way according to the need/greed? There used to be the old school thought of ecologists that the ecosystem should be restored to the way it was without any alteration or modification, and only then can it be considered ecosystem restoration. Ecological restoration basically assists the recovery of scratched, destructed, or besmirched ecosystems. The prime goal of ecosystem restoration is to bring the ecosystem to its historic trajectory and not compulsorily to its historic conditions [65].

Some contemporary factors, such as climate change or global warming, can change the ecosystem's trajectory, just like they might have changed the trajectory of the adjacent ecosystem. So, it might be possible that the ecosystem restored is not the same as in the case of structure and function, but our main objective is to restore it to as superlative a condition as possible. To achieve this, we cannot halt our progress, but we can ensure sustainable progress by removing the causative damaging agents from that particular ecosystem and leaving it to self-natural restoration. So, it can establish a self-organized ecosystem on the trajectory of repossession [66]. We can monitor the recovery at different time interludes without disturbing the restoration or recovery progression. So we can develop a healthy ecological relationship with nature. The ecosystem has been divided into polar, temperate, and tropical due to different temperatures and climatic conditions. In this section, we will be primarily focusing on temperate ecosystem restoration and tropical ecosystem restoration.

A temperate ecosystem is chiefly an environment where the biotic communities interact with their non-living environment in a region of moderate climate or intermediate tropical and polar zones having discrete warm or hot summer periods and cool or cold winter periods, respectively. The temperate biome includes the temperate forest present in the zones of North America, Asia, and Europe. The temperate zone covers an enormous area of diversity because of the presence of diverse forests broadly disseminated from temperate deciduous forests to pine wood, grasslands, and geographically circumscribed temperate rainforests [67]. On the contrary, the topical ecosystem lies in the equatorial belt and experiences a warm and humid climate. It receives abundant rainfall because of the convection of the wind. Along with excessive precipitation, it receives a considerable amount of sunlight, thus providing an ideal condition for the growth of luxuriant shrubbery.

The primary regions included under the tropical biome are the Amazon basin in Brazil, the Congo basin in West Africa, Indonesia, *etc*. Both temperate and tropical ecosystems are rich in diversification and contribute significantly to maintaining the ecosystem balance and sustaining human requirements. However, many natural and human activities have damaged this precious ecosystem, such as deforestation, urbanization, overgrazing of animals, biodiversity loss because of the invasion of foreign species, logging, *etc*. So, the primary demand of the scenario is to bring back the dented ecosystem to the recovery trajectory. Microbes have been consistently leading in achieving this objective after removing primary anthropogenic causative agents. Microbes are the most positive and practical approach to ecosystem restoration and sustaining a healthy relationship between nature and humans [68]. Various microbes such as bacteria, fungi, protozoans, and viruses are involved in effective ecosystem restoration as key players. They indulge in many natural activities, such as carbon sinking by excreting mucins that aggregate large particles and act as reservoirs. They also help in nitrogen fixation and recycling nutrients into the soil so that appropriate flora and fauna can be reputable [69]. They also help in the formation of soil and maintain the pH of the soil so that the degraded temperate and tropical ecosystems can be restored.

The knowledge of the diversity of the microbes in a particular ecosystem and comparing them with the existing diversity acquaintance through metagenomic tools can aid us in knowing their role in the refurbishment of temperate and tropical ecosystems. Microbes underwrite in upholding the atmospheric carbon dioxide and nitrogen that are worn out by the greenhouse effect and hence help in the restorations of temperate and tropical forest ecosystems. About 50% of oxygen and carbon dioxide are produced by bacteria. They act as a primary successor in the ecosystem succession pyramid and change the environment in the course of succession to evolve a newer ecosystem or restore the prior ecosystem [70]. The list of microbes and their role in ecosystem restoration is mentioned in Table **2**.

Table 2. Imperative group of microbes and their role in ecological refurbishment [71].

Microbes	Beneficial Progressions	Role in Ecological Refurbishment
Archaea	Involved in elemental alteration like nitrification, methanogenesis, *etc.*	Regulates nutrient cycle, atmospheric and environmental conditions
Chemoautotrophic bacteria	Helps in sulfate reduction and iron oxidation	Purifies water manages the climatic regulation and nutrient cycling
Heterotrophic bacteria	Cessation of organic matter, mineralization, polymer production, *etc.*	Carbon sequestration regulates decomposition and environmental cleaning
Photoautotrophic bacteria	Photosynthesis	Oxygen and carbon dioxide regulation, maintaining flora and fauna
Fungi, Arbuscular mycorrhizal fungi	Mineralization, nutrient cycling, and consumption of organic matter	Regulates soil formation and nutrient cycling and helps in the repair of the soil ecosystem
Protozoans	Ingesting of pathogenic microbes and wastewater management	Disintegration of organic matters, nutrient cycling
Viruses	lysis	Increases availability of nutrients

Metagenomics and microbial diversity play crucial roles in restoring ecosystems. They provide valuable insight into microbial community's composition, function, and interactions for ecosystem functioning and resilience [72 - 74]. Metagenomics involves sequencing and analyzing DNA extracted from the ecosystem to characterize microbial communities and their genetic potential [72, 75]. This approach helps to assess disturbances due to ecosystem degradation and microbial community dynamics [73, 76, 77]. Understanding microbial diversity and interactions is crucial for promoting ecosystem functioning, nutrient cycling, and resilience. The role of metagenomics approaches and microbial diversity in ecosystem restoration is presented in Fig. (**5**).

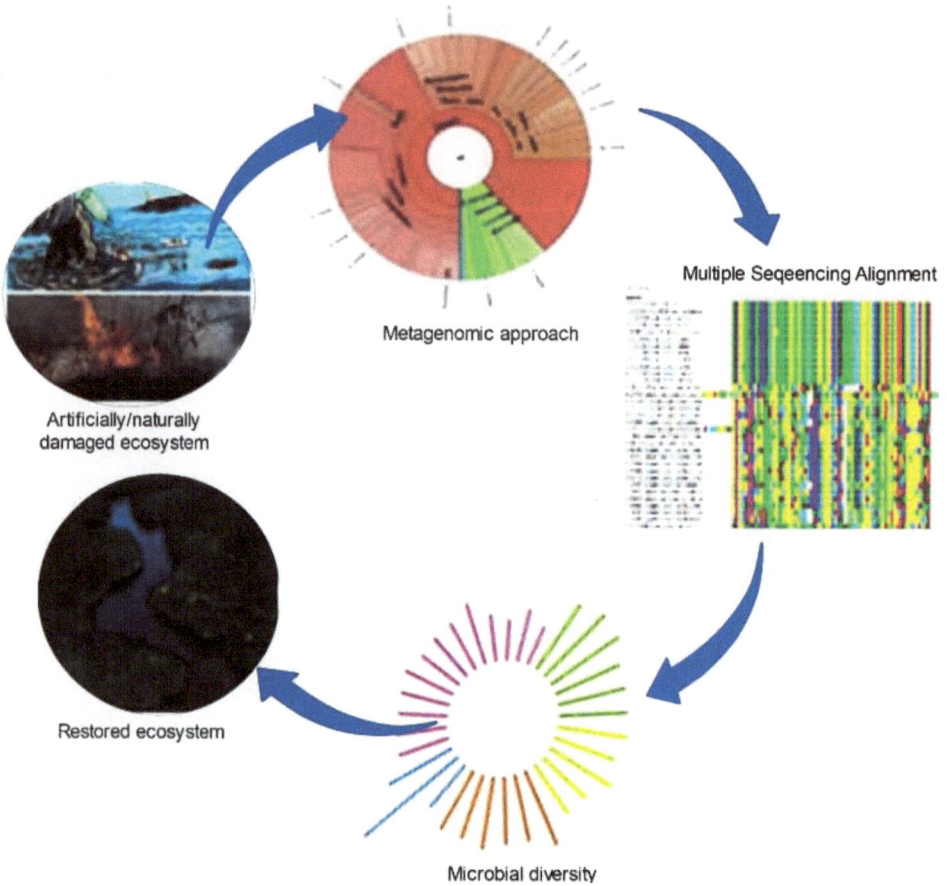

Metagenomic approach

Multiple Seqeencing Alignment

Artificially/naturally damaged ecosystem

Restored ecosystem

Microbial diversity

Fig. (5). Role of metagenomics approaches and microbial diversity in ecosystem restoration.

CONCLUSION

Restoration of the ecosystem is one of the most imperative segments for present ecological conditions to mitigate the changing environment and obtain the utmost ecosystem assistance by upholding a balanced and mutual relationship with the ecosystem for overall sustainable development. Ecologists have carried out many artificial projects to restore the ecosystem for decades, but all the projects have either attained limited triumph or been botched completely. Hence, to break this gigantic delinquent, a potential solution is understanding the natural science of restoration of ecology. Microbes are ubiquitous, providing many indispensable elements to the ecosystem by providing sustainable plant throughput and maintaining the temperate and tropical ecosystem. They also stabilize the environment to ensure that it is optimum for human life. The rich soil microbial diversity plays a central role in the functioning of the ecosystem and in preserving the soil environment, which is essential for nourishing plant diversity in temperate and tropical forest ecosystems. Exploring this beneficial microbiome is essential by using our advanced metagenomic tools, which act as a boon to reconnoiter microbial diversity and their connection with the ecosystem. The information on the microbial community will help in reinventing the diverse inherent plant communities in temperate and tropical ecosystems. It will also help in soil formation, reinstating plant coteries within phonological cycles, and establishing late-succession plant communities that were extinct or endangered during temperate/tropical ecosystem dilapidation.

ACKNOWLEDGEMENTS

The facility and support from the Department of Bioengineering and Biotechnology, Birla Institute of Technology, Mesra, are duly acknowledged.

REFERENCES

[1]   Reid W, Mooney H, Cropper A, Capistrano D. Ecosystem and human well-being: synthesis: A Millennium ecosystem assessment. Washington, DC: Island Press 2005.

[2]   Freedman B. Resources and Sustainable Development. In: Freedman B, Ed. Environmental Science a Canadian prespective. Dalhousie University Libraries Digitial Editions 2018; pp. 1-37.

[3]   Nuss P, Günther J, Kosmol J, Golde M, Müller F, Frerk M. Monitoring framework for the use of natural resources in Germany. Resour Conserv Recycling 2021; 175(105858): 105858.
[http://dx.doi.org/10.1016/j.resconrec.2021.105858]

[4]   Wassie SB. Natural resource degradation tendencies in Ethiopia: a review. Environ Syst Res 2020; 9(1): 33.
[http://dx.doi.org/10.1186/s40068-020-00194-1]

[5]   McLauchlan KK, Craine JM, Oswald WW, Leavitt PR, Likens GE. Changes in nitrogen cycling during the past century in a northern hardwood forest. Proc Natl Acad Sci USA 2007; 104(18): 7466-70.
[http://dx.doi.org/10.1073/pnas.0701779104] [PMID: 17446271]

[6]   Kareiva P, Watts S, McDonald R, Boucher T. Domesticated nature: shaping landscapes and ecosystems for human welfare. Science 2007; 316(5833): 1866-9.
[http://dx.doi.org/10.1126/science.1140170] [PMID: 17600209]

[7]   Zoomers, Z., Wrathall, D., & van der Geest, K. Loss and damage to ecosystem services. UNU_EHS Working Paper Series, No.2, Bonn, United Nations Institute For Environment and Human security.2014:1-22.

[8]   Gill JC, Malamud BD. Anthropogenic processes, natural hazards, and interactions in a multi-hazard framework. Earth Sci Rev 2017; 166: 246-69.
[http://dx.doi.org/10.1016/j.earscirev.2017.01.002]

[9]   McGuire B, Maslin M. Climate forcing of geological Hazards. John Wiley& Sons 2012.

[10]   Bradford, AAcid rain: Causes, effects and solution. Ben Biggs. 2022.

[11]   Dey S, Mehta N. Automobile pollution control using catalysis. Resources. Environment and Sustainability 2020; 2: 100006.

[12]   Vitousek PM, Porder S, Houlton BZ, Chadwick OA. Terrestrial phosphorus limitation: mechanisms, implications, and nitrogen–phosphorus interactions. Ecol Appl 2010; 20(1): 5-15.
[http://dx.doi.org/10.1890/08-0127.1] [PMID: 20349827]

[13]   Pajares S, Bohannan B, Souza V. The role of microbial communities in tropical ecosystem. Front Microbiol 1805; 2016(7): 1-3.

[14]   Hamonts K, Bissett A, Macdonald BCT, Barton PS, Manning AD, Young A. Effects of ecological restoration on soil microbial diversity in a temperate grassy woodland. Appl Soil Ecol 2017; 117-118: 117-28.
[http://dx.doi.org/10.1016/j.apsoil.2017.04.005]

[15]   Gould S. Full House. New York: Three Rivers Press 1996.
[http://dx.doi.org/10.4159/harvard.9780674063396]

[16]   Woese C. The quest for Darwin's grail. ASM News-American Society of Microbiology 1999; 65(5): 260-5.

[17]   Dunlap P. Microbial diversity. Elsevier Inc. 2001; pp. 191-205.

[18]   Willey J, Sherwood M, Woolverton C, Prescott LM. Prescott, Harley: and Klein's Microbiology. New York: McGraw-Hill Higher Education 2008.

[19]   Madigan M, Martino J. Brock Biology of Microorganisms. Pearson 2006.

[20]   Horiike T, Miyata D, Hamada K, *et al.* Phylogenetic construction of 17 bacterial phyla by new method and carefully selected orthologs. Gene 2009; 429(1-2): 59-64.
[http://dx.doi.org/10.1016/j.gene.2008.10.006] [PMID: 19000750]

[21]   Stetter KO. History of discovery of the first hyperthermophiles. Extremophiles 2006; 10(5): 357-62.
[http://dx.doi.org/10.1007/s00792-006-0012-7] [PMID: 16941067]

[22]   Koffler H. Protoplasmic difference between mesophiles and thermophiles. Bacteriol Rev 1957; 21(4): 227-40.

[23]   Zuberer DA, Zibilske LM. Composting: the microbiological processing of organic wastes.InPrinciples and Application of soil. Microbiology 2021; 655-79.

[24]   Clarke A, Morris GJ, Fonseca F, Murray BJ, Acton E, Price HC. A low temperature limit for life on earth. PLoS One 2013; 8(6): e66207.
[http://dx.doi.org/10.1371/journal.pone.0066207] [PMID: 23840425]

[25]   D'Amico S, Collins T, Marx JC, Feller G, Gerday C, Gerday C. Psychrophilic microorganisms: challenges for life. EMBO Rep 2006; 7(4): 385-9.
[http://dx.doi.org/10.1038/sj.embor.7400662] [PMID: 16585939]

[26]   Panikov NS, Flanagan PW, Oechel WC, Mastepanov MA, Christensen TR. Microbial activity in soils frozen to below −39°C. Soil Biol Biochem 2006; 38(4): 785-94.
[http://dx.doi.org/10.1016/j.soilbio.2005.07.004]

[27]   Feller G, Gerday C. Psychrophilic enzymes: hot topics in cold adaptation. Nat Rev Microbiol 2003; 1(3): 200-8.
[http://dx.doi.org/10.1038/nrmicro773] [PMID: 15035024]

[28]   Erimban S, Daschakraborty S. Cryostabiliztion of the cell membrane a psychrotolerant bacteria *via* home viscous adaptation. J Phys Chem Lett 2020; 11(18): 7709-16.
[http://dx.doi.org/10.1021/acs.jpclett.0c01675] [PMID: 32840376]

[29]   Chattopadhyay MK. Mechanism of bacterial adaptation to low temperature. J Biosci 2006; 31(1): 157-65.
[http://dx.doi.org/10.1007/BF02705244] [PMID: 16595884]

[30]   De Maayer P, Anderson D, Cary C, Cowan DA. Some like it cold: understanding the survival strategies of psychrophiles. EMBO Rep 2014; 15(5): 508-17.

[http://dx.doi.org/10.1002/embr.201338170] [PMID: 24671034]

[31]   Lu Y, Lai Q, Zhang C, *et al.* Characteristics of hydrogen and methane production from cornstalks by an augmented two- or three-stage anaerobic fermentation process. Bioresour Technol 2009; 100(12): 2889-95.
[http://dx.doi.org/10.1016/j.biortech.2009.01.023] [PMID: 19231169]

[32]   Tang YQ, Ji P, Hayashi J, Koike Y, Wu XL, Kida K. Characteristic microbial community of a dry thermophilic methanogenic digester: its long-term stability and change with feeding. Appl Microbiol Biotechnol 2011; 91(5): 1447-61.
[http://dx.doi.org/10.1007/s00253-011-3479-9] [PMID: 21789494]

[33]   Kumar P, Bhim J, Paliwal A, *et al.* Biotechnology and microbial standpoint choot in bioremediation InSmart Bioremediation Technology. Academic Press 2019; pp. 137-58.

[34]   Thierry, A., Falentin, H., & Deutsch, G. Jan. Bacteria, Beneficial *Propionibacterium* spp., pp 403-411.Encyclopedia of Dairy Science. Elsevier, London, United Kingdom. 2011.

[35]   Roginski, H. Fermented Milks: Nordic fermented milks. Encyclopedia of dairy science. 2011 496-502.

[36]   Verma, A., Banerjee, R., Dwivedi, H., & Juneja, V.Bacteriocins: Potential in food preservation. Encyclopedia of Food Microbiology.2014;108-186.

[37]   Alquéres SMC, Almeida RV, Clementino MM, *et al.* Exploring the biotechnologial applications in the archaeal domain. Braz J Microbiol 2007; 38(3): 398-405.
[http://dx.doi.org/10.1590/S1517-83822007000300002]

[38]   Antranikian G, Vorgias C, Bertoldo C. Extreme environments as a resource for microorganisms and novel biocatalysts. Advance Biochemical Engineering and Biotechnology 2005; pp. 219-62.
[http://dx.doi.org/10.1007/b135786]

[39]   Arima K, Liu WH, Beppu T. Studies on the lipase of thermophilic fungus *Humicola lanuginosa*. Agric Biol Chem 1972; 36(5): 893-5.
[http://dx.doi.org/10.1080/00021369.1972.10860340]

[40]   Wyss M, Brugger R, Kronenberger A, *et al.* Biochemical characterization of fungal phytases (myo-inositol hexakisphosphate phosphohydrolases): catalytic properties. Appl Environ Microbiol 1999; 65(2): 367-73.
[http://dx.doi.org/10.1128/AEM.65.2.367-373.1999] [PMID: 9925555]

[41]   Synowiecki J. Thermostable enzyme in food processing, in recent research development in food biotechnology enzyme as additives or processing aids. Kerala: Research Signpost 2008.

[42]   Jones MD, Foulkes NS. Reverse transcription of mRNA by *Thermus aquaticus* DNA polymerase. Nucleic Acids Res 1989; 17(20): 8387-8.
[http://dx.doi.org/10.1093/nar/17.20.8387] [PMID: 2478963]

[43]   Banerjee UC, Sani RK, Azmi W, Soni R. Thermostable alkaline protease from Bacillus brevis and its characterization as a laundry detergent additive. Process Biochem 1999; 35(1-2): 213-9.
[http://dx.doi.org/10.1016/S0032-9592(99)00053-9]

[44]   Perler FB, Comb DG, Jack WE, *et al.* Intervening sequences in an Archaea DNA polymerase gene. Proc Natl Acad Sci USA 1992; 89(12): 5577-81.
[http://dx.doi.org/10.1073/pnas.89.12.5577] [PMID: 1608969]

[45]   Liao WY, Shen CN, Lin LH, *et al.* Asperjinone, a nor-neolignan, and terrein, a suppressor of ABCG2-expressing breast cancer cells, from thermophilic Aspergillus terreus. J Nat Prod 2012; 75(4): 630-5.
[http://dx.doi.org/10.1021/np200866z] [PMID: 22360613]

[46]   Georlette D, Jónsson ZO, Van Petegem F, *et al.* A DNA ligase from the psychrophile *Pseudoalteromonas haloplanktis* gives insights into the adaptation of proteins to low temperatures. Eur J Biochem 2000; 267(12): 3502-12.
[http://dx.doi.org/10.1046/j.1432-1327.2000.01377.x] [PMID: 10848966]

[47] Hauksson JB, Andrésson ÓS, Ásgeirsson B. Heat-labile bacterial alkaline phosphatase from a marine Vibrio sp. Enzyme Microb Technol 2000; 27(1-2): 66-73.
[http://dx.doi.org/10.1016/S0141-0229(00)00152-6] [PMID: 10862903]

[48] Takasawa T, Sagisaka K, Yagi K, *et al.* Polygalacturonase isolated from the culture of the psychrophilic fungus *Sclerotinia borealis*. Can J Microbiol 1997; 43(5): 417-24.
[http://dx.doi.org/10.1139/m97-059] [PMID: 9165700]

[49] Yokoigawa K, Okubo Y, Kawai H, Esaki N, Soda K. Structure and function of psychrophilic alanine racemase. J Mol Catal, B Enzym 2001; 12(1-6): 27-35.
[http://dx.doi.org/10.1016/S1381-1177(00)00200-9]

[50] Ramana K, Singh L, Dhaked R. Biotechnological application of psychrophiles and their habitable to low temperature. J Sci Ind Res (India) 2000; 87-101.

[51] Margesin R, Miteva V. Diversity and ecology of psychrophilic microorganisms. Res Microbiol 2011; 162(3): 346-61.
[http://dx.doi.org/10.1016/j.resmic.2010.12.004] [PMID: 21187146]

[52] Atlas R. Diversity of microbial communities In advance in microbial ecology. New York: Springer 1984; pp. 1-47.

[53] Mendes LW, Braga LPP, Navarrete AA, Souza DG, Silva GGZ, Tsai SM. Using metagenomics to connect microbial community biodiversity and functions. Curr Issues Mol Biol 2017; 24(1): 103-18.
[http://dx.doi.org/10.21775/cimb.024.103] [PMID: 28686570]

[54] Handelsman J. Metagenomics: application of genomics to uncultured microorganisms. Microbiol Mol Biol Rev 2004; 68(4): 669-85.
[http://dx.doi.org/10.1128/MMBR.68.4.669-685.2004] [PMID: 15590779]

[55] Pagani I, Liolios K, Jansson J, *et al.* The Genomes OnLine Database (GOLD) v.4: status of genomic and metagenomic projects and their associated metadata. Nucleic Acids Res 2012; 40(D1): D571-9.
[http://dx.doi.org/10.1093/nar/gkr1100] [PMID: 22135293]

[56] Maiden MCJ, Bygraves JA, Feil E, *et al.* Multilocus sequence typing: A portable approach to the identification of clones within populations of pathogenic microorganisms. Proc Natl Acad Sci USA 1998; 95(6): 3140-5.
[http://dx.doi.org/10.1073/pnas.95.6.3140] [PMID: 9501229]

[57] Urwin R, Maiden MCJ. Multi-locus sequence typing: a tool for global epidemiology. Trends Microbiol 2003; 11(10): 479-87.
[http://dx.doi.org/10.1016/j.tim.2003.08.006] [PMID: 14557031]

[58] Maiden MCJ, van Rensburg MJJ, Bray JE, *et al.* MLST revisited: the gene-by-gene approach to bacterial genomics. Nat Rev Microbiol 2013; 11(10): 728-36.
[http://dx.doi.org/10.1038/nrmicro3093] [PMID: 23979428]

[59] Zhang W, Jayarao BM, Knabel SJ. Multi-virulence-locus sequence typing of Listeria monocytogenes. Appl Environ Microbiol 2004; 70(2): 913-20.
[http://dx.doi.org/10.1128/AEM.70.2.913-920.2004] [PMID: 14766571]

[60] Schloss PD, Westcott SL, Ryabin T, *et al.* Introducing mothur: open-source, platform-independent, community-supported software for describing and comparing microbial communities. Appl Environ Microbiol 2009; 75(23): 7537-41.
[http://dx.doi.org/10.1128/AEM.01541-09] [PMID: 19801464]

[61] Kuczynski J, Stombaugh J, Walter W, *et al.* Using QIIME to analyze 16S rRNA gene sequence from microbial communities. Curr Protoc Bioinformatics 2012; 27(1): 1-20.

[62] Caporaso JG, Kuczynski J, Stombaugh J, *et al.* QIIME allows analysis of high-throughput community sequencing data. Nat Methods 2010; 7(5): 335-6.
[http://dx.doi.org/10.1038/nmeth.f.303] [PMID: 20383131]

[63]   Staley JT, Konopka A. Measurement of *in situ* activities of nonphotosynthetic microorganisms in aquatic and terrestrial habitats. Annu Rev Microbiol 1985; 39(1): 321-46.
[http://dx.doi.org/10.1146/annurev.mi.39.100185.001541] [PMID: 3904603]

[64]   Angly F, Rodriguez-Brito B, Bangor D, *et al.* PHACCS, an online tool for estimating the structure and diversity of uncultured viral communities using metagenomic information. BMC Bioinformatics 2005; 6(1): 41.
[http://dx.doi.org/10.1186/1471-2105-6-41] [PMID: 15743531]

[65]   Gann GD, McDonald T, Walder B, *et al.* International principles and standards for the practice of ecological restoration. Second edition. Restor Ecol 2019; 27(S1): S1-S46.
[http://dx.doi.org/10.1111/rec.13035]

[66]   Walker LR, Walker J, Hobbs R. Linking restoration and ecological succession. Springer-Verlag New York 2007; p. 188.
[http://dx.doi.org/10.1007/978-0-387-35303-6]

[67]   Gilliam FS. Forest ecosystems of temperate climatic regions: from ancient use to climate change. New Phytol 2016; 212(4): 871-87.
[http://dx.doi.org/10.1111/nph.14255] [PMID: 27787948]

[68]   Suding K, Higgs E, Palmer M, *et al.* Committing to ecological restoration. Science 2015; 348(6235): 638-40.
[http://dx.doi.org/10.1126/science.aaa4216] [PMID: 25953995]

[69]   Ducklow H. Microbial services: challenges for microbial ecologists in a changing world. Aquat Microb Ecol 2008; 53(1): 13-9.
[http://dx.doi.org/10.3354/ame01220]

[70]   Kling, G. Microbes: Transformers of matter and material. Ann Arbor, MI: University of Michigan: Ecology and Evolutionary Biology. 2010.

[71]   Singh A, Sisodia A, Sisodia V, Padhi M. New and Future Development In Microbial Biotechnology and Bioengineering Role of microbes in restoration ecology and ecosystem services. Elsevier publications 2019; pp. 57-68.

[72]   Hohenlohe PA, Funk WC, Rajora OP. Population genomics for wildlife conservation and management. Mol Ecol 2021; 30(1): 62-82.
[http://dx.doi.org/10.1111/mec.15720] [PMID: 33145846]

[73]   Stach TL, Sieber G, Shah M, *et al.* Temporal disturbance of a model stream ecosystem by high microbial diversity from treated wastewater. MicrobiologyOpen 2023; 12(2): e1347.
[http://dx.doi.org/10.1002/mbo3.1347] [PMID: 37186231]

[74]   Graham EB, Knelman JE. Implication of microbial community assembly for ecosystemresoration: patterns, process,potential. EcoEvoRxiv 2022; 1-34.

[75]   Zhang L, Chen F, Zeng Z, *et al.* Advances in metagenomics and its application in environmental microorganisms. Front Microbiol 2021; 12: 766364.
[http://dx.doi.org/10.3389/fmicb.2021.766364] [PMID: 34975791]

[76]   Dong C, Liu X, Tang F, Qiu S. How upstream innovativeness of ecosystems affects firms' innovation: The contingent role of absorptive capacity and upstream dependence. Technovation 2023; 124: 102735.
[http://dx.doi.org/10.1016/j.technovation.2023.102735]

[77]   Fan X, Hao X, Hao H, Zhang J, Li Y. Comprehensive Assessment Indicator of Ecosystem Resilience in Central Asia. Water 2021; 13(2): 124.
[http://dx.doi.org/10.3390/w13020124]

CHAPTER 10

Basic and Traditional Microbial Techniques in Ecosystem Restoration

R.V. Akil Prasath[1,*], S. Akila[2], M. Shankar[3], R. Raveena[4], M. Prasanthrajan[5], K. Boomiraj[4], S. Karthika[4] and Selvaraj Keerthana[4]

[1] *Department of Environmental Science and Management, Bharathidasan University, Tiruchirappalli–620024, India*

[2] *National Agro Foundation, Research & Development Centre, Anna University Taramani Campus, Taramani, Chennai, Tamil Nadu-600113, India*

[3] *Bureau of Plant Genetic Resources, ICAR- Indian Agricultural Research Institute, New Delhi, India*

[4] *Department of Environmental Sciences, Tamil Nadu Agricultural University, Coimbatore, Tamil Nadu-641003, India*

[5] *Centre for Agricultural Nanotechnology, Tamil Nadu Agricultural University, Coimbatore, Tamil Nadu-641003, India*

Abstract: The onset of the anthropogenic destruction of ecosystems is one of the ongoing problems that can threaten the existence of organisms, including humans. The emerging problem can be effectively addressed through restoration ecology, a nature-based solution that promises to be cost-effective. Microorganisms, including bacteria, fungi, and viruses, are omnipresent and provide numerous benefits to the ecosystem, such as sustainable plant productivity, enriched soil nutrients, increased soil carbon pool, decomposition, and a stable environment for human life. Soil microorganisms also play a fundamental role in ecosystem functioning and conserving plant diversity. Exploring voluminous beneficial microorganisms and promoting the reestablishment of these beneficial microbes in the soil will preserve Earth's diverse native plant populations, which, in turn, will help in improving soil and be a vital player in enhancing ecosystem primary productivity, food chain, and locking away atmospheric carbon into its plant body and soil. Microbial restoration can be achieved by basic and traditional methods, *i.e.*, (i) by treating the soil with organic matter-rich manure harvested from bio piles, (ii) composting, (iii) graze manuring, (iv) natural manuring, and (v) plant-assisted microbial restoration technique. Regenerative/carbon farming can also be practiced in parallel to enhance the restoration rate and protect beneficial microbial life in the soil. However, the increasing use of microbial inoculants is also raising several queries about their effectiveness and their impacts on autochthonous soil microorganisms, which should be cautiously considered before introducing bioinoculants for restoration. Even if bioinoculants restore the microbial community,

---

* **Corresponding author R.V. Akil Prasath:** Department of Environmental Science and Management, Bharathidasan University, Tiruchirappalli–620024, India; E-mail: akilprasath13@gmail.com

Shiv Prasad, Govindaraj Kamalam Dinesh, Murugaiyan Sinduja, Velusamy Sathya, Ramesh Poornima & Sangilidurai Karthika (Eds.)

they have the following shortcomings: (i) prolonged persistence of microbial colonies and detection in soil; (ii) the monitoring of the impact of the introduced bioinoculants on native soil microbial communities, which needs to be monitored examined periodically. This chapter delves into fundamental and conventional techniques and approaches that can be employed to maintain soil microbial populations. Furthermore, the chapter investigates the possibility of creating protocols for regulatory or commercial objectives, emphasizing the significance of ecological restoration by using bioinoculants or microbial colonies in degraded sites.

Keywords: Ecosystem restoration, Ecosystem functioning, Microorganisms, Microbial colonies.

INTRODUCTION

Land degradation has become a global environmental issue, affecting both natural and human systems. Restoration of degraded land is now indispensable since land degradation adversely impacts soil productivity, resulting in a decline in soil organic matter, soil microbial diversity, and water quality, which consequently affects ecosystemic functions and processes [1]. At the same time, microorganisms play a pivotal role in the mineralization, oxidation, reduction, and immobilization of both organic and mineral materials within the soil matrix. Basic and traditional microbial techniques are now envisaged as a viable solution in ecosystem restoration involving microorganisms to improve soil health and the functioning of degraded ecosystems [2]. Microbes, including bacteria and fungi, are crucial components of healthy, productive ecosystems, playing a pivotal role in soil formation, nutrient cycling, and plant growth. Introducing specific beneficial microorganisms into degraded soils can help improve soil quality, increase nutrient availability, and enhance plant growth [3]. Basic and traditional microbial techniques, such as composting, bio piles, graze manure-based restoration, inoculation with beneficial symbiotic bacteria, regenerative agriculture/restoration, and mycorrhizal inoculation, have been practiced by civilizations for centuries to improve soil fertility [4]. These techniques are largely feasible, low-cost, low-input, and environmentally friendly, altogether making them ideal for use in areas with limited resources or where chemical inputs are undesirable. The current chapter aids in understanding the ecological principles underlying basic and traditional microbial techniques, their conceivable applications in ecosystem restoration, and their importance for endorsing more active, sustainable restoration practices.

By increasing soil qualities, microorganisms can play a key role in recovering degraded lands and facilitating ecosystem restoration. Studies suggest that microorganisms can lessen soil degradation and increase soil characteristics, including fertility [5]. This will encourage plant growth, subsequent ecosystem restoration, and new organic matter produced by bacteria. Collaboration between the domains of microbiology and soil, which has not been very common up to this point, is necessary for such a restoration plan to understand soil microbial dynamics. We accept "ecological restoration" as the most inclusive and enduring phrase to describe the science and practice of repairing harm, deterioration, and destruction at all scales, from a single organism to entire landscapes. Soil microbial communities are an essential part of ecosystems and play crucial roles in ecological processes related to the cycling of carbon (C), nitrogen (N), and phosphorus (P) [6]. Evaluating ecosystem restoration requires an investigation of microbial diversity. The impacts of biochar on microbial biomass and microbial communities, including bacteria, archaea, and fungi, have been the subject of numerous studies. Ecological functions can only be partially understood by studying microbial communities alone.

The ecosystem services delivered by soil, the living skin of Earth, are vital for life: The majority of our antibiotics come from bacteria that live in soil, which also serves as a filter and water storage system, a habitat for a wide variety of creatures, and a medium for plants and heterotrophs to develop in. Climate change, population growth, and land degradation, such as carbon loss, biodiversity decrease, and erosion, present a growing set of challenges for humanity [7]. For example, soil hydrological function is decreased by land degradation, which also affects various other ecosystem services. Changes in hydraulic operation, infiltration and soil moisture storage, carbon cycling, biological activity, movement of nutrients and pollutants, and plant development all contribute to these effects [8]. As soil ecosystem degradation has a growing impact on biodiversity, food production, climate regulation, and human livelihoods, it is essential to understand its effects and take steps to reverse them.

Most soil life consists of microorganisms that control nutrient cycling, break down organic matter, control soil-borne plant diseases, define soil structure, and increase plant productivity [9]. Soil microorganisms facilitate the alteration of the soil environment. On the other hand, research to date has concentrated chiefly on how microbial communities adapt to environmental change, and microbial community structure and diversity can also be employed as indicators of soil health. It was proposed in an article on restoration ecology written 13 years ago that microbial populations, when adequately maintained to support the restoration process and system health, serve as markers of change and support the regeneration of degraded ecosystems. Since then, many researchers have considered the possibility of using microorganisms as ecosystem regulators, particularly to increase crop productivity and engineer dryland restoration. These studies show that microorganisms have the power to change soil functions fundamentally.

BASIC AND TRADITIONAL MICROBIAL TECHNIQUE

Using microorganisms to remove various polluting wastes is a technique known as bioremediation [10]. It is both environmentally acceptable and economically advantageous. *Ex-situ* and *in-situ* bioremediation services are used to treat heavy metal waste, petroleum hydrocarbons, agro-industry, dyestuff, agrochemicals, organic and volatile compounds, lignocellulose biomass, and nuclear waste. These services include biopiling, composting, land farming, bioventing, biosparging, biostimulation, and bioaugmentation. Remediation techniques using microbes are presented in Fig. (**1**). The bioremediation process uses a variety of microorganisms (natural, exotic, and created), each of which has a unique metabolic capacity and the ability to produce a variety of enzymes that fall into one of six categories: oxidoreductases, transferases, hydrolases, lyases, isomerases, and ligases (synthetases) [11].

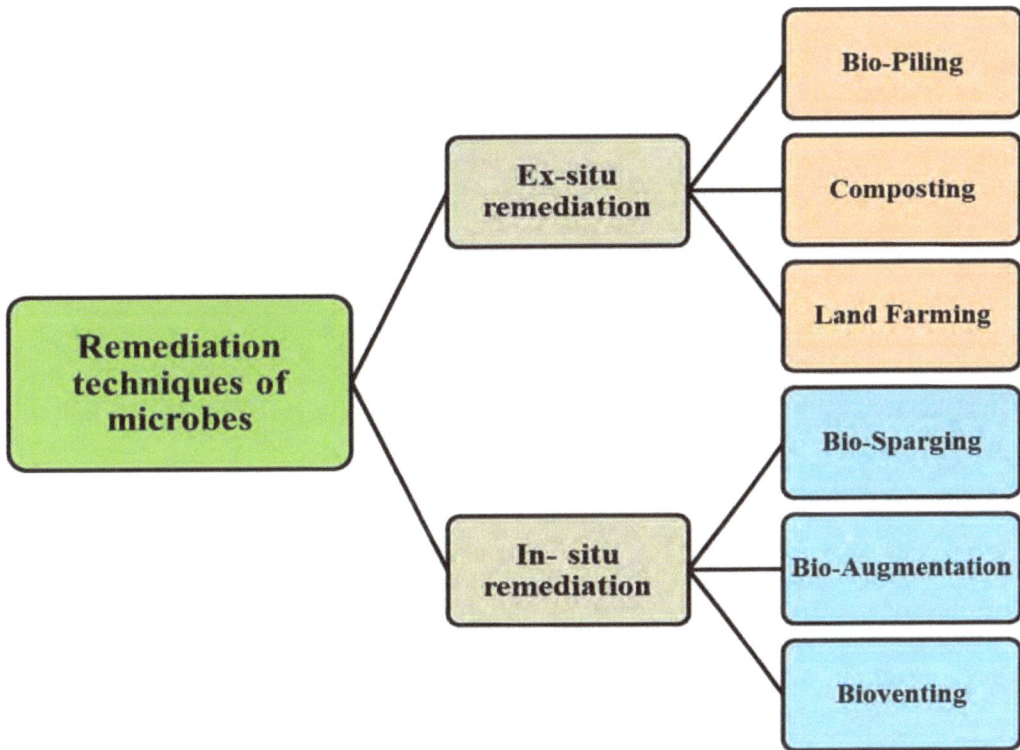

Fig. (1). Remediation techniques using microbes.

Solid-phase traditional microbial restoration techniques use microbial inoculants in a solid medium to restore degraded ecosystems [12]. These techniques have proven to be successful over centuries in soil fertility enhancement and continue to be an essential tool in modern restoration ecology. Solid phase techniques involve the use of a carrier material, such as bio pile carbon-rich amendments, compost, or vermiculite, to support the growth and survival of the inoculated microbes [13, 14]. Microbial inoculants are composed of beneficial microorganisms, such as bacteria (symbiotic nitrogen fixers, plant growth promoters) and fungi (symbiotic and free-living mycorrhiza), that are introduced into the soil to improve soil structure and nutrient cycling and to suppress pathogenic organisms [15]. This technique effectively restores a wide range of ecosystems, including degraded agricultural lands, mine sites, and polluted soils. Solid-phase traditional microbial restoration techniques have the potential to play a critical part in the sustainable management of ecosystems and the conservation of biodiversity [16].

Biopiles - Carbon-rich Soil Amendments

Biopiles are a composting method that is increasingly used in agriculture and waste management. Biopiles consist of large piles of organic waste, such as agricultural residues, food waste, and yard waste, that are piled up in a specific location and allowed to decompose [17]. The decomposition process is facilitated by various microorganisms, including bacteria, fungi, and other decomposers, that break down the organic matter and convert it into a nutrient-rich soil amendment. Bio piles have several advantages over traditional composting methods [18]. One advantage is that bio piles can handle a much larger volume of organic waste than traditional composting systems, making it ideal for use in large-scale agricultural operations and for managing municipal solid waste. Second, bio piles are a cost-effective way to manage organic waste, as they require minimal equipment and can be managed with relatively low labor costs [19].

Biopiles can also be referred to as compost piles, bio-cells, bio-heaps, and bio-mounds. *Ex situ* bioremediation techniques like this one are utilized to reduce pollution concentrations in excavated soils while biodegradation occurs. In biopiles, dirt is excavated, moved, and piled into heaps [20]. A bottom inert liner creates a protective layer on top of which the soil is put. In this procedure, the bio-pile system is fed with air *via* a piping and pump system that either pushes air into the pile at positive pressure or pulls air through the pile at negative pressure. The breakdown of adsorbed pollutants increased as a result of the increased microbial activity brought on by microbial respiration [21].

Specific Conditions to Maintain Bio Piles

Bio piles are typically managed using a set of best practices to ensure efficient decomposition and minimize the risk of odors and other environmental impacts

[22]. One important best practice is to maintain a balance between carbon-rich and nitrogen-rich materials in the pile. Carbon-rich materials, such as dry leaves and straw, provide the energy source for the microorganisms that break down the organic matter, while nitrogen-rich materials, such as food waste and manure, provide the nutrients that the microorganisms need to grow and multiply. The fundamental bio pile system comprises a treatment bed, an aeration system, an irrigation/nutrient system, and a leachate collection system [23]. Moisture, heat, nutrition, oxygen, and pH should all be under control for effective breakdown. Underground irrigation equipment uses hooves to provide nutrients and air. The soil is coated with plastic to avoid runoff, preventing evaporation and volatilization and encouraging sun heating [24]. The biopile therapy technique can take 20 days to 3 months to finish.

Influence of Physiochemical Parameters on Microbes

Maintaining a proper balance between these two types of materials is critical to the success of the bio pile. Another important best practice is to maintain proper moisture levels in the pile. Microorganisms need moisture to survive and to break down organic matter [25]. However, too much moisture can lead to anaerobic conditions in the pile, producing odors and other environmental impacts. It is important to monitor moisture levels in the pile and to add water as needed to maintain a proper balance. Temperature is also an important factor in the decomposition process. As microorganisms break down the organic matter, they generate heat. This heat can raise the temperature of the pile to a level that kills the microorganisms. Thus, it is vital to monitor the temperature of the pile and to turn the pile as needed to ensure even decomposition and to maintain a proper temperature range [26].

In conclusion, biopile is a promising tool for managing organic waste and improving soil health. By following best practices for managing bio piles, we can ensure feasible decomposition and minimize the risk of environmental impacts. Bio piles have several advantages over traditional composting systems and can be used in a variety of applications [19]. With further research and development, bio piles have the potential to play an essential role in sustainable agriculture and waste management.

COMPOST

Composting is the practice of breaking down organic materials such as leftover food scraps, yard waste, and other organic matter into a nutrient-rich soil amendment [27]. The resulting compost is a valuable resource for gardeners, farmers, and landscapers, as it can be used to improve soil health, increase crop yields, and reduce the need for synthetic fertilizers and pesticides. Composting is an aerobic microbiological process that is facilitated by bacteria and fungi. Composting is also a method to produce fertilizer or soil conditioner [28]. Composting is exothermic bio-oxidation in which the organic substrate is biodegraded by a mixed population of microorganisms, including bacteria, archaea, and fungi. During this process, the temperature in the composted material rises to 50–60°C or more at a relatively low moisture content (approximately 60 percent at the beginning of the process and 40–50 percent at the end of compost maturation). During decomposition, microorganisms consume oxygen, which creates heat [29]. This heat helps to speed up the composting process and can raise the temperature inside the compost pile to over 140 degrees Fahrenheit. Both bacteria and fungi are present and active in a typical composting process [30].

Earlier studies have revealed that major bacterial groups at the beginning of the composting process are mesophilic organic acid-producing bacteria such as *Lactobacillus* spp. and *Acetobacter* spp. Later, at the thermophilic stage, gram-positive bacteria, such as *Bacillus* spp. and Actinobacteria, become dominant [13]. However, it has been observed that the most efficient composting process is achieved by mixed communities of bacteria and fungi [31].

Microbial System of Compost

A high concentration of *Lactobacillus* spp. as well as numerous *Acetobacter* spp. sequences are symptomatic of low pH and mesophilic temperatures at the beginning of the composting process. However, a relatively high concentration of Bacillus spp. sequences in samples from the feeding end of the drum suggest that the decomposition of proteins and amino acids had started [32]. Also, higher numbers of Actinobacteria, compared to the compost of equal age from the full-scale feeding end, indicate the beginning of the decomposition of slowly degradable material like lignin and cellulose. The composting process under changing temperature conditions involves various microorganisms, primarily representatives of four bacterial phyla—Firmicutes, Proteobacteria, Bacteroidetes, and Actinobacteria. These account for 85% of all identified and classified microorganisms in the compost samples studied [13]. The first mesophilic composting stage is performed by mesophilic bacteria and fungi, which are replaced with thermophiles as the temperature of the compost increases. Representatives of the large family Enterobacteriales (Proteobacteria) are facultative anaerobes that inhabit the soil and the gastrointestinal tract of humans and animals and are active during the early stages of composting. Moderately thermophilic bacteria of the family Lactobacillales (Firmicutes) are active at the beginning of compost heating and after repeated loosening of the compost material at the beginning of the cooling stage [33].

Various bacilli of the genus Bacillus (Firmicutes) may make up more than 80% of all bacteria at the thermophilic stage of composting (45–60°C). The three species of Bacillus that are most frequently found are *B. subtilis, B. licheniformis,* and *B. circulans.* Thermus species break down different macromolecules and can grow between 65 and 82°C [34]. Bacteria from the genera Thermobifida and Bacillus and tiny fungi from the genera Thermomyces and Aspergillus are the most prevalent members of the stabilized compost layers [33]. Actinomycetes (Actinomycetales) and bacteria from the genera of families Enterobacteriales and Pseudomonadales, which support more thorough decomposition of leftover organic material, become active at the third stage (cooling) and the compost maturation stage. The list of effective microorganisms that play a vital role at various stages of composting is presented in Table **1**.

Table 1. Effective microorganisms at various stages of compost.

S. No.	Composting Phase	Microorganisms	Refs.
1	Thermophilic phase (Day 45)	*Curtobacterium citreum, Stenotrophomonas rhizophila, Stenotrophomonas maltophilia, Microbacterium foliorum, Xanthomonas oryzae, Pseudoxanthomonas taiwanensis, Bacillus ginsengihumi, Serratia marcescens, Serratia odorifera Rhabditidae spp., Panagolamidae sp. Diplogasteridae Sp., Cephalobidae sp., Mononchoides sp., Ditylenchus filimus*	[3]
2	Mesophilic phase (Day 139)	*Xenophilus azovorans, Bacillus licheniformis, Pseudomonas mendocina, Rhodococcus rhodochrous Bacillus sp., Paenibacillius sp, Actinomycetes, Aspergillus fumigatus, Feacal coliforms, Pseudomonas sp., Streptococcus sp., Proteus sp., Serratia sp.*	[14]
3	Psychophilic phase	*Asprgillus fumigatus, Emericella sp., Aspergillus ochraceus, Aspergillus terreus, Penicillium oxalicum*	[35]

Effect of Microbes on Compost Quality

Nitrogen fixers, or diazotrophs, develop in the compost and play a crucial role in ensuring soil fertility owing to their ability to fix molecular nitrogen from the atmosphere due to the presence of the nitrogenase enzyme complex. Diazotrophs that are representatives of the bacterial genera Azotobacter, Klebsiella, Pseudomonas, Xanthomonas, Alcaligenes, Stenotrophomonas, Caulobacter, Achromobacter, and Clostridium, which are involved in the degradation of organic matter of the compost, were isolated from the compost mass. The nitrogen fixation ability was also shown for some archaea present in the compost. Due to nitrogen fixation, nitrogen, the key nutrient of plants whose deficiency is usually observed in soils, is supplied from the atmosphere [20].

Benefits of Composting

Reduce carbon footprint: By composting food scraps and yard waste, the amount of waste that ends up in landfills can be reduced [36]. This not only saves space in landfills but also reduces the production of greenhouse gases, which contribute to climate change.

A microbial community's dynamics or succession during the composting process reflects how well they can break down the compost mixture [3]. The composition of the raw materials and nutrient supplements, environmental conditions (ambient or experimental), and interactions between all these parameters play a significant role in the changes produced in a microbiome along the process. The most prevalent and quickly growing microorganisms during composting here are bacteria and fungus. The substrates used and the bacteria active during the process significantly impact the compost's quality [13] by releasing a variety of substrate-based hydrolytic enzymes that break down the intricately structured molecules and produce water-soluble chemicals. Besides metabolizing the organics, they produce simple, usable compounds that enhance agricultural possibilities and stabilize the natural ecosystem when added to the soil.

The importance of soil organisms and have pioneered and promoted the use of numerous practices that increase microbial activity and diversity. Management practices employed by organic growers, such as the use of composts and manures, cover crops, and diverse crop rotations, have been reported to increase biologically available forms of OM and increase the activities of beneficial soil microbes, including invertebrates.

Economic feasibility: Composting can save money by reducing the need for synthetic fertilizers and pesticides. Instead, compost can be used to naturally fertilize plants and reduce the risk of pests and diseases.

GRAZE MANURE-BASED RESTORATION

Graze manure-based restoration is an innovative land restoration approach using livestock to improve soil fertility and restore ecosystem functions. Graze manure-based restoration seeks to mimic this natural cycle by using livestock to graze on degraded land and deposit manure, thereby restoring soil fertility and promoting plant growth. Livestock, such as cattle, sheep, and goats, are strategically grazed on degraded land, depositing manure that provides soil nutrients [37]. The manure also enhances soil structure, water-holding capacity, and microbial activity, which are critical for restoration. Over time, the restoration process leads to the development of healthy and functional ecosystems that support a range of ecological services [38].

Graze Manure-Based Restoration: Sustainable Land Management

Measurements of functional diversity can provide information about the functioning of soils. In recent years, microbial diversity, activity, and resilience have been linked to crops and/or forests' sustainability in various ecosystems [3]. A sustainable approach to land management that adheres to the ideals of sustainable agriculture is graze manure-based restoration. A closed nutrient cycle that is aided by the use of cattle in restoration lessens the need for synthetic fertilizers and improves the ecological resilience of ecosystems. Additionally, by storing carbon in the soil, grazing manure-based restoration can aid in mitigating climate change. The restoration process promotes the growth of perennial vegetation, which has the potential to sequester more carbon than annual crops or bare soil [39].

A Successful Case Study - The Restoration of Degraded Rangelands in Mongolia

A sustainable ecosystem depends on functional soil microbial communities, which play an essential role in nutrient cycling, plant health, and growth, as well as the degradation of toxic substances. Thus, the microbial community is an essential component of soil quality and may indicate changes in soil quality and health. Besides soil organic carbon, changes in pH, soil water content management, soil type plant diversity and composition, and soil mineral nutrient availability have all been shown to influence soil microbial communities. While land reclamation and cultivation are known to alter plant species and associated soil properties, the mechanisms by which changes in land use shape the underlying microbial communities are less well known [40].

Graze manure-based restoration has been successfully implemented in various ecosystems worldwide. One example is the restoration of degraded rangelands in Mongolia. Graze manure-based restoration has been used to restore degraded rangelands by strategically grazing livestock and planting trees [41]. This has helped to improve soil fertility, increase plant diversity, and support local livelihoods in the region. Another example is the restoration of degraded grasslands in Brazil. Graze manure-based restoration has been used to restore grasslands degraded by overgrazing and intensive agriculture by strategically grazing cattle on degraded land. The cattle deposit manure, which adds nutrients to the soil and promotes the growth of new plant communities of nutrient transfer and recycling in ecosystems. This process creates a self-sustaining cycle of nutrient cycling, which maintains soil fertility and promotes biodiversity [40].

MUTUALISTIC - PLANT AND MICROBE-ASSISTED ECO-RESTORATION TECHNIQUES

Roots and Microbes: Mutualistic Rhizosphere

Plants became essential for life on Earth by providing us with food, oxygen, and a range of other valuable resources. However, plants cannot survive and thrive alone, and they rely on a complex network of interactions with other organisms in the soil, including microbes, for their growth. The rhizosphere is the narrow zone of soil surrounding plant roots. It is a highly active and dynamic environment that is rich in nutrients and microbial life. In particular, roots and microbes have a mutualistic relationship that is critical in promoting plant health, growth, and productivity. A plant's root system is a complex structure that provides a physical and chemical interface between the plant and the soil. The rhizosphere is the site for a complex and dynamic network of interactions between the plant and soil microorganisms (Table **2**), which can have both beneficial and harmful effects on plant health [42, 43].

Table 2. Plant growth regulators and their explicit role in regulating plant growth by symbiotic bacteria (Adapted from [35]).

Symbiotic Bacteria	PGRs	Crops	Responses	Refs.
Kluyvera ascorbata SUD 165	Siderophores, indole-3-acetic acid	Canola, tomato	Strains exhibited reduced plant growth inhibition caused by heavy metals such as nickel, lead, and zinc.	[46]
Rhizobium leguminosarum	Indole-3-acetic acid	Rice	Inoculation of rice seedlings with *R. leguminosarum* resulted in notable growth-promoting effects.	[47]
Rhizobium leguminosarum	Indole-3-acetic acid	Rice	Observations revealed that inoculating axenically grown rice seedlings had growth-promoting effects.	[48]
Azotobacter sp.	Indole-3-acetic acid	Maize	Maize seedlings showed significant growth-promoting effects when inoculated with a strain that efficiently produces IAA.	[49]
Rhizobacterial isolates	Auxins	Wheat, rice	Wheat and rice exhibited notable growth-promoting effects when inoculated with rhizobacterial isolates.	[35]
Rhizobacteria (unidentified)	Indole-3-acetic acid	Brassica	The modified jar experiments revealed a significant correlation between the auxin production by PGPR *in vitro* and the growth promotion of rapeseed seedlings that were inoculated.	[50]

(Table 2) cont.....

Symbiotic Bacteria	PGRs	Crops	Responses	Refs.
Rhizobacteria (unidentified)	Indole-3-acetic acid	Wheat, rice	Inoculated seedlings showed relatively more positive effects when exposed to rhizobacterial strains that were active in producing IAA.	[35]
Pseudomonas fluorescens	Siderophores, indole-3-acetic acid	Groundnut	The involvement of ACC deaminase and siderophore production facilitated the promotion of nodulation and yield of groundnut.	[51]
Rhizobacteria (Unidentified)	Auxin, indole--acetic acid, acetamide	Wheat	The strain that produced the highest amount of auxin in non-sterilized soil resulted in the maximum increase in growth yield.	[35]
Azospirillum brasilense A3, A4, A7, A10, CDJA	Indole-3-acetic acid,	Rice	All bacterial strains caused an increase in rice grain yield compared to the uninoculated control.	[52]
Bacillus circulans P2, *Bacillus* sp.				
P3,*Bacillus magaterium* P5, *Bacillus*. Sp. Psd7				
Streptomyces anthocysnicus				
Pseudomonas aeruginosa Psd5				
Pseudomonas pieketti Psd6, *Pseudomonas fluorescens*				
MTCC103				
Azospirillum lipoferum strains 15		Wheat	In a growth chamber pot experiment, the development of the wheat root system was promoted even under crude oil contamination.	[53]
Pseudomonas denitrificans	Auxin	Wheat, maize	In pot experiments, all bacterial strains were found to enhance the growth of wheat and maize plants.	[54]
Pseudomonas rathonis				
Azotobacter sp.	Indole-3-acetic acid	Sesbenia, mung bean	An increase in tryptophan concentration from 1 mg/ml to 5 mg/ml led to a decline in growth in both crops.	[54]
Pseudomonas sp.				
Pseudomonas sp.	Indole-3-acetic acid	Wheat	The nutritional quality of wheat grain was enhanced by a combined bio-inoculation of diacetyl-phloroglucinol-producing PGPR and AMF.	[55]

(Table 2) cont.....

Symbiotic Bacteria	PGRs	Crops	Responses	Refs.
Bacillus cereus RC 18,				
Bacillus licheniformis RC08,	Indole-3-acetic acid	Wheat, spinach	All bacterial strains were proficient in producing indole acetic acid (IAA), leading to a significant increase in the growth of wheat and spinach.	[56]
Bacillus megaterium RC07				

The relationship between roots and microbes in the rhizosphere is a mutualistic relation at most, given that both the plant and the microbes benefit from the interaction. The plant provides carbon compounds in the form of sugars and other organic molecules to the microbes, which use them as an energy source for growth and metabolism. In return, the microbes provide a range of benefits to the plant.

Nutrient Acquirement: Microbes in the rhizosphere can help plants acquire essential nutrients, such as nitrogen, phosphorus, and potassium, that are present in the soil in forms that are not readily available to the plant [44].

Disease Suppression: Some microbes in the rhizosphere can suppress soil-borne pathogens that can cause disease in plants, thereby improving plant health and yield.

Stress Tolerance: Microbes in the rhizosphere can help plants tolerate environmental stresses, such as drought, salinity, and extreme temperatures, by improving root growth and nutrient uptake [45].

Sustainable Restoration Practice Using Symbiotic Microbes

The mutualistic relationship between roots and microbes in the rhizosphere has important implications for sustainable restoration practice. Promoting a healthy and diverse microbial community in the rhizosphere can restore a degraded ecosystem, improving soil fertility, reducing the need for synthetic fertilizers and pesticides, and even increasing crop yields. This can lead to a range of benefits, including:

Improved Soil Health: By promoting a diverse and healthy microbial community in the rhizosphere, farmers can improve soil health and reduce soil degradation.

Reduced Environmental Impacts: By reducing the need for synthetic fertilizers and pesticides, farmers can reduce the environmental impacts of agriculture, such as soil and water pollution [57].

Increased Resilience: By improving crop yields and plant health, farmers can increase the resilience of their farming systems to environmental stresses, such as drought, pests, and disease [58].

Microbial Effect as the Symbiotic Association

The symbiotic association between roots and microorganisms within the rhizosphere has been effectively incorporated into diverse agricultural systems globally. One example is using cover crops in sustainable agriculture/ecological restoration. Cover crops are planted between cash crops to improve soil health and fertility. They promote a diverse and healthy microbial community in the rhizosphere by bedding an organic litter layer, which can improve nutrient uptake and reduce the need for synthetic fertilizers [59]. Another example is the use of biofertilizers in organic agriculture. Biofertilizers are composed of live microbial inoculants that are applied to the soil to improve soil fertility and plant health [60].

REGENERATIVE ECO-RESTORATION/CARBON FARM TECHNIQUE

The effects of climate change are becoming more apparent each year, with rising temperatures, more frequent extreme weather events, and changes in ecosystems. While global efforts to reduce greenhouse gas emissions are crucial, it is also essential to consider ways to remove carbon dioxide from the atmosphere. One promising approach is regenerative eco-restoration, a carbon farming technique that restores degraded land and enhances its capacity to store carbon. Regenerative eco-restoration is a technique that involves restoring degraded ecosystems, such as degraded farmland or grasslands, to their natural state through the use of ecological principles. This technique involves a combination of practices that focus on restoring the ecosystem's natural balance, such as planting native species, restoring soil health, and enhancing biodiversity. The goal of regenerative eco-restoration is to create a resilient and self-sustaining ecosystem that can provide ecosystem services such as carbon storage, water infiltration, and habitat for wildlife.

Diverse Manifestations of Carbon and Chemotaxis of Microbes

In order to trap carbon from the atmosphere and store it in the soil and vegetation, regenerative eco-restoration first establishes a healthy ecosystem. This is done by combining several techniques that improve soil health and natural carbon sequestration. Planting native tree species, restoring the health of the soil, and enhancing biodiversity are some of the primary techniques utilized in regenerative eco-restoration [61]. The biosphere's essential ingredient, carbon, is crucial to every living thing's metabolism. The most numerous and diverse class of organisms on the Earth, microbes have developed a range of ways to recognize and react to carbon sources in their surroundings. One such process that enables microorganisms to find and use carbon sources is chemotaxis, or the capacity to move towards or away from chemical gradients [62].

CARBON IN THE ENVIRONMENT

Carbon can be found in the environment in several ways, such as carbon dioxide, organic matter, and dissolved carbon molecules. Depending on elements, including soil type, temperature, and pH, carbon availability in various forms can vary greatly. Although most microorganisms depend on plants as their primary carbon source, they have evolved various methods, including photosynthesis, respiration, and fermentation. Increased soil microbial activity and biomass, improved soil water-stable aggregation, and recalcitrant C are just a few ways that soil organic additions might improve soil C storage [63].

Microbes in the Carbon World

In vivo, turnover, which results in the deposition of C generated from microorganisms, is the microorganism-mediated SOM transformation/formation mechanism. The amount of microbial-derived carbon and the ratio of microbial-to plant-derived carbon in the soil C pool is directly impacted by changes in the anabolic capability and activity rates of bacteria [64]. Not only does microbial anabolism play a crucial role in the creation of biomass, but it also plays a role in the digestion of readily available organic molecules and their resynthesis into the unique forms found in microbial biomass and necromass. By creating stable or stabilized SOM, soil bacteria that primarily degrade SOM can also promote C sequestration. The balance between microbial catabolic activity, which releases carbon dioxide (CO_2), and anabolic activity, which contributes to senesced microbial biomass, determines how much carbon is stored in soil [63].

Chemotaxis in Microbes

Microbes can move towards or away from a chemical gradient through a mechanism called chemotaxis. In this process, the microorganisms sense chemical cues, prompting a directed migration towards or away from the chemical source. Microbes may find and use nutrients, including carbon sources, through a process called chemotaxis [65]. Chemotaxis in microorganisms can be either positive or negative. Negative chemotaxis refers to a microorganism's movement in the opposite direction of a chemical gradient, whereas positive chemotaxis describes migration in the direction of a chemical gradient. The movement is focused on or away from a particular chemical stimulus source [66].

There are many different types of microbial motility, including swimming, swarming, gliding, twitching, and even surfing. Transmembrane chemoreceptors, frequently grouped at the cell poles, are used by motile cells to continuously measure specific chemical concentrations to conduct chemotaxis [67]. Because one detection event can amplify nearby chemoreceptors, clustering enables bacteria to respond to very modest relative changes in particular chemicals. The cytoplasm then receives information from the chemoreceptors, which activates a signaling mechanism that affects the rotation of the flagellar motor or motors, changing the direction of swimming [68]. Even though motility and chemotaxis have mainly been studied in a small number of model species, such as Escherichia coli and Bacillus subtilis, a fundamental sensing pathway has remained intact. The well-studied instance is E. coli, which splays out its flagella from the cell body, interrupting its run and causing a fall. However, many marine bacteria only have one flagellum, making the E. coli swimming method ineffective [54].

Even though eukaryotes have been linked to the great majority of microbial symbionts discovered to date, there is mounting evidence that prokaryotic symbioses are equally common. Tight metabolic coupling benefits aggregating microorganisms, and the employment of motility and chemotaxis can help get around rate restrictions and short chemical diffusion distances brought on by the small size of both partners. In many settings, chemotaxis frequently mediates the formation and upkeep of highly organized microbial consortia [69]. For instance, the thick-walled, nitrogen-fixing heterocysts of filamentous cyanobacteria like Anabaena spp. produce particular signaling chemicals that draw Pseudomonas spp. and improve nitrogen fixation rates. Sulfate-reducing *Thioploca spp.*, another filamentous bacterial taxon dwelling at the interface of sulfide-rich sediments, is colonized by *Desulfonema* spp. using gliding motility, enabling complete sulfate reduction and reoxidation among these organisms.

Chemotaxis is a behavior displayed by bacteria like *E. coli* towards glucose and other simple carbohydrates [70]. These sugars are an essential supply of carbon for bacteria, and their existence depends on their ability to find and use them effectively. Other microorganisms, such as methanotrophs, use methane as a carbon source. Methane is a simple chemical that is prevalent in some habitats, such as marshes and ruminant stomachs [71]. Methanotrophs use a sophisticated chemotaxis system that controls the expression of numerous genes and the coordination of numerous cellular pathways in order to detect and use methane. Microbes also show chemotaxis toward other chemical cues in the environment, such as oxygen, pH, and temperature, in addition to carbon sources. The chemotaxis of microbes towards the root for the symbiotic relationship is presented in Fig. (**2**).

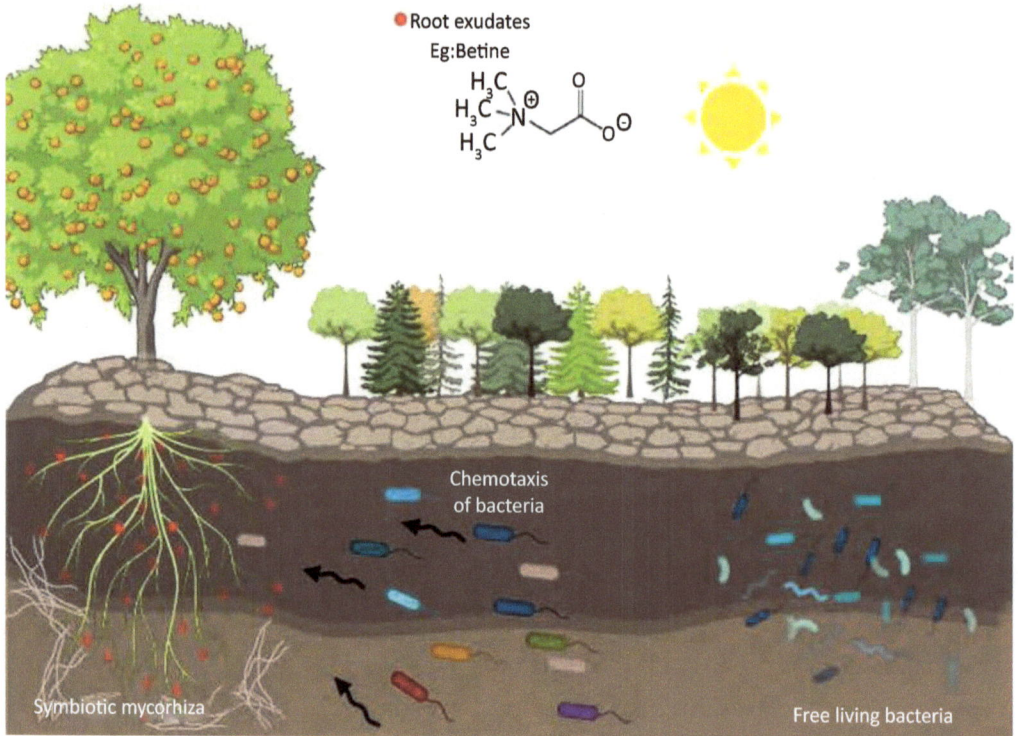

Fig. (2). Chemotaxis of microbes towards the root for the symbiotic relationship.

For example, some bacteria are able to sense the pH of their environment and move towards or away from areas of high or low pH. Thus, plant-microbe symbioses are achieved for mutual existence [59].

CONCLUSION

The ecosystem services delivered by soil microbes, the wealth of soil on Earth, are vital for life. The majority of our antibiotics come from bacteria that live in soil, which also serves as a filter, a habitat for a wide variety of creatures, and a medium for plants and heterotrophs to develop in. Climate change, population growth, and land degradation, such as carbon loss, biodiversity decrease, and erosion, present a growing set of challenges for humanity. The soil function is decreased by land degradation, which also affects various other ecosystem services. Changes in hydraulic operation, infiltration and soil moisture storage, carbon cycling, biological activity, movement of nutrients and pollutants, and plant development all contribute to these effects. As soil ecosystem degradation has a growing impact on biodiversity, food production, climate regulation, and human livelihoods, it is essential to understand its effects and take steps to reverse them with the microbes as a converter.

Basic and traditional microbial techniques play a crucial role in ecological restoration by promoting the growth and establishment of plant communities. The use of techniques such as bio-piling, composting, promotion of symbiotic mycorrhiza, and regenerative agriculture can improve soil quality and enhance nutrient cycling, leading to increased plant growth and biodiversity by using microbial techniques, even phytoremediation, *i.e.*, restoring degraded sites by removing contaminants and pollutants from the soil. While inoculating bioinoculants in the degraded soil by practicing regenerative restoration techniques, the microbial community can be retained in the soil for successive generations. While more sophisticated techniques have emerged in modern years, basic and traditional microbial techniques remain a vital part of ecological restoration. They ought to persist to be utilized in combination with new and innovative methods to ensure the success of restoration efforts.

REFERENCES

[1]   Johnson DB, Williamson JC. Conservation of mineral nitrogen in restored soils at opencast coal mine sites: I. Results from field studies of nitrogen transformations following restoration. Eur J Soil Sci 1994; 45(3): 311-7.
[http://dx.doi.org/10.1111/j.1365-2389.1994.tb00514.x]

[2]   Liang JL, Liu J, Jia P, *et al.* Novel phosphate-solubilizing bacteria enhance soil phosphorus cycling following ecological restoration of land degraded by mining. ISME J 2020; 14(6): 1600-13.
[http://dx.doi.org/10.1038/s41396-020-0632-4] [PMID: 32203124]

[3]   Miransari M. Soil microbes and the availability of soil nutrients. Acta Physiol Plant 2013; 35(11): 3075-84.
[http://dx.doi.org/10.1007/s11738-013-1338-2]

[4]   Stamenković S, Beškoski V, Karabegović I, Lazić M, Nikolić N. Microbial fertilizers: A comprehensive review of current findings and future perspectives. Span J Agric Res 2018; 16(1): e09R01.
[http://dx.doi.org/10.5424/sjar/2018161-12117]

[5]   Ding Y, Liu Y, Liu S, *et al.* Biochar to improve soil fertility. A review. Agron Sustain Dev 2016; 36(2): 36.
[http://dx.doi.org/10.1007/s13593-016-0372-z]

[6]   Lladó S, López-Mondéjar R, Baldrian P. Forest soil bacteria: diversity, involvement in ecosystem processes, and response to global change. Microbiol Mol Biol Rev 2017; 81(2): e00063-16.
[http://dx.doi.org/10.1128/MMBR.00063-16] [PMID: 28404790]

[7]   Karlen D, Rice C. Soil degradation: Will humankind ever learn? Sustainability (Basel) 2015; 7(9): 12490-501.
[http://dx.doi.org/10.3390/su70912490]

[8]   Robinson DA, Hopmans JW, Filipovic V, *et al.* Global environmental changes impact soil hydraulic functions through biophysical feedbacks. Glob Change Biol 2019; 25(6): 1895-904.
[http://dx.doi.org/10.1111/gcb.14626] [PMID: 30900360]

[9] Shafique HA, Sultana V, Ehteshamul-Haque S, Athar M. Management of soil-borne diseases of organic vegetables. J Plant Prot Res 2016; 56(3): 221-30.
[http://dx.doi.org/10.1515/jppr-2016-0043]

[10] Sharma I. Bioremediation techniques for polluted environment: concept, advantages, limitations, and prospects Trace Met Environ approaches Recent Adv. IntechOpen 2020.

[11] Tegene BG, Tenkegna TA. Mode of action, mechanism and role of microbes in bioremediation service for environmental pollution management. Journal of Biotechnology & Bioinformatics Research 2020; 2: 1-18.
[http://dx.doi.org/10.47363/JBBR/2020(2)116]

[12] Tripathi V, Edrisi SA, Chen B, *et al.* Biotechnological advances for restoring degraded land for sustainable development. Trends Biotechnol 2017; 35(9): 847-59.
[http://dx.doi.org/10.1016/j.tibtech.2017.05.001] [PMID: 28606405]

[13] Beesley L, Moreno-Jiménez E, Gomez-Eyles JL, Harris E, Robinson B, Sizmur T. A review of biochars' potential role in the remediation, revegetation and restoration of contaminated soils. Environ Pollut 2011; 159(12): 3269-82.
[http://dx.doi.org/10.1016/j.envpol.2011.07.023] [PMID: 21855187]

[14] Lu H, Yan M, Wong MH, *et al.* Effects of biochar on soil microbial community and functional genes of a landfill cover three years after ecological restoration. Sci Total Environ 2020; 717: 137133.
[http://dx.doi.org/10.1016/j.scitotenv.2020.137133] [PMID: 32062262]

[15] Ahirwal J, Maiti SK, Satyanarayana Reddy M. Development of carbon, nitrogen and phosphate stocks of reclaimed coal mine soil within 8 years after forestation with Prosopis juliflora (Sw.) Dc. Catena 2017; 156: 42-50.
[http://dx.doi.org/10.1016/j.catena.2017.03.019]

[16] Kumar V, Shahi SK, Singh S. Bioremediation: an eco-sustainable approach for restoration of contaminated sites. Microb Bioprospecting Sustain Dev 2018; pp. 115-36.

[17] Sadik MW, El Shaer HM, Yakot HM. Recycling of agriculture and animal farm wastes into compost using compost activator in Saudi Arabia. J Int Environ Appl Sci 2010; 5: 397.

[18] Hemidat S, Jaar M, Nassour A, Nelles M. Monitoring of composting process parameters: a case study in Jordan. Waste Biomass Valoriz 2018; 9(12): 2257-74.
[http://dx.doi.org/10.1007/s12649-018-0197-x]

[19] Philp JC, Atlas RM. Bioremediation of contaminated soils and aquifers. Bioremediation Appl Microb Solut Real□World Environ Cleanup 2005; pp. 139-236.

[20] Uddin MJ, Sagar G, Jagdeeshwar J. Soil Pollution and Soil Remediation Techniques. International Journal of Advance Research, Ideas and Innovations in Technology, 2017, 3.

[21] Labud V, Garcia C, Hernandez T. Effect of hydrocarbon pollution on the microbial properties of a sandy and a clay soil. Chemosphere 2007; 66(10): 1863-71.
[http://dx.doi.org/10.1016/j.chemosphere.2006.08.021] [PMID: 17083964]

[22] Jouhara H, Czajczyńska D, Ghazal H, *et al.* Municipal waste management systems for domestic use. Energy 2017; 139: 485-506.
[http://dx.doi.org/10.1016/j.energy.2017.07.162]

[23] Cristorean C, Micle V, Sur IM. A critical analysis of *ex-situ* bioremediation technologies of hydrocarbon polluted soils. ECOTERRA J Environ Res Prot 2016; 13: 17-29.

[24] Shukla KP, Singh NK, Sharma S. Bioremediation: developments, current practices and perspectives. Genet Eng Biotechnol J 2010; 3: 1-20.

[25] Hoorman JJ. The role of soil bacteria. Columbus: Ohio State Univ Extension 2011; pp. 1-4.

[26] Cooperband L. The art and science of composting. Cent Integr Agric Syst 2002.

[27]   Pant A, Wang K. Recycle Organic Waste through Vermicomposting 2015.

[28]   Partanen P, Hultman J, Paulin L, Auvinen P, Romantschuk M. Bacterial diversity at different stages of the composting process. BMC Microbiol 2010; 10(1): 94.
[http://dx.doi.org/10.1186/1471-2180-10-94] [PMID: 20350306]

[29]   Azim K, Soudi B, Boukhari S, Perissol C, Roussos S, Thami Alami I. Composting parameters and compost quality: a literature review. Org Agric 2018; 8(2): 141-58.
[http://dx.doi.org/10.1007/s13165-017-0180-z]

[30]   D'Imporzano G, Re I, Spina F, *et al.* Optimizing bioremediation of hydrocarbon polluted soil by life cycle assessment (LCA) approach. Environ Eng Manag J 2019; (18): 2155-62.

[31]   Du T, Wang D, Bai Y, Zhang Z. Optimizing the formulation of coal gangue planting substrate using wastes: The sustainability of coal mine ecological restoration. Ecol Eng 2020; 143: 105669.
[http://dx.doi.org/10.1016/j.ecoleng.2019.105669]

[32]   Prenafeta-Boldú FX, Fernández B, Viñas M, *et al.* Effect of Bacillus spp. direct-fed microbial on slurry characteristics and gaseous emissions in growing pigs fed with high fibre-based diets. Animal 2017; 11(2): 209-18.
[http://dx.doi.org/10.1017/S1751731116001415] [PMID: 27412081]

[33]   Nozhevnikova AN, Mironov VV, Botchkova EA, Litti YV, Russkova YI. Composition of a microbial community at different stages of composting and the prospects for compost production from municipal organic waste. Appl Biochem Microbiol 2019; 55(3): 199-208.
[http://dx.doi.org/10.1134/S0003683819030104]

[34]   Zeikus JG. Thermophilic bacteria: Ecology, physiology and technology. Enzyme Microb Technol 1979; 1(4): 243-52.
[http://dx.doi.org/10.1016/0141-0229(79)90043-7]

[35]   Hayat R, Ali S, Amara U, Khalid R, Ahmed I. Soil beneficial bacteria and their role in plant growth promotion: a review. Ann Microbiol 2010; 60(4): 579-98.
[http://dx.doi.org/10.1007/s13213-010-0117-1]

[36]   Brown S. Greenhouse gas accounting for landfill diversion of food scraps and yard waste. Compost Sci Util 2016; 24(1): 11-9.
[http://dx.doi.org/10.1080/1065657X.2015.1026005]

[37]   Papanastasis VP. Restoration of degraded grazing lands through grazing management: can it work? Restor Ecol 2009; 17(4): 441-5.
[http://dx.doi.org/10.1111/j.1526-100X.2009.00567.x]

[38]   Chadwick D, Wei J, Yan'an T, Guanghui Y, Qirong S, Qing C. Improving manure nutrient management towards sustainable agricultural intensification in China. Agric Ecosyst Environ 2015; 209: 34-46.
[http://dx.doi.org/10.1016/j.agee.2015.03.025]

[39]   Yang Y, Tilman D, Furey G, Lehman C. Soil carbon sequestration accelerated by restoration of grassland biodiversity. Nat Commun 2019; 10(1): 718.
[http://dx.doi.org/10.1038/s41467-019-08636-w] [PMID: 30755614]

[40]   Olukoye GA, Wamicha WN, Kinyamario JI. An ecosystem modelling approach to rehabilitating semi-desert rangelands of North Horr, Kenya Pastor Syst Marg Environ. Wageningen, Netherlands: Wageningen Academic Publishers 2005.

[41]   Bekunda M, Sanginga N, Woomer PL. Restoring soil fertility in sub-Sahara Africa. Adv Agron 2010; 108: 183-236.
[http://dx.doi.org/10.1016/S0065-2113(10)08004-1]

[42]   Venturi V, Keel C. Signaling in the Rhizosphere. Trends Plant Sci 2016; 21(3): 187-98.
[http://dx.doi.org/10.1016/j.tplants.2016.01.005] [PMID: 26832945]

[43]   Schulz B. Mutualistic interactions with fungal root endophytes. Microb Root Endophytes 2006; pp. 261-79.

[44]   Li E, de Jonge R, Liu C, *et al.* Rapid evolution of bacterial mutualism in the plant rhizosphere. Nat Commun 2021; 12(1): 3829.
[http://dx.doi.org/10.1038/s41467-021-24005-y] [PMID: 34158504]

[45]   Bhatia NP, Adholeya A, Sharma A. Biomass production and changes in soil productivity during longterm cultivation of Prosopis juliflora (Swartz) DC inoculated with VA mycorrhiza and Rhizobium spp. in a semi-arid wasteland. Biol Fertil Soils 1998; 26(3): 208-14.
[http://dx.doi.org/10.1007/s003740050369]

[46]   Burd GI, Dixon DG, Glick BR. Plant growth-promoting bacteria that decrease heavy metal toxicity in plants. Can J Microbiol 2000; 46(3): 237-45.
[http://dx.doi.org/10.1139/w99-143] [PMID: 10749537]

[47]   Biswas JC, Ladha JK, Dazzo FB, Yanni YG, Rolfe BG. Rhizobial inoculation influences seedling vigor and yield of rice. Agron J 2000; 92(5): 880-6.
[http://dx.doi.org/10.2134/agronj2000.925880x]

[48]   Biswas JC, Ladha JK, Dazzo FB. Rhizobia inoculation improves nutrient uptake and growth of lowland rice. Soil Sci Soc Am J 2000; 64(5): 1644-50.
[http://dx.doi.org/10.2136/sssaj2000.6451644x]

[49]   Zahir ZA, Abbas SA, Khalid M, Arshad M. Substrate dependent microbially derived plant hormones for improving growth of maize seedlings. Pak J Biol Sci 2000; 3(2): 289-91.
[http://dx.doi.org/10.3923/pjbs.2000.289.291]

[50]   H A, Z Z, M A, A K. Relationship between *in vitro* production of auxins by rhizobacteria and their growth-promoting activities in Brassica juncea L. Biol Fertil Soils 2002; 35(4): 231-7.
[http://dx.doi.org/10.1007/s00374-002-0462-8]

[51]   Dey R, Pal KK, Bhatt DM, Chauhan SM. Growth promotion and yield enhancement of peanut (Arachis hypogaea L.) by application of plant growth-promoting rhizobacteria. Microbiol Res 2004; 159(4): 371-94.
[http://dx.doi.org/10.1016/j.micres.2004.08.004] [PMID: 15646384]

[52]   Thakuria D, Talukdar NC, Goswami C, Hazarika S, Boro RC, Khan MR. Characterization and screening of bacteria from rhizosphere of rice grown in acidic soils of Assam. Curr Sci 2004; 86(7): 978-85.

[53]   Tekle Abegaz S. We are IntechOpen, the world's leading publisher of open-source books. Built by scientists for scientists TOP 1%. University of Gondar Institutional Repository; Gondar, Ethiopia: 2021.

[54]   Egamberdiyeva D. Plant☐growth☐promoting rhizobacteria isolated from a Calcisol in a semi☐arid region of Uzbekistan: biochemical characterization and effectiveness. J Plant Nutr Soil Sci 2005; 168(1): 94-9.
[http://dx.doi.org/10.1002/jpln.200321283]

[55]   Roesti D, Gaur R, Johri B, *et al.* Plant growth stage, fertiliser management and bio-inoculation of arbuscular mycorrhizal fungi and plant growth promoting rhizobacteria affect the rhizobacterial community structure in rain-fed wheat fields. Soil Biol Biochem 2006; 38(5): 1111-20.
[http://dx.doi.org/10.1016/j.soilbio.2005.09.010]

[56]   Cakmakci R, Dönmez MF, Erdoğan Ü. The effect of plant growth promoting rhizobacteria on barley seedling growth, nutrient uptake, some soil properties, and bacterial counts. Turk J Agric For 2007; 31: 189-99.

[57]   De Vries FT, Griffiths RI, Knight CG, Nicolitch O, Williams A. Harnessing rhizosphere microbiomes for drought-resilient crop production. Science (80-) 2020; 368: 270–4.
[http://dx.doi.org/10.1126/science.aaz5192]

[58]   Malviya D, Singh UB, Singh S, Sahu PK, Pandiyan K, Kashyap AS, *et al.* Microbial interactions in the rhizosphere contributing crop resilience to biotic and abiotic stresses. Rhizosph Microbes Soil Plant Funct 2020; pp. 1-33.
[http://dx.doi.org/10.1007/978-981-15-9154-9_1]

[59]   Drinkwater LE, Snapp SS. Understanding and managing the rhizosphere in agroecosystems Rhizosph. Elsevier 2007; pp. 127-53.

[60]   Pathak DV, Kumar M. Microbial inoculants as biofertilizers and biopesticides Microb Inoculants Sustain Agric Product. Res Perspect 2016; Vol. 1: pp. 197-209.

[61]   Juwarkar AA, Singh L, Kumar GP, Jambhulkar HP, Kanfade H, Jha AK. Biodiversity promotion in restored mine land through plant-animal interaction. J Ecosyst Ecography 2016; 6: 2.

[62]   Tomar OS, Minhas PS, Sharma VK, Singh YP, Gupta RK. Performance of 31 tree species and soil conditions in a plantation established with saline irrigation. For Ecol Manage 2003; 177(1-3): 333-46.
[http://dx.doi.org/10.1016/S0378-1127(02)00437-1]

[63]   Cienciala E, Centeio A, Blazek P, Cruz Gomes Soares M, Russ R. Estimation of stem and tree level biomass models for Prosopis juliflora/pallida applicable to multi-stemmed tree species. Trees (Berl) 2013; 27(4): 1061-70.
[http://dx.doi.org/10.1007/s00468-013-0857-1]

[64]   Fan Z, Liang C. Significance of microbial asynchronous anabolism to soil carbon dynamics driven by litter inputs. Sci Rep 2015; 5(1): 9575.
[http://dx.doi.org/10.1038/srep09575] [PMID: 25849864]

[65]   Colin R, Ni B, Laganenka L, Sourjik V. Multiple functions of flagellar motility and chemotaxis in bacterial physiology. FEMS Microbiol Rev 2021; 45(6): fuab038.
[http://dx.doi.org/10.1093/femsre/fuab038] [PMID: 34227665]

[66]   Gupta Sood S. Chemotactic response of plant-growth-promoting bacteria towards roots of vesicular-arbuscular mycorrhizal tomato plants. FEMS Microbiol Ecol 2003; 45(3): 219-27.
[http://dx.doi.org/10.1016/S0168-6496(03)00155-7] [PMID: 19719591]

[67]   Raina JB, Fernandez V, Lambert B, Stocker R, Seymour JR. The role of microbial motility and chemotaxis in symbiosis. Nat Rev Microbiol 2019; 17(5): 284-94.
[http://dx.doi.org/10.1038/s41579-019-0182-9] [PMID: 30923350]

[68]   Karmakar R. State of the art of bacterial chemotaxis. J Basic Microbiol 2021; 61(5): 366-79.
[http://dx.doi.org/10.1002/jobm.202000661] [PMID: 33687766]

[69]   Ibrar M, Khan S, Hasan F, Yang X. Biosurfactants and chemotaxis interplay in microbial consortium-based hydrocarbons degradation. Environ Sci Pollut Res Int 2022; 29(17): 24391-410.
[http://dx.doi.org/10.1007/s11356-022-18492-9] [PMID: 35061186]

[70]   Bi S, Sourjik V. Stimulus sensing and signal processing in bacterial chemotaxis. Curr Opin Microbiol 2018; 45: 22-9.
[http://dx.doi.org/10.1016/j.mib.2018.02.002] [PMID: 29459288]

[71]   Cicerone RJ, Oremland RS. Biogeochemical aspects of atmospheric methane. Global Biogeochem Cycles 1988; 2(4): 299-327.
[http://dx.doi.org/10.1029/GB002i004p00299]

CHAPTER 11

Molecular Techniques in Ecosystem Restoration

R. Shivakumar[1,*] and **B. Balaji**[2,3]

[1] *Department of Biotechnology, Centre for Plant Molecular Biology and Bioinformatics, Tamil Nadu Agricultural University, Coimbatore-641003, India*

[2] *ICAR-National Institute for Plant Biotechnology, LBS Centre, Pusa Campus, New Delhi, India*

[3] *Post Graduate School, Indian Agricultural Research Institute, Pusa, New Delhi-110012, India*

Abstract: A damaged ecosystem must be rebuilt to its original form, or a new ecosystem must be created in a degraded area. Ecosystem restoration is a complex procedure. Researchers can now investigate the structure and function of ecosystems at the molecular level thanks to the development of molecular techniques as a potent tool for ecosystem restoration. This chapter examines the application of molecular methods to ecosystem regeneration. The various available molecular methods and how they have been applied to monitor ecosystem health, identify microbial communities in ecosystems, and comprehend interactions between microbes and plants are discussed. The chapter also examines the application of molecular methods to the restoration of ecosystems that have been damaged, including the use of plant-microbe interactions to promote plant development in contaminated soils. The chapter emphasizes the significance of molecular methods in ecosystem restoration and their potential to offer a more precise and thorough comprehension of ecosystem processes. The conclusion highlights the importance of ongoing investigation into the use of molecular methods for ecosystem restoration, especially in creating novel methods and their incorporation with existing restoration techniques. In the end, applying molecular methods can help develop practices for ecological restoration that are more efficient and long-lasting.

Keywords: Bio-sensors, Environmental DNA, Heavy metal pollution check, Metagenomics, Ocean acidification, 16s rRNA sequencing.

INTRODUCTION

Ecosystem restoration is the process of repairing and restoring damaged or degraded ecosystems to their original or functional state. It entails various activities, including habitat restoration, reforestation, soil restoration, and water restoration. Because of growing concerns about climate change, biodiversity loss,

---

[*] **Corresponding author R. Shivakumar:** Department of Biotechnology, Centre for Plant Molecular Biology and Bioinformatics, Tamil Nadu Agricultural University, Coimbatore-641003, India; E-mail: shivakumar.r.writer@gmail.com

Shiv Prasad, Govindaraj Kamalam Dinesh, Murugaiyan Sinduja, Velusamy Sathya, Ramesh Poornima &
Sangilidurai Karthika (Eds.)

and natural resource depletion, ecosystem restoration has become increasingly important in recent years. Molecular techniques have transformed the field of ecosystem restoration by allowing scientists and practitioners to study and comprehend the underlying mechanisms of ecosystem functioning and design effective ecosystem restoration strategies.

This chapter provides an overview of molecular techniques used in ecosystem restoration, such as DNA-, RNA-, and protein-based techniques. The chapter begins by discussing the significance of molecular techniques in ecosystem restoration and the problems that these techniques can help to solve. It then details the various molecular techniques used in ecosystem restoration, their advantages and disadvantages, and their applications in various contexts.

IMPORTANCE OF MOLECULAR TECHNIQUES

For various reasons, molecular techniques have become an indispensable tool for ecosystem restoration. For starters, molecular techniques allow scientists to investigate and comprehend the underlying mechanisms of ecosystem function. Identifying the key species and processes that are critical to ecosystem functioning, as well as the interactions between these species and processes, is part of this. Scientists can design more effective ecosystem restoration strategies and predict the outcomes of restoration activities if they understand the underlying mechanisms of ecosystem functioning.

Second, molecular techniques can assist in addressing some of the challenges associated with ecosystem restoration. Human activities, for instance, exploit a large number of ecosystems, altering their biodiversity and composition. When creating reintroduction plans for native species that were once part of the ecosystem, molecular techniques can be used to identify these species. The identification of genes and traits essential to the survival of native species can also be aided by molecular techniques, as can the selection of individuals with these traits for reintroduction. Monitoring the success of ecosystem restoration efforts can be aided by molecular techniques. By observing changes in species abundance and diversity as well as changes in ecosystem processes, scientists can evaluate the efficacy of ecosystem restoration efforts and make necessary adjustments.

VARIOUS MOLECULAR APPROACHES TO ECOSYSTEM RESTORATION

Identification of species is critical in ecosystem restoration, especially when dealing with threatened or endangered species. Molecular techniques such as DNA barcoding and sequencing have become valuable tools for identifying species and detecting cryptic or rare species that are difficult to identify using traditional methods [1]. These techniques have also been used to investigate population genetic diversity and the origins and evolutionary history of species. Another crucial area of study in ecosystem restoration is population genetics. Microsatellites and single nucleotide polymorphisms (SNPs) are examples of molecular markers that have been used to study population genetic structure and monitor individual movement both within and between populations [2]. These markers can also be used to estimate population sizes, gene flow, inbreeding, and genetic drift.

Conservation biology is a branch of biology that seeks to protect and conserve ecosystem biodiversity. Molecular techniques have been critical in this field, particularly in identifying and tracking endangered species and understanding the genetic and ecological factors influencing species survival. Microsatellites and mitochondrial DNA have been used to determine the source populations for translocation programs and monitor their effectiveness. The interactions of species with their environment have also been studied using molecular methods. For instance, transcriptomics and proteomics have been used to study how species react to environmental stressors like pollution and climate change. By shedding light on the molecular processes underlying species' responses to environmental stressors, these techniques can aid in the identification of potential conservation targets.

Overall, molecular methods have emerged as a crucial tool for restoring ecosystems, providing scientists with a potent arsenal of instruments for investigating the genetic diversity, population dynamics, and ecological interactions of species in an ecosystem. This chapter will examine the various molecular approaches that have been used in ecosystem restoration, as well as their advantages, drawbacks, and potential applications in the future.

PCR-based Approaches

Most molecular-based studies begin with DNA extraction from a particular organism, followed by the amplification of particular DNA segments using the polymerase chain reaction (PCR). The fact that only minute amounts of DNA are needed makes PCR worthwhile [3]. This is especially helpful when scientists need numerous samples or cannot obtain large amounts of tissue, as in population genetic studies. DNA was taken from people in various populations and used in a PCR-based genetic diversity survey. This kind of investigation can provide insights into the historical processes that led to variations in the genetic makeup of populations dispersed across a variety of environments and geographies. A schematic representation of the polymerase chain reaction is shown in Fig. (1).

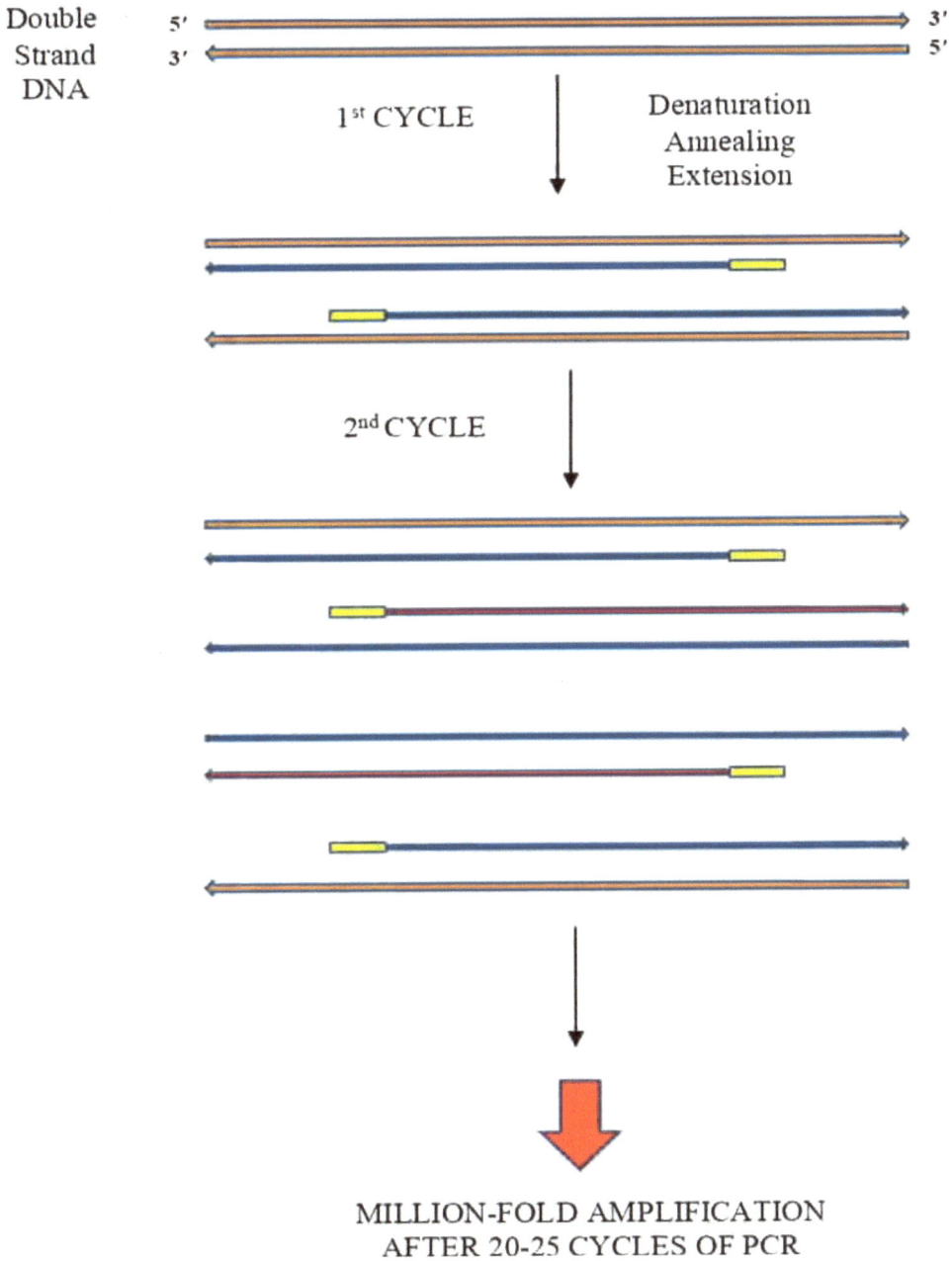

Fig. (1). Schematic representation of the polymerase chain reaction. Dotted lines in each cycle indicate the newly synthesized DNA. Solid rectangles indicate oligonucleotide primers. Each DNA strand is marked with an arrow indicating the 5' to 3'.

In order to reconstruct the phylogenetic history of each species within the complex, PCR is used to amplify specific coding or non-coding regions of DNA from different species. Once established, the phylogenetic tree resulting from this study can reveal details about the species complex's diversity and which species are most closely related to one another [4]. This, in turn, can shed light on the ecological and behavioral elements that have contributed to the diversity of a species complex.

Marker-based Approaches

Marker-based approaches have been used in ecological studies for several analyses. Such studies use a variety of DNA markers. Microsatellites, mini satellites, restriction fragment length, and DNA sequence information are a few of the significant markers. In contrast to mini-satellites, which function similarly to microsatellites but have a longer sequence than the former, microsatellites are highly repetitive sequences that can heritably function independently and distinguish between different individuals [5]. RFLPs are the specific sites in the DNA that are cut by the specific nucleases varying in the differently sized fragments between individuals. These various techniques could be detected by agarose gel electrophoresis by creating discrete bands [6]. That is also detected visually by chemifluorescence, which could detect the fluorescent emission from the designed primers. The basic approaches are described further,

Amplified Fragment Length Polymorphisms

A technique known as amplified fragment length polymorphisms (AFLPs) can produce markers that are anonymous. A schematic flow chart showing the principle of the AFLP method is shown in Fig. (**2**).

This technique uses PCR and restriction enzymes to produce millions of unique fragments that can be used to genetically identify people within or between species within the same genus [7]. Because it does not require prior knowledge of an organism's genome, the AFLP method is advantageous. The researcher does not know which genome regions are targeted by this technique. This approach frequently produces knowledge regarding the fundamental genetic differentiation and diversity levels. As a result, using AFLP markers as a starting point when analyzing population or species differences is common. Only one allele is amplified by AFLP fragments, which represent distinct restriction sites that are either present or absent in each individual, limiting the amount of data that can be gathered. The dominant markers produced by a different technique, known as random amplified polymorphic DNA (RAPD), are typically visualized using agarose gel electrophoresis. However, the AFLP method, which typically uses chemifluorescence and a genetic analyzer for visualization, has largely replaced this approach.

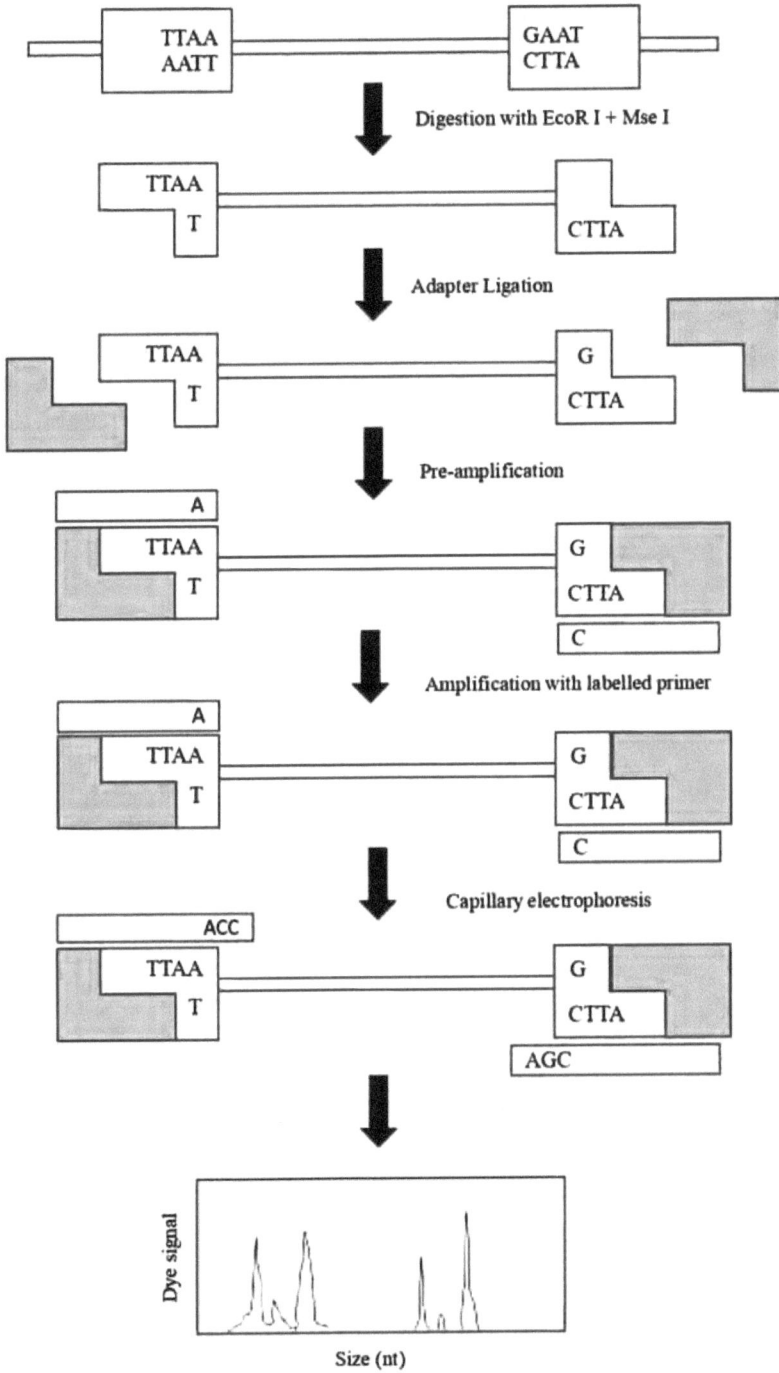

Fig. (2). Schematic flow chart showing the principle of the AFLP method.

Sequence-Tagged Site (STS) Markers

STS markers provide an alternative method for describing genetic diversity within and between species. A sequence-tagged site is a brief (200–500 bp) nucleotide sequence located in a specific region of the genome and targeted using PCR and primers created by the researcher [8]. One type of STS marker (VNTR) is microsatellites (MSATs), also known as simple sequence repeats (SSRs) or variable number tandem repeats. Contrary to AFLPs, these markers do call for some familiarity with specific regions containing tandemly repeated nucleotide motifs, like "ATC" or "GAG," which are frequently found in non-coding regions of DNA. When used with PCR amplification and primers specifically created to target these sites, the STS method offers a much finer level of discrimination between individuals. Codominant markers can show whether a person is heterozygous at a specific locus because both alleles are amplified during the PCR process [9].

DNA Sequencing

Understanding the target genes or gene regions of the researcher is essential for DNA sequencing. This approach gives molecular ecologists the most precise and fundamental level of genetic detail currently available when combined with PCR and well-designed primers. This is because a wide range of taxonomic levels, from phyla to species, and, depending on levels of variation, even among individuals within a population, can be cross-comparison analyzed to determine the exact nucleotide sequence. The genetic basis of adaptive trait loci (for example, genes involved in responses to day length in plants), historical patterns of animal species migration and expansion (for example, from the Pleistocene to the present day), and the evolution of specific traits involved in taxonomic diversification can all be inferred from an organism's evolutionary history using DNA sequencing [10]. A typical workflow of next-generation sequencing methods is shown in Fig. (3).

DISTINCT APPROACH OF MARKER

The discovery of molecular markers has spawned a deluge of research using them to address issues such as the degree of genetic variation within a species, behavioral patterns, how gene expression patterns can differ among populations that are closely related to one another, and many other aspects of organismal variation. In one of the earliest molecular ecological studies, they found that the genetic diversity of cheetahs in South Africa was incredibly low. In fact, O'Brien *et al.* (1985) found that because different cheetahs were genetically similar, their immune systems did not reject the tissue grafts when they transplanted skin between them [11]. A population of organisms with the same genetic makeup will react to environmental changes similarly. This environmental change is likely to have an adverse effect on all genetically similar individuals in the population, increasing the likelihood of extinction if an organism's capacity to survive or reproduce is hampered. However, this is not always the case, as some populations with closely related genes appear to flourish. Therefore, the importance of genetic diversity for population survival is still debatable.

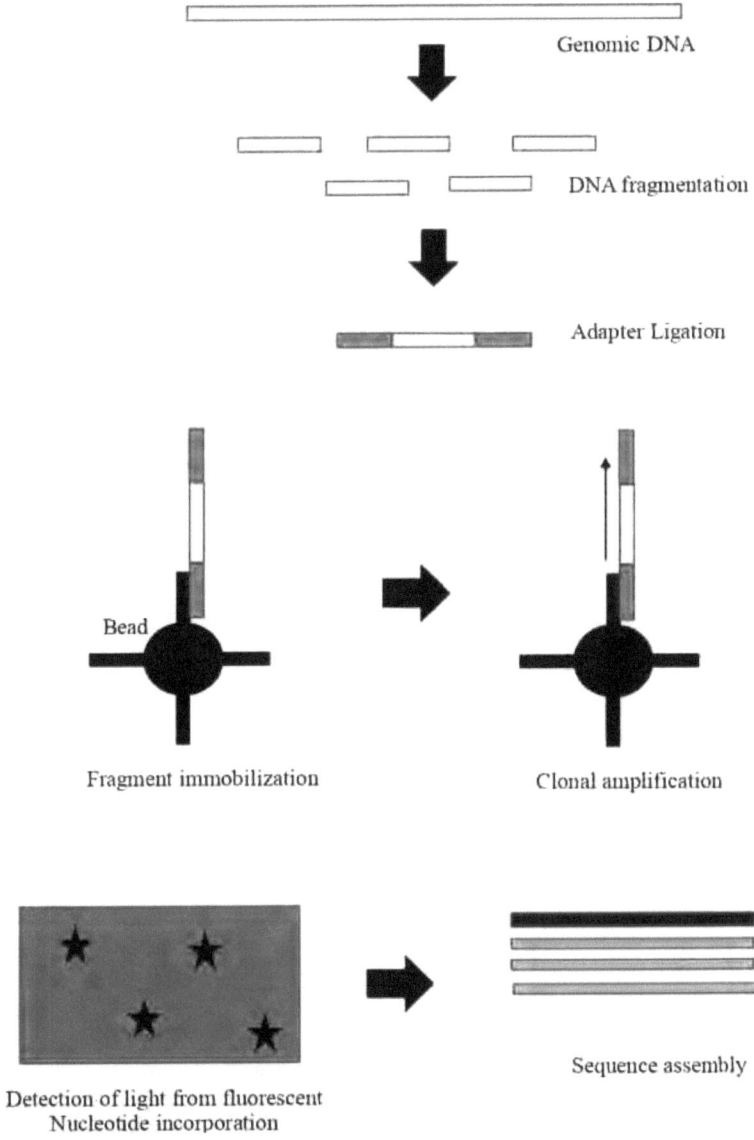

Fig. (3). A typical workflow of next-generation sequencing methods.

Biosensors – DNA Probe

Biosensors make use of biological elements as a recognition component. The selectivity, sensitivity, repeatability, durability, and cost-effectiveness of the transduction process are all improved by the nanotechnology component of a nanobiosensor [12]. Only a few cases mention nanomaterial as a recognizable and permeating component. Nanobiosensors called DNA nanosensors use nonmaterials for transducing and DNA for recognition. Numerous DNA-based nanosensors have been created for use in a wide range of human endeavors, including detecting noxious environmental substances.

Biosensors use a biological recognition component, such as an enzyme, antibody, or nucleic acid, to specifically bind to a target molecule. For example, an electrical or optical signal that can be detected and quantified is produced when the recognition element interacts with the target molecule. A wide range of targets, including pathogens, organic pollutants, and heavy metals, can be programmed into biosensors.DNA elements like ssDNA, dsDNA, mismatch DNA, CA rich, C rich, T rich, G-quadruplex, and aptamer have been used for pollutant detection. Nanomaterials that are frequently used include metallic, metal oxides, quantum dots (QDs), platinum (Pt), copper (Cu), magnetic, tungsten disulfide (WS2), mesoporous silica (MSN), graphdiyne, graphene, and graphene oxide [13].

Since each pollutant has a unique toxic response, the interaction between DNA and pollutants is unique. Various analytes can be sensed thanks to DNA's versatility in changing specificity as its sequence and structure change. A wide range of applications can be created by combining DNA with the unique qualities of nanomaterials. There is potential for a wide range of detection applications with fewer restrictions due to the propensity and response of DNA interaction with biological or chemical molecules. The color, UV-visible absorption intensity, and wavelength of nanomaterials change when ssDNA or dsDNA binds to an analyte in a DNA nanosensor. The two nanomaterials most frequently used in creating DNA nanosensors are AuNPs and AgNPs [14]. One type of biosensor used for environmental assessment is the microbial biosensor. Microbial biosensors use microorganisms like yeast or bacteria as their recognition component. These microorganisms create a detectable signal when interacting with a particular target molecule. For instance, heavy metals like lead and cadmium have been detected in water and soil using microbial biosensors. The microorganisms used in these biosensors produce a fluorescent or bioluminescent signal when exposed to the target metal, enabling researchers to quickly and easily detect the presence of these pollutants.

Enzyme-based Biosensors

The enzyme-based biosensor is another type of biosensor used for environmental assessment. Enzymes are used as the recognition element in enzyme-based biosensors. Proteins that catalyze specific chemical reactions are known as enzymes. An enzyme catalyzes a reaction that yields a measurable signal, such as an increase in electrical conductivity or a change in pH, when it comes into contact with its target molecule [15]. Biosensors based on enzymes have been used to detect various targets, including glucose, cholesterol, and pesticides.

A schematic representation of typical biosensor system architecture is shown in Fig. (**4**). The target analyte is detected by the bioreceptor and translated into a signal for analysis with the aid of a transducer. (A) Sample. (B) Biosensor electrode: composed of the bioreceptor, the immobilization surface, and the transducer element. (C) Physicochemical reaction. (D) Data analysis [16].

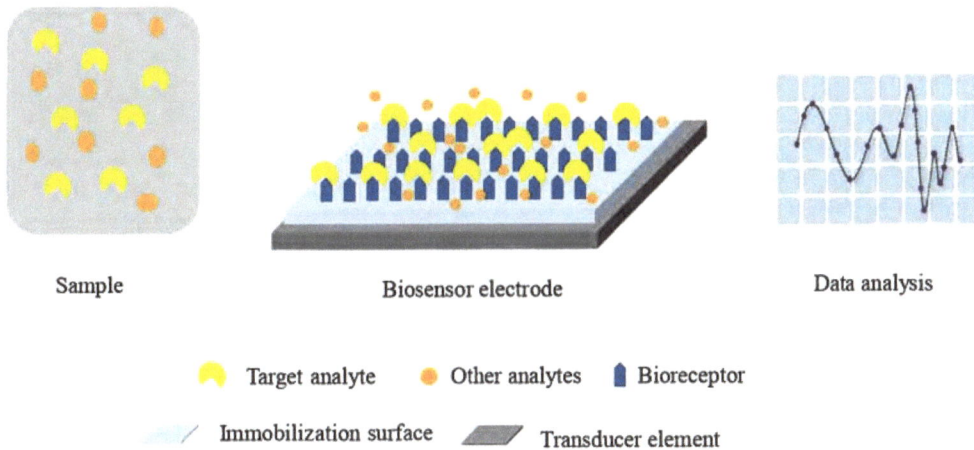

Fig. (4). Schematic of a typical biosensor system architecture.

DNA-based Biosensors

DNA probes are short DNA segments that are designed to bind to specific DNA or RNA sequences. FISH is an excellent tool for monitoring microbial populations in water, soil, and other environmental samples. A schematic representation for nucleic acid RNA/DNA-based biosensor development using an Au-coated three-electrode system based on other reported biosensors ssDNA: single-stranded DNA, Au: gold [17] is presented in Fig. (**5**).

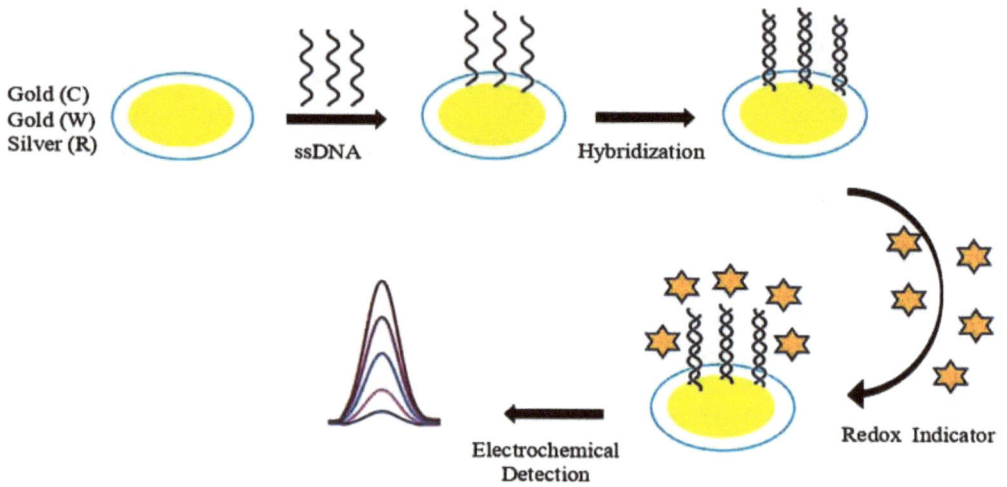

Fig. (5). Schematic representation for nucleic acid RNA/DNA-based biosensor development using Au-coated three-electrode system based on other reported biosensors ssDNA: single-stranded DNA, Au: gold [17].

DNA probes frequently detect and identify specific microorganisms in environmental samples, such as bacteria or viruses. DNA probes can be programmed to target specific genes or regions of the genome that are specific to a given organism. Another example of a DNA probe used for environmental assessment is the fluorescent *in situ* hybridization (FISH) method. In the FISH method, fluorescently labeled DNA probes are used to bind to particular DNA sequences in a sample. Because probes can target particular microorganisms, researchers can use them to identify and quantify microbial populations in environmental samples [18].

Various Detections by Biosensors

Pathogen detection: Contaminants in water V. cholera can be recognized by selective binding of the O1 OmpW gene with two DNA probes. The magnetic NP-probe1-O1 OmpW-fluorescein amidite (FAM) probe2-AuNP complex only forms in test samples that contain V. cholera. The FAM probe can be isolated and measured using fluorescence [19].

Antibiotic detection: A DNA nanosensor can identify various antibiotics in water and a sophisticated biological medium. Aptamer-AuNPs remained stable in salt solution (OFL) without ofloxacin. AuNP aggregated due to the presence of OFL in water and synthetic urine samples, which was evident as a red to purple/blue color shift. Salt causes AuNPs to aggregate because aptamer preferentially binds to complementary strands that have been FAM-labeled in solution. The solution appears blue because FAM emits much fluorescence [20].

Pesticide detection: The main classes detected by DNA-based nanosensors include triazines, neonicotinoids, carbamates, and organophosphorus. In order to find acetamiprid in celery and green tea leaves, a colorimetric DNA nanosensor has been created. The solution turns purple due to salt-induced aggregation of the aptamer binding to ssDNA-AuNPs [21]. As a result of the formation of a red aptamer-acetamiprid complex in the presence of acetamiprid, ssDNA-AuNPs become soluble.

Heavy metals detection: Sulphide ions (S2) were found using the fluorescence quenching technique in hot spring and seawater samples. Fluorescence quenching is brought on by a conformational change in the DNA probe located over the Au/Ag nanoclusters brought on by S2 in the test sample [22]. The sensor needs sodium peroxydisulfate to prevent non-specific interaction with interfering iodide ions (I). I-specific DNA template functionalized Au/Ag nanoclusters experience fluorescence quenching and color change from colorless transparent to purple-red when (I) is present in the test sample.

Toxic pollutants detection: DNA nanosensors have reportedly been used to detect contaminant dyes, explosives, and toxins. Aptamers release Fluorogenic Rhodamine B dye trapped in MSN pores. A fluorescence signal is produced when the aptamer binds to bisphenol A in tap water samples. Dopamine can photo-induced electron transfer with CuNPs to quench the fluorescence of dsDNA-CuNP. The dsDNA template containing AgNPs releases DNA when dopamine is present in the test sample, making that DNA available for the intercalating dye GeneFinder to produce fluorescence [23].

ENVIRONMENTAL DNA TECHNOLOGY

Environmental DNA, or "eDNA," which can be extracted from environmental samples and used to identify species, populations, and communities, can potentially address many of the problems with biodiversity monitoring. Significant technological and scientific advancements have been made throughout the past ten years due to the discovery that higher creatures' DNA can remain in the environment and may thus be sampled, retrieved, and studied. Species constantly release DNA into their surroundings as they interact with the environment. For higher organisms, this DNA can come from expelled cells or tissues like skin, hair, urine, feces, and hair, as well as plainly from deceased individuals releasing the genetic material.

Collection of e-DNA Samples

Regions and various habitats, including glaciers, permafrost/tundra, aquatic sediments, lakes and streams, terrestrial habitats, and oceans, could be used to collect samples [24]. The first three are from ancient environments dating back to prehistoric times, while the remaining ones are from the present.

Ancient deposits were the starting point for studies on soil eDNA, but more recent work has focused on current ecosystems. Both intracellular and extracellular DNA comprises the total amount of DNA in soil, with extracellular DNA most likely making up the majority. As a result, metabarcoding of eDNA from surface soil has been successfully used as a proxy for plant taxonomic diversity in several terrestrial environments. In aquatic species, the genetic material, preferably the DNA, that the aquatic animals shed into the environment. This basically includes the skin particles, feces, and urine. This DNA content is, therefore, used to determine the presence of the species or their relative abundance in the particular habitat. The DNA is transported with the flow and remains suspended in the water column over time, making waterways particularly appropriate for eDNA sampling. This technique has great potential for finding species that are hard to locate and take a lot of time, such as cryptic, rare species (like galaxiids, crayfish, and freshwater mussels) in far-off places. The capacity to recognize and respond to pest species outbreaks will also improve with the use of eDNA.

Techniques Involved in e-DNA Technology

Two crucial approaches have been implied for the detection of species through this technology were single-species detection by PCR and multi-species detection by NGS (Next generation sequencing). A schematic representation of environmental DNA metabarcoding workflow [25] is presented in Fig. (**6**).

Single species detection by PCR: Single species detection by PCR is essential and efficient when one target species exists. The probes and primers thus developed are species-specific. Despite having excellent specificity, sensitivity, and quantification capabilities, the method is less effective because it can only detect one target organism at a time. This approach quickly becomes inefficient and expensive for increasingly complex systems due to the dearth of DNA extract for many reactions.

DNA metabarcoding approach: This technique is efficient and cost-effective, and mass DNA sequencing technology keeps improving, including next generation sequencing. The main issue with using generic primers for metabarcoding is that some sequences of specific species amplify less effectively due to primer affinity bias. This will then limit the results to the species that have the highest affinity for the primer or to the species that are already known to exist locally, allowing for the development of customized primers. These limitations will gradually become less important as more effective and published generic primers for metabarcoding research are developed.

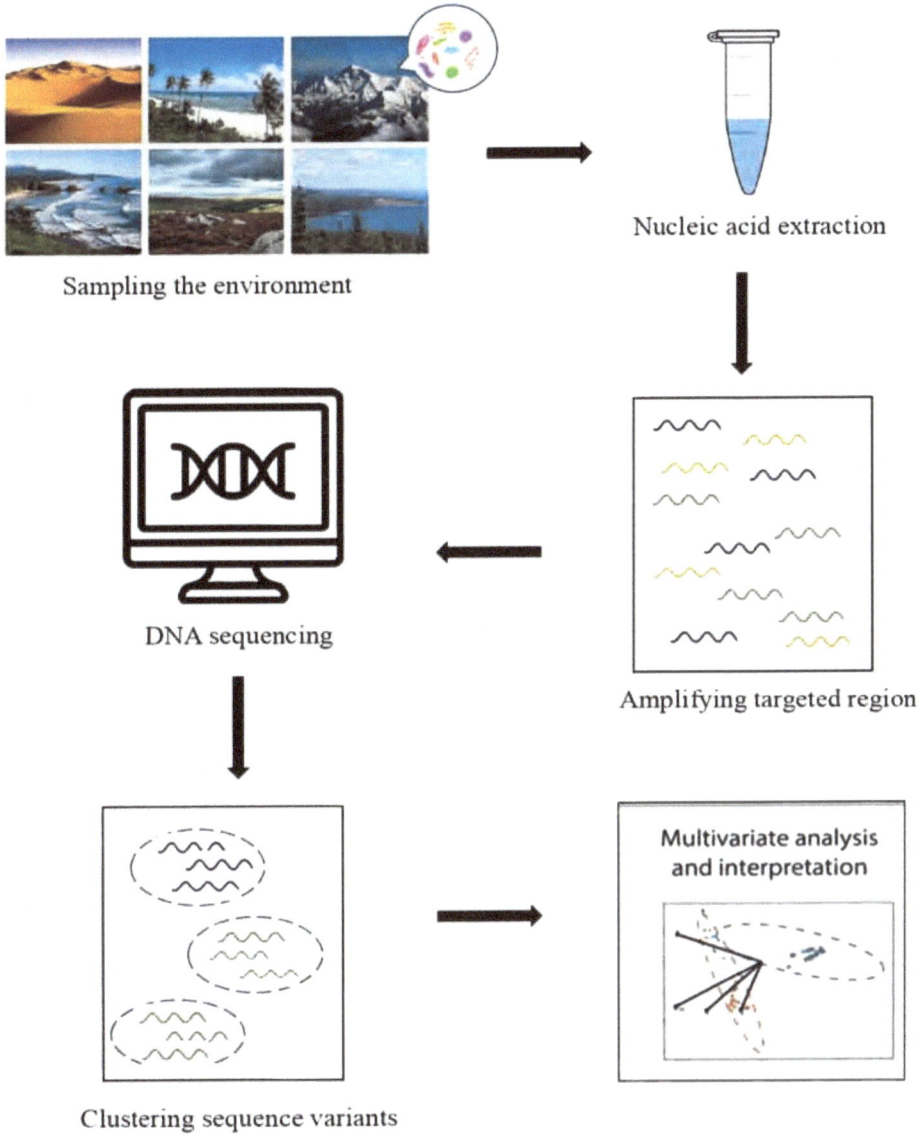

Fig. (6). Schematic representation of environmental DNA metabarcoding workflow.

Concerns and Future Perspectives of e-DNA Technology

Interpretation of results is a greater concern in distinguishing the living and dead organisms; it is challenging to determine particular stages of the life cycle of the organisms and variate the hybrid species. Hence this could be done with the help of mitochondrial markers, which are also used to detect the maternal path of the hybrid species. The exploitation of the entire DNA obtained from a particular habitat for the study also has temporal and special constraints. The lower sediments of any habitat, say aquatic, may have the DNA existence of the ancient species that restrict the studies conducted for contemporary organisms. Some terrestrial bacteria have the capacity to naturally transform mammoth DNA down to just 20 bp in age and degradation. Therefore, during the evolution of life on Earth, the vast pool of DNA in the environment may have influenced the evolution of bacteria by being incorporated into their genomes [26]. These altered eDNA fragments may also be redeposited in the environment after cell death, making eDNA fragments continuously accessible thousands or millions of years after they were first shed from their original source.

- This technology is developing with the evolving technology; hence, a few points are to be noted for future developments.
- Efficient monitoring of the spatial and temporal distribution of eDNA in different habitats
- Studies to devise knowledge outcomes about eDNA concentration and species abundance
- To determine the source of DNA from where it has been obtained, like, epithelial, intestinal cells, feces, and urine, and also to know about the variation in life stages of the target organism.
- Studies about temperature, pH, and salinity are physio-chemical variables that affect the availability and degradation of eDNA.

16S RRNA SEQUENCING FOR POLLUTION CHECK

Microorganisms are important creatures in all types of ecosystems. Any changes in the environment would affect the microbial population. The microbial studies under ecology have been improved due to next-generation sequencing, including 16S rRNA sequencing [27]. The 16S RNAs are small sequences present in all organisms. They are species-specific due to the hypervariable regions (V1 to V9), which made it a fast and accurate tool for studying the ecological changes highly influenced by microbial distribution and diversity [28 - 30].

Further, the 16S rRNA sequencing method is more accurate than traditional identification methods, which are primarily based on phenotypes and are highly mutable [31]. Since most of the microorganisms are non-culturable, 16S RNA sequencing is highly helpful in metagenomics and ecological studies such as environmental barcoding and metabarcoding [32, 33]. Because of the high resolution and easy accessibility, 16S rRNA sequencing is used to monitor and investigate the pollution sites and their effects on microbial populations. The phylogenetic analysis through hypervariable regions in the 16s rRNA sequence from the sites also shows the distribution of related microorganisms, such as phylotypes, over the sites. Observations and identification of 16s rRNA sequence act as genetic markers, valid to monitor and check the environmental quality and also help to detect the cause of pollution [34]. Ecotoxicological studies show fluctuation in the microbial population, diversity, and activities that can be seen in the soil-polluted sites, including heavy metals [35, 36].

Heavy Metal Pollution Check

Heavy metals such as As, Cd, Cr, Cu, Hg, Ni, Pb, and Zn are highly toxic and show bioaccumulation and biomagnification. These metals persist in the environment and are hard to degrade [37, 38].16s rRNA study on the heavy metal contaminated site in Slovakia shows proteobacteria were well distributed and also possess heavy metal resistant genes like czcA and nccA [39]. 16S rRNA sequencing of Chromium polluted sites shows Firmicutes (30-50%), Gammaproteobacteria (18%), Actinobacteria (13-14.5%), Bacteriodetes (7.75-9.5%), Alphaproteobacteria (7%), Betaproteobacteria (8%), Deltaproteobacteria (9.5%), Epsilonproteobacteria (3-4%) are widely distributed [40]. In the soil of metal accumulating plants, Thlaspicaerulescens, Actinobacteria, Acidobacteria, α-Proteobacteria, β-Proteobacteria, and Planctomycetes were found through 16s rDNA and 16S rRNA libraries. Among them, Actinobacteria were prevalent and could be metabolically active in heavy metal-contaminated sites [41]. There are certain bacteria promoting plant growth in heavy metal-contaminated areas. Plant growth promotion happens through solubilizing nutrients, accumulating toxic metals, and growth hormone production, thereby helping the plants to tolerate the effects of heavy metals [42].

Application on Fecal Contamination Water

Fecal pollution negatively affects sanitary status, water quality, and the aquatic ecosystem. The fecal pollution monitoring through the fecal pollution indicators such as coliform bacteria, fecal enterococci, *etc.*, have no information about the source of fecal contamination [43]. Molecular approaches, including 16S rRNA, were used to identify the source of fecal contamination. Fecal pollution sources include human waste, animal feces and effluents, and manure runoff. 16S rRNA sequence can be used as genetic markers for identifying and distinguishing the source for fecal contamination, *i.e.*, human and cow fecal, through the difference in the composition of Bifidobacterium and Bacteroides-Prevotella [44]. The 16S rRNA gene libraries are made to understand the phylogenetic relationship among the microbes from the fecal polluted sites. The fecal-specific markers can be made from the host-specific clusters obtained in the phylogenetic analysis and serve as a differentiating tool for fecal contaminant sources in the contaminated sites. Real-time PCR of these markers can also quantify the source of fecal contamination [45].

ANTIBIOTIC-RESISTANCE GENES IN UNTREATED WASTEWATER

Strategies to reduce the Antibiotic-resistance genes in wastewater are presented in Fig. (**7**). Global health is threatened by antibiotic-resistant genes (ARGs) and antibiotic-resistant bacteria (ARB). Since antibiotics are frequently used in the environment, finding antibiotic-resistance genes in untreated wastewater and hospital waste is common. Through horizontal gene transfer, these genes can also move between various microbe species, leading to the development of highly resistant microorganisms with ARGs [46, 47]. Transformation of ARGs into disease-causing pathogens will end in risk to ecological sustainability. The presence of ARGs in water can be detected by normal PCR using antibiotic gene primers, and the quantity of ARGs can be identified through qPCR, and the particular organism holding the resistance is identified through the culturing and antibiogram test [48, 49]. Hospital wastewater is an important source of antibiotic-resistance genes such as the β-lactam resistance gene bla_{GES-1}, quinolone resistance gene *qnr*, and integrin *intI1*. *VanA* and bla_{KPC}, the resistance genes of Vancomycin and Kanamycin, respectively, were found abundant in hospital wastewater. There have been 25% more ARGs from hospital wastewater inferred through the qPCR of antibiotic genes [50]. *Escherichia coli* isolates from the wastewater and river water were found to be 43% multi-antibiotic-resistant isolates in the untreated wastewater, and genes responsible for tetracycline *tet*A, *tet*Band*tet*K were noted. Virulence genes were also noted in the isolated wastewater as *bfp*A, *est*A, and *eae* as 65%, 56%, and 39%, respectively [51].

Several incidences show that antibiotic-resistant genes can occur in treated water [52, 53]. The unmanaged wastewater treatment units create a suitable environment for developing ARBs through selective pressure on the bacteria. Bacitracin and Rifamycin resistance genes were abundant in treated wastewater due to high microbial loads in the treatment, adaptability, and nutrient availability [54]. There are health risks in wastewater treatment plants to workers exposed to concentrated multidrug-resistant pathogens. Their pathogens were transmitted through bioaerosols from the plant sludge [55]. Municipal wastewater treatments are inefficient in removing antibiotic resistance genes. Several strategies can reduce the ARGs in wastewater treatment, as shown in (Fig. **2**). Aerobic digestion and anaerobic digestion can reduce the antibiotic-resistant bacteria and fecal indicator bacteria. This can effectively remove *tet*, *bla*, *qnr,* and *er*genes [56, 57]. Installing a membrane bioreactor removes the antibiotics and ARBs selectively. Combining membrane bioreactors can somewhat reduce the number of antibiotics and ARBs [58].

Chemical approaches such as chlorination are efficient in controlling ARG-bearing microbial communities. Ozonation and UV treatment at secondary and tertiary wastewater treatment are not promising in reducing the ARGs and other microbes, as several regrowths have been observed [59, 60]. Ionizing radiation treatment such as γ-rays affects microbial activities and reduces antibiotic resistance genes [61]. Treating with reactive oxygen species damages the cells of microbes and disinfects the ARBs. Using TiO_2 as photo-catalysis generates superoxide radicals. These reactive oxygen species damage the cell structure and oxidize the genetic materials [62]. Reducing the immoderate use of antibiotics in clinical treatment is imperative and needs to be followed, and adapting different strategies for the disease treatment must be encouraged. By combining all of the strategies, antibiotic-resistant bacteria and gene levels can be reduced in water treatment, and the causes of environmental gene pollution can be avoided.

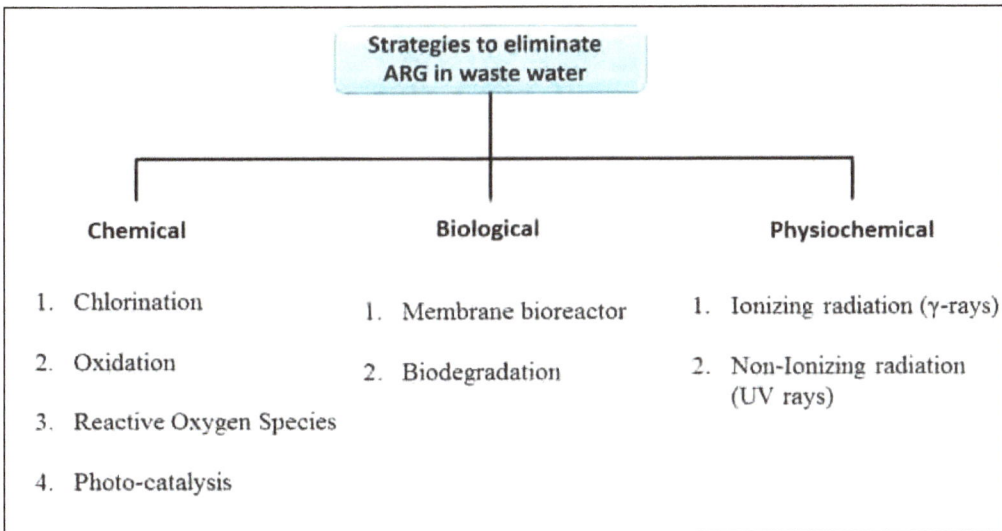

Fig. (7). Strategies to reduce antibiotic-resistance genes in wastewater.

OCEAN ACIDIFICATION ANALYSIS-METAGENOMICS

The ocean is comprised of diverse ecological units that interact with each other. Ocean acidification (OA) is the increase in acidity or the decrease in the average pH of the ocean. CO_2 in the atmosphere dissolves in the ocean water and turns into carbonic and bicarbonate ions, increasing the proton (H^+) concentration. This process is a natural process. However, due to anthropogenic activities, there has been an increase in CO_2 concentration, resulting in OA and air pollution [63, 64].

The OA has adverse effects on the aquatic ecosystem. In understanding the ocean ecosystem under acidification, metagenomics is a good approach. Metagenomic studies conducted in sites such as ocean volcanoes and submarine CO_2 vents where the concentration of CO_2 is relatively higher than in other ocean ecosystems give more understanding. Such a type of metagenomics study shows Firmicutes, Proteobacteria, and several unclassified bacteria were found in the sites. Firmicutes, including *Bacillus vulcani and B. Aeolus,* were prevalent, and Proteobacteria included Gamma-Proteobacteria and Epsilon-proteobacteria. There was an unusual distribution of microbes than the actual distribution of marine ecosystems due to low pH and high CO_2 concentration. The ocean ecosystem has a low diversity index at low pH and high CO_2 [65]. In the ocean and coastal areas, corals, algae, and sponges can serve as habitats for diverse species and one of the key organisms to maintain the marine ecosystem. These key organisms are affected by OA, and it creates a substantial negative impact on the other organism's habitats. Rhodoliths are calcareous algae, photosynthesizers, and vital members of coastal ecosystems providing shelter for diverse organisms. OA has a direct role in these holobionts. Metagenomics of the rhodolith microbiome under OA shows host-microbes interactions under adverse environmental conditions [66].

Corals are one of the key organisms vulnerably affected under OA. Corals under low pH due to OA show delayed calcium carbonate growth rates, resulting in coral bleaching. Bleaching occurs when there is a distribution in the symbiosis between the corals and zooxanthellae. Metagenomics and functional metagenomics of Corals *Balanophylliaeuropaea*using 16S rRNA under acidification showed that the microbiome has a quantitative shift in the physiological and biogeochemical process. There has been a considerable change in the Nitrogen metabolism to acclimatize to the acidified environment metabolism such as Nitrogen. Microbiome change affects carbohydrate metabolism through gluconeogenesis and glycolysis [67]. The coral *Stylophorapistillata* metagenomic studies using 16s rRNA showed that active microbial communities were affected under OA. Relative change in the microbial community infers diseased corals. Healthy corals can be identified through abundant coral-associated microbes such as archaea, bacteria, and protista, and diseased corals have fungi, fusobacteria, and bacteriodetes [68].

Similarly, variations in the microbial communities were seen in sponges under OA. The metagenome of plastic biofilms under OA shows an abundance of diatoms, and antibiotic genes are relatively less in those biofilms [69]. Metagenomics is a reliable resource for studying ocean ecology and understanding the community structure and interaction among different organisms under adverse conditions.

CONCLUSION

The potential of molecular methods in ecosystem restoration efforts is enormous. These methods add a fresh perspective to conventional methods, enabling a more in-depth comprehension of an ecosystem's genetic diversity and ecological functions. The use of molecular methods in ecosystem restoration has increased monitoring effectiveness, decreased time and resource requirements, and increased the success rates of restoration efforts. Despite the encouraging outcomes, issues and constraints still need to be resolved if molecular techniques are to realize their potential in ecosystem restoration fully. In order to overcome these difficulties and maximize the use of molecular techniques in ecosystem restoration, additional study and development is required. With continued advancement, molecular techniques will unquestionably play a more significant part in efforts to restore ecosystems, protect and preserve biodiversity, and maintain healthy, functional ecosystems for future generations.

ACKNOWLEDGEMENTS

The authors thank the Department of Biotechnology, Centre for Plant Molecular Biology and Bioinformatics, Tamil Nadu Agricultural University, Coimbatore, CAR-National Institute for Plant Biotechnology, LBS Centre, Pusa Campus New Delhi, Post Graduate School, Indian Agricultural Research Institute, Pusa, New Delhi for their continual encouragement and unflinching support.

REFERENCES

[1]   Valentini A, Pompanon F, Taberlet P. DNA barcoding for ecologists. Trends Ecol Evol 2009; 24(2): 110-7.
        [http://dx.doi.org/10.1016/j.tree.2008.09.011] [PMID: 19100655]

[2]   Morin PA, Luikart G, Wayne RK. SNPs in ecology, evolution and conservation. Trends Ecol Evol 2004; 19(4): 208-16.
        [http://dx.doi.org/10.1016/j.tree.2004.01.009]

[3]   Herman JG, Graff JR, Myöhänen S, Nelkin BD, Baylin SB. Methylation-specific PCR: a novel PCR assay for methylation status of CpG islands. Proc Natl Acad Sci USA 1996; 93(18): 9821-6.
        [http://dx.doi.org/10.1073/pnas.93.18.9821] [PMID: 8790415]

[4]   Sharma P, Johnson MA, Mazloom R, *et al.* Meta-analysis of the *Ralstonia solanacearum* species complex (RSSC) based on comparative evolutionary genomics and reverse ecology. Microb Genom 2022; 8(3): 000791.

[http://dx.doi.org/10.1099/mgen.0.000791] [PMID: 35297758]

[5]   Kashi Y, King D. Simple sequence repeats as advantageous mutators in evolution. Trends Genet 2006; 22(5): 253-9.
         [http://dx.doi.org/10.1016/j.tig.2006.03.005] [PMID: 16567018]

[6]   Semagn K, Bjørnstad Å, Ndjiondjop MN. An overview of molecular marker methods for plants. Afr J Biotechnol 2006; 5(25)

[7]   Agarwal M, Shrivastava N, Padh H. Advances in molecular marker techniques and their applications in plant sciences. Plant Cell Rep 2008; 27(4): 617-31.
         [http://dx.doi.org/10.1007/s00299-008-0507-z] [PMID: 18246355]

[8]   Jones N, Ougham H, Thomas H, Pašakinskienė I. Markers and mapping revisited: finding your gene. New Phytol 2009; 183(4): 935-66.
         [http://dx.doi.org/10.1111/j.1469-8137.2009.02933.x] [PMID: 19594696]

[9]   Thomas MR, Scott NS. Microsatellite repeats in grapevine reveal DNA polymorphisms when analysed as sequence-tagged sites (STSs). Theor Appl Genet 1993; 86(8): 985-90.
         [http://dx.doi.org/10.1007/BF00211051] [PMID: 24194007]

[10]   Ellegren H. Genome sequencing and population genomics in non-model organisms. Trends Ecol Evol 2014; 29(1): 51-63.
         [http://dx.doi.org/10.1016/j.tree.2013.09.008] [PMID: 24139972]

[11]   O'Brien SJ, Roelke ME, Marker L, *et al.* Genetic basis for species vulnerability in the cheetah. Science 1985; 227(4693): 1428-34.
         [http://dx.doi.org/10.1126/science.2983425] [PMID: 2983425]

[12]   Naresh V, Lee N. A review on biosensors and recent development of nanostructured materials-enabled biosensors. Sensors (Basel) 2021; 21(4): 1109.
         [http://dx.doi.org/10.3390/s21041109] [PMID: 33562639]

[13]   Kumar V, Guleria P. Application of DNA-nanosensor for environmental monitoring: recent advances and perspectives. Curr Pollut Rep 2020; 10(4): 765-85.
         [http://dx.doi.org/10.1007/s40726-020-00165-1] [PMID: 33344145]

[14]   Wang Z, Li P, Cui L, Qiu JG, Jiang B, Zhang C. Integration of nanomaterials with nucleic acid amplification approaches for biosensing. Trends Analyt Chem 2020; 129: 115959.
         [http://dx.doi.org/10.1016/j.trac.2020.115959]

[15]   Park M, Tsai SL, Chen W. Microbial biosensors: engineered microorganisms as the sensing machinery. Sensors (Basel) 2013; 13(5): 5777-95.
         [http://dx.doi.org/10.3390/s130505777] [PMID: 23648649]

[16]   Campaña AL, Florez SL, Noguera MJ, *et al.* Enzyme-based electrochemical biosensors for microfluidic platforms to detect pharmaceutical residues in wastewater. Biosensors (Basel) 2019; 9(1): 41.
         [http://dx.doi.org/10.3390/bios9010041] [PMID: 30875946]

[17]   Ravina , Dalal A, Mohan H, Prasad M, Pundir CS. Detection methods for influenza A H1N1 virus with special reference to biosensors: a review. Biosci Rep 2020; 40(2): BSR20193852.
         [http://dx.doi.org/10.1042/BSR20193852] [PMID: 32016385]

[18]   Franks AH, Harmsen HJM, Raangs GC, Jansen GJ, Schut F, Welling GW. Variations of bacterial populations in human feces measured by fluorescent *in situ* hybridization with group-specific 16S rRNA-targeted oligonucleotide probes. Appl Environ Microbiol 1998; 64(9): 3336-45.
         [http://dx.doi.org/10.1128/AEM.64.9.3336-3345.1998] [PMID: 9726880]

[19]   Narmani A, Kamali M, Amini B, Kooshki H, Amini A, Hasani L. Highly sensitive and accurate detection of Vibrio cholera O1 OmpW gene by fluorescence DNA biosensor based on gold and magnetic nanoparticles. Process Biochem 2018; 65: 46-54.
         [http://dx.doi.org/10.1016/j.procbio.2017.10.009]

[20] Ng S, Lim HS, Ma Q, Gao Z. Optical aptasensors for adenosine triphosphate. Theranostics 2016; 6(10): 1683-702.
[http://dx.doi.org/10.7150/thno.15850] [PMID: 27446501]

[21] Mitra T, Sahoo SK, Banerjee A, Upadhyay AK, Hasan KN. Potential of Nanobiosensors for Environmental Pollution Detection: Nanotechnology Combined with Enzymes, Antibodies, and Microorganisms. Nanotechnology For Environmental Pollution Decontamination 2022 Nov 30 (pp. 189-218). Apple Academic Press.

[22] Chen WY, Lan GY, Chang HT. Use of fluorescent DNA-templated gold/silver nanoclusters for the detection of sulfide ions. Anal Chem 2011; 83(24): 9450-5.
[http://dx.doi.org/10.1021/ac202162u] [PMID: 22029551]

[23] Arumugasamy SK, Chellasamy G, Gopi S, Govindaraju S, Yun K. Current advances in the detection of neurotransmitters by nanomaterials: An update. Trends Analyt Chem 2020; 123: 115766.
[http://dx.doi.org/10.1016/j.trac.2019.115766]

[24] Mahon AR, Jerde CL, Galaska M, *et al.* Validation of eDNA surveillance sensitivity for detection of Asian carps in controlled and field experiments. PLoS One 2013; 8(3): e58316.
[http://dx.doi.org/10.1371/journal.pone.0058316] [PMID: 23472178]

[25] Yap W, Switzer AD, Gouramanis C, *et al.* Environmental DNA signatures distinguish between tsunami and storm deposition in overwash sand. Communications Earth & Environment 2021; 2(1): 129.
[http://dx.doi.org/10.1038/s43247-021-00199-3]

[26] Thomsen PF, Willerslev E. Environmental DNA – An emerging tool in conservation for monitoring past and present biodiversity. Biol Conserv 2015; 183: 4-18.
[http://dx.doi.org/10.1016/j.biocon.2014.11.019]

[27] Dueholm MS, Andersen KS, McIlroy SJ, *et al.* Generation of comprehensive ecosystem-specific reference databases with species-level resolution by high-throughput full-length 16S rRNA gene sequencing and automated taxonomy assignment (AutoTax). MBio 2020; 11(5): e01557-20.
[http://dx.doi.org/10.1128/mBio.01557-20] [PMID: 32963001]

[28] Chaudhary N, Sharma AK, Agarwal P, Gupta A, Sharma VK. 16S classifier: a tool for fast and accurate taxonomic classification of 16S rRNA hypervariable regions in metagenomic datasets. PLoS One 2015; 10(2): e0116106.
[http://dx.doi.org/10.1371/journal.pone.0116106] [PMID: 25646627]

[29] Han XY, Pham AS, Tarrand JJ, Sood PK, Luthra R. Rapid and accurate identification of mycobacteria by sequencing hypervariable regions of the 16S ribosomal RNA gene. Am J Clin Pathol 2002; 118(5): 796-801.
[http://dx.doi.org/10.1309/HN44-XQYM-JMAQ-2EDL] [PMID: 12428802]

[30] Liu WT, Marsh TL, Cheng H, Forney LJ. Characterization of microbial diversity by determining terminal restriction fragment length polymorphisms of genes encoding 16S rRNA. Appl Environ Microbiol 1997; 63(11): 4516-22.
[http://dx.doi.org/10.1128/aem.63.11.4516-4522.1997] [PMID: 9361437]

[31] Petti CA, Polage CR, Schreckenberger P. The role of 16S rRNA gene sequencing in identification of microorganisms misidentified by conventional methods. J Clin Microbiol 2005; 43(12): 6123-5.
[http://dx.doi.org/10.1128/JCM.43.12.6123-6125.2005] [PMID: 16333109]

[32] Bukin YS, Galachyants YP, Morozov IV, Bukin SV, Zakharenko AS, Zemskaya TI. The effect of 16S rRNA region choice on bacterial community metabarcoding results. Sci Data 2019; 6(1): 190007.
[http://dx.doi.org/10.1038/sdata.2019.7] [PMID: 30720800]

[33] Hajibabaei M, Shokralla S, Zhou X, Singer GAC, Baird DJ. Environmental barcoding: a next-generation sequencing approach for biomonitoring applications using river benthos. PLoS One 2011; 6(4): e17497.

[http://dx.doi.org/10.1371/journal.pone.0017497] [PMID: 21533287]

[34]   Brookes PC. The use of microbial parameters in monitoring soil pollution by heavy metals. Biol Fertil Soils 1995; 19(4): 269-79.
[http://dx.doi.org/10.1007/BF00336094]

[35]   Bourhane Z, Lanzén A, Cagnon C, *et al.* Microbial diversity alteration reveals biomarkers of contamination in soil-river-lake continuum. J Hazard Mater 2022; 421: 126789.
[http://dx.doi.org/10.1016/j.jhazmat.2021.126789] [PMID: 34365235]

[36]   Gołębiewski M, Deja-Sikora E, Cichosz M, Tretyn A, Wróbel B. 16S rDNA pyrosequencing analysis of bacterial community in heavy metals polluted soils. Microb Ecol 2014; 67(3): 635-47.
[http://dx.doi.org/10.1007/s00248-013-0344-7] [PMID: 24402360]

[37]   Doyi I, Essumang D, Gbeddy G, Dampare S, Kumassah E, Saka D. Spatial distribution, accumulation and human health risk assessment of heavy metals in soil and groundwater of the Tano Basin, Ghana. Ecotoxicol Environ Saf 2018; 165: 540-6.
[http://dx.doi.org/10.1016/j.ecoenv.2018.09.015] [PMID: 30223167]

[38]   Erasmus JH, Malherbe W, Zimmermann S, *et al.* Metal accumulation in riverine macroinvertebrates from a platinum mining region. Sci Total Environ 2020; 703: 134738.
[http://dx.doi.org/10.1016/j.scitotenv.2019.134738] [PMID: 31731169]

[39]   Karelová E, Harichová J, Stojnev T, Pangallo D, Ferianc P. The isolation of heavy-metal resistant culturable bacteria and resistance determinants from a heavy-metal-contaminated site. Biologia (Bratisl) 2011; 66(1): 18-26.
[http://dx.doi.org/10.2478/s11756-010-0145-0]

[40]   Desai C, Parikh RY, Vaishnav T, Shouche YS, Madamwar D. Tracking the influence of long-term chromium pollution on soil bacterial community structures by comparative analyses of 16S rRNA gene phylotypes. Res Microbiol 2009; 160(1): 1-9.
[http://dx.doi.org/10.1016/j.resmic.2008.10.003] [PMID: 18996186]

[41]   Gremion F, Chatzinotas A, Harms H. Comparative 16S rDNA and 16S rRNA sequence analysis indicates that *Actinobacteria* might be a dominant part of the metabolically active bacteria in heavy metal□contaminated bulk and rhizosphere soil. Environ Microbiol 2003; 5(10): 896-907.
[http://dx.doi.org/10.1046/j.1462-2920.2003.00484.x] [PMID: 14510843]

[42]   Kumar V, Singh S, Singh J, Upadhyay N. Potential of plant growth promoting traits by bacteria isolated from heavy metal contaminated soils. Bull Environ Contam Toxicol 2015; 94(6): 807-14.
[http://dx.doi.org/10.1007/s00128-015-1523-7] [PMID: 25782590]

[43]   Field KG, Bernhard AE, Brodeur TJ. Molecular approaches to microbiological monitoring: fecal source detection.

[44]   Bernhard AE, Field KG. Identification of nonpoint sources of fecal pollution in coastal waters by using host-specific 16S ribosomal DNA genetic markers from fecal anaerobes. Appl Environ Microbiol 2000; 66(4): 1587-94.
[http://dx.doi.org/10.1128/AEM.66.4.1587-1594.2000] [PMID: 10742246]

[45]   Mieszkin S, Yala JF, Joubrel R, Gourmelon M. Phylogenetic analysis of *Bacteroidales* 16S rRNA gene sequences from human and animal effluents and assessment of ruminant faecal pollution by real□time PCR. J Appl Microbiol 2010; 108(3): 974-84.
[http://dx.doi.org/10.1111/j.1365-2672.2009.04499.x] [PMID: 19735325]

[46]   Dantas G, Sommer MOA, Oluwasegun RD, Church GM. Bacteria subsisting on antibiotics. Science 2008; 320(5872): 100-3.
[http://dx.doi.org/10.1126/science.1155157] [PMID: 18388292]

[47]   Wang RN, Zhang Y, Cao ZH, *et al.* Occurrence of super antibiotic resistance genes in the downstream of the Yangtze River in China: Prevalence and antibiotic resistance profiles. Sci Total Environ 2019; 651(Pt 2): 1946-57.

[http://dx.doi.org/10.1016/j.scitotenv.2018.10.111] [PMID: 30321718]

[48]   Schwartz T, Kohnen W, Jansen B, Obst U. Detection of antibiotic-resistant bacteria and their resistance genes in wastewater, surface water, and drinking water biofilms. FEMS Microbiol Ecol 2003; 43(3): 325-35.
[http://dx.doi.org/10.1111/j.1574-6941.2003.tb01073.x] [PMID: 19719664]

[49]   Tascini C, Sozio E, Viaggi B, Meini S. Reading and understanding an antibiogram. Ital J Med 2016; 10(4): 289-300.
[http://dx.doi.org/10.4081/itjm.2016.794]

[50]   Paulus GK, Hornstra LM, Alygizakis N, Slobodnik J, Thomaidis N, Medema G. The impact of on-site hospital wastewater treatment on the downstream communal wastewater system in terms of antibiotics and antibiotic resistance genes. Int J Hyg Environ Health 2019; 222(4): 635-44.
[http://dx.doi.org/10.1016/j.ijheh.2019.01.004] [PMID: 30737165]

[51]   Osińska A, Korzeniewska E, Harnisz M, Niestępski S. The prevalence and characterization of antibiotic-resistant and virulent Escherichia coli strains in the municipal wastewater system and their environmental fate. Sci Total Environ 2017; 577: 367-75.
[http://dx.doi.org/10.1016/j.scitotenv.2016.10.203] [PMID: 27816226]

[52]   Czekalski N, Gascón Díez E, Bürgmann H. Wastewater as a point source of antibiotic-resistance genes in the sediment of a freshwater lake. ISME J 2014; 8(7): 1381-90.
[http://dx.doi.org/10.1038/ismej.2014.8] [PMID: 24599073]

[53]   LaPara TM, Burch TR, McNamara PJ, Tan DT, Yan M, Eichmiller JJ. Tertiary-treated municipal wastewater is a significant point source of antibiotic resistance genes into Duluth-Superior Harbor. Environ Sci Technol 2011; 45(22): 9543-9.
[http://dx.doi.org/10.1021/es202775r] [PMID: 21981654]

[54]   Zieliński W, Hubeny J, Buta-Hubeny M, *et al.* Metagenomics analysis of probable transmission of determinants of antibiotic resistance from wastewater to the environment – A case study. Sci Total Environ 2022; 827: 154354.
[http://dx.doi.org/10.1016/j.scitotenv.2022.154354] [PMID: 35259375]

[55]   Zieliński W, Korzeniewska E, Harnisz M, Drzymała J, Felis E, Bajkacz S. Wastewater treatment plants as a reservoir of integrase and antibiotic resistance genes – An epidemiological threat to workers and environment. Environ Int 2021; 156: 106641.
[http://dx.doi.org/10.1016/j.envint.2021.106641] [PMID: 34015664]

[56]   Burch TR, Sadowsky MJ, LaPara TM. Aerobic digestion reduces the quantity of antibiotic resistance genes in residual municipal wastewater solids. Front Microbiol 2013; 4: 17.
[http://dx.doi.org/10.3389/fmicb.2013.00017] [PMID: 23407455]

[57]   Yao S, Ye J, Yang Q, *et al.* Occurrence and removal of antibiotics, antibiotic resistance genes, and bacterial communities in hospital wastewater. Environ Sci Pollut Res Int 2021; 28(40): 57321-33.
[http://dx.doi.org/10.1007/s11356-021-14735-3] [PMID: 34089156]

[58]   Sun Y, Shen Y, Liang P, Zhou J, Yang Y, Huang X. Multiple antibiotic resistance genes distribution in ten large-scale membrane bioreactors for municipal wastewater treatment. Bioresour Technol 2016; 222: 100-6.
[http://dx.doi.org/10.1016/j.biortech.2016.09.117] [PMID: 27716561]

[59]   Sousa JM, Macedo G, Pedrosa M, *et al.* Ozonation and UV$_{254nm}$ radiation for the removal of microorganisms and antibiotic resistance genes from urban wastewater. J Hazard Mater 2017; 323(Pt A): 434-41.
[http://dx.doi.org/10.1016/j.jhazmat.2016.03.096] [PMID: 27072309]

[60]   Zhang Y, Zhuang Y, Geng J, *et al.* Inactivation of antibiotic resistance genes in municipal wastewater effluent by chlorination and sequential UV/chlorination disinfection. Sci Total Environ 2015; 512-513: 125-32.
[http://dx.doi.org/10.1016/j.scitotenv.2015.01.028] [PMID: 25616228]

[61]   Chu L, Wang J, He S, Chen C, Wojnárovits L, Takács E. Treatment of pharmaceutical wastewater by ionizing radiation: Removal of antibiotics, antimicrobial resistance genes and antimicrobial activity. J Hazard Mater 2021; 415: 125724.
[http://dx.doi.org/10.1016/j.jhazmat.2021.125724] [PMID: 34088196]

[62]   Gelover S, Gómez LA, Reyes K, Teresa Leal M. A practical demonstration of water disinfection using TiO2 films and sunlight. Water Res 2006; 40(17): 3274-80.
[http://dx.doi.org/10.1016/j.watres.2006.07.006] [PMID: 16949121]

[63]   Guinotte JM, Fabry VJ. Ocean acidification and its potential effects on marine ecosystems. Ann N Y Acad Sci 2008; 1134(1): 320-42.
[http://dx.doi.org/10.1196/annals.1439.013] [PMID: 18566099]

[64]   Hönisch B, Ridgwell A, Schmidt DN, *et al.* The geological record of ocean acidification. Science 2012; 335(6072): 1058-63.
[http://dx.doi.org/10.1126/science.1208277] [PMID: 22383840]

[65]   Aguayo P, Campos VL, Henríquez C, Olivares F. De IaIglesia R, Ulloa O, Vargas CA. The influence of p CO2-driven ocean acidification on open ocean bacterial communities during a short-term microcosm experiment in the eastern tropical South Pacific (ETSP) off northern Chile. Microorganisms 2020; 8(12): 1924.
[http://dx.doi.org/10.3390/microorganisms8121924] [PMID: 33291533]

[66]   Cavalcanti GS, Shukla P, Morris M, *et al.* Rhodoliths holobionts in a changing ocean: host-microbes interactions mediate coralline algae resilience under ocean acidification. BMC Genomics 2018; 19(1): 701.
[http://dx.doi.org/10.1186/s12864-018-5064-4] [PMID: 29291715]

[67]   Palladino G, Caroselli E, Tavella T, *et al.* Metagenomic shifts in mucus, tissue and skeleton of the coral *Balanophyllia europaea* living along a natural CO2 gradient. ISME Communications 2022; 2(1): 65.
[http://dx.doi.org/10.1038/s43705-022-00152-1] [PMID: 37938252]

[68]   Thurber RV, Willner-Hall D, Rodriguez-Mueller B, *et al.* Metagenomic analysis of stressed coral holobionts. Environ Microbiol 2009; 11(8): 2148-63.
[http://dx.doi.org/10.1111/j.1462-2920.2009.01935.x] [PMID: 19397678]

[69]   Kerfahi D, Harvey BP, Kim H, Yang Y, Adams JM, Hall-Spencer JM. Whole community and functional gene changes of biofilms on marine plastic debris in response to ocean acidification. Microb Ecol 2022; 1-3.
[PMID: 35378620]

SUBJECT INDEX

A

Abiotic 20, 21, 47, 114, 120, 123, 125, 128, 130, 131
  diseases 123
  stresses 20, 21, 47, 114, 120, 123, 125, 128, 130, 131
Actinomycetes 24, 136, 163, 170, 229
Aerobic microbiological process 228
AFLP method 248, 249
AMPK protein kinases 124
Anthropogenic 13, 199, 200
  stress 13
Antibacterial agent 203
Antibiotic 126, 209, 260, 261
  -resistant bacteria (ARBs) 209, 260, 261
  -resistant genes (ARGs) 126, 260, 261
Antifreeze proteins (APs) 126
Arbuscular mycorrhizal fungi (AMF) 46, 50, 51, 58, 104, 105, 107, 147, 215, 233
Auxiliary metabolic genes (AMGs) 126

B

Bacteria 115, 259
  coliform 259
  microscopic 115
Bacterial endophyte 59
Biofertilizers, microbe-based 23
Biogas, methane-rich 166
Biomass 50, 60, 100, 140, 150, 160, 167, 169, 225, 236
  lignocellulose 225
  lignocellulosic 160, 167
  production and nutrient removal 150
Bioremediation 19, 71, 74, 77, 171
  techniques 19, 74
  technologies 71, 77, 171
Bioremediation processes 83, 84, 85, 188, 225
  anaerobic 84
  microbial-mediated 83

C

Carbon 23, 47, 114, 115, 117, 120, 121, 149, 150, 159, 162, 183, 186, 187, 222, 235, 236, 237
  atmospheric 23, 222
  emissions 117, 149
  farming technique 235
  fixation 47, 183
  microbial-derived 236
Cheese production 203
Chemical 22, 92, 93, 94, 96, 102, 107, 108, 109, 163, 170
  decomposition 163
  fertilizers 22, 92, 93, 94, 96, 102, 107, 108, 109, 170
Chemosynthesis 101
Chemotaxis system 237
Contaminants 6, 47, 55, 76, 77, 83, 84, 85, 108, 137, 138, 145, 163, 168, 171, 180
  emerging 137
  environmental 180
  organic 6, 77
  toxic 145
Contaminated 6, 48, 70, 71, 75, 76, 77, 79, 80, 81, 82, 84, 85, 101, 148
  environments 6, 79, 85
  soil 48, 70, 71, 75, 76, 77, 80, 81, 84, 101, 148
  water 77, 82
Contamination 30, 130, 157, 233
  crude oil 233
  risks 30
CRISPR 188, 189
  gene editing technology 188
  technology 188, 189
Crop(s) 92, 117, 124, 129
  infectious 124
  production 92, 117, 129
Cytochrome oxidase 210